Genetics

3rd Edition

by Tara Rodden Robinson, PhD
and Lisa Cushman Spock, PhD, CGC

Genetics For Dummies®, 3rd Edition

Published by **John Wiley & Sons, Inc.,** 111 River Street, Hoboken, NJ 07030-5774, www.wiley.com

Published simultaneously in Canada

For general information on our other products and services, please contact our Customer Care Department within the U.S. at 877-762-2974, outside the U.S. at 317-572-3993, or fax 317-572-4002. For technical support, please visit www.wiley.com/techsupport.

Wiley publishes in a variety of print and electronic formats and by print-on-demand. Some material included with standard print versions of this book may not be included in e-books or in print-on-demand. If this book refers to media such as a CD or DVD that is not included in the version you purchased, you may download this material at http://booksupport.wiley.com. For more information about Wiley products, visit www.wiley.com.

Library of Congress Control Number: 2019954514

ISBN: 978-1-119-63303-7; ISBN (ePDF): 978-1-119-63308-2; ISBN (ePub): 978-1-119-63312-9

Manufactured in the United States of America

V10016415_121919

Contents at a Glance

Table of Contents

Introduction

Genetics affects all living things. Although sometimes complicated and always diverse, all genetics comes down to basic principles of *heredity* — how traits are passed from one generation to the next — and how DNA is put together. As a science, genetics is a fast-growing field because of its untapped potential — for good and for bad. Despite its complexity, genetics can be surprisingly accessible. Genetics is a bit like peeking behind a movie's special effects to find a deceptively simple and elegant system running the whole show.

About This Book

Genetics For Dummies, 3rd Edition, is an overview of the entire field of genetics. Our goal is to explain every topic so that anyone, even someone without any genetics background at all, can follow the subject and understand how it works. As in the first and second editions, we include many examples from the frontiers of research. We also make sure that the book has detailed coverage of some of the hottest topics that you hear about in the news, including gene therapy, pharmacogenetics, and gene editing. And we address the practical side of genetics: how it affects your health and the world around you. In short, this book is designed to be a solid introduction to genetics basics and to provide some details on the subject.

Genetics is a fast-paced field; new discoveries are coming out all the time. You can use this book to help you get through your genetics course or for self-guided study. *Genetics For Dummies*, 3rd Edition, provides enough information for you to get a handle on the latest press coverage, understand the genetics jargon that mystery writers like to toss around, and translate information imparted to you by medical professionals. The book is filled with stories of key discoveries and "wow" developments. Although we try to keep things light and inject some humor when possible, we also make every effort to be sensitive to whatever your circumstances may be.

This book is a great guide if you know nothing at all about genetics. If you already have some background, then you're set to dive into the details of the subject and expand your horizons.

Conventions Used in This Book

It would be very easy for us to use specialized language that you'd need a translator to understand, but what fun would that be? Throughout this book, we try to avoid jargon as much as possible, but at the same time, we use and carefully define terms that scientists actually use. After all, it may be important for you to understand some of these multisyllabic jawbreakers in the course of your studies or your or a loved one's medical treatment.

To help you navigate through this book, we use the following typographical conventions:

- We use *italic* for emphasis and to highlight new words or terms that we define in the text.
- We use **boldface** to indicate keywords in bulleted lists or the action parts of numbered steps.
- We use `monofont` for websites and email addresses.

Foolish Assumptions

It's a privilege to be your guide into the amazing world of genetics. Given this responsibility, you were in our thoughts often while we were writing this book. Here's how we imagine you, our reader:

- You're a student in a genetics or biology class.
- You're curious to understand more about the science you hear reported in the news.
- You're an expectant or new parent or a family member who's struggling to come to terms with what doctors have told you.
- You're affected by cancer or some hereditary disease, wondering what it means for you and your family.

If any of these descriptions fit, you've come to the right place.

How This Book Is Organized

We designed this book to cover background material in the first two parts and then all the applications in the rest of the book. We think you'll find it quite accessible.

Part 1: The Lowdown on Genetics: Just the Basics

Part 1 explains how trait inheritance works. The first chapter introduces you to the field of genetics and what genetics professionals may do in their day-to-day work lives. The second chapter gives you a handle on how genetic information gets divvied up during cell division; these events provide the foundation for just about everything else that has to do with genetics. From there, we explain simple inheritance of one gene and then move on to more complex forms of inheritance.

Part 2: DNA: The Genetic Material

Part 2 covers what's sometimes called *molecular genetics.* Don't let the word "molecular" scare you off. We give you nitty-gritty details, but we break them down so that you can easily follow along. We track the progress of how your genes work from start to finish: how your DNA is put together, how it gets copied, and how the building plans for your body are encoded in the double helix. To help you understand how scientists explore the secrets stored in your DNA, we also cover how DNA is sequenced. In the process, we relate the fascinating story behind the Human Genome Project.

Part 3: Genetics and Your Health

Part 3 is intended to help you see how genetics affects your health and well-being. We cover the subjects of genetic counseling; inherited diseases; genetics and cancer; and chromosome disorders such as Down syndrome. We also include a chapter on gene therapy, a practice that may hold the key to cures or treatments for many of the disorders we describe in this part of the book.

Part 4: Genetics and Your World

Part 4 explains the broader impact of genetics and covers some hot topics that are often in the news. We explain how various technologies work and highlight both the possibilities and the perils of each. We delve into population genetics (of both humans, past and present, and endangered animal species), evolution, DNA and forensics, genetically modified plants and animals, and the issue of ethics, which is raised on a daily basis as scientists push the boundaries of the possible with cutting-edge technology.

Part 5: The Part of Tens

In Part 5, you get our lists of ten milestone events and important people that have shaped genetics history, and ten of the next big things in the field.

Icons Used in This Book

All *For Dummies* books use icons to help readers keep track of what's what. Here's a rundown of the icons we use in this book and what they all mean.

This icon points out stories about the people behind the science and accounts of how discoveries came about.

This icon flags information that's critical to your understanding or that's particularly important to keep in mind.

These details are useful but not necessary to know. If you're a student, though, these sections may be especially important to you.

Points in the text where we provide added insight on how to get a better handle on a concept are found here. We draw on our personal experience for these tips and alert you to other sources of information you can check out.

Beyond This Book

We've included a ton of extra content on the website that accompanies this book. To find it, simply open your favorite web browser, go to `www.dummies.com`, and search for **Genetics For Dummies** to find the following:

* **Cheat Sheet:** We've created Cheat Sheet pages that review basic genetics terminology; the structure of cells, chromosomes, and DNA; the laws of inheritance; solving genetics problems; and the basics of transcription and translation.
* **Bonus chapters:** While the book covers a lot of the hottest topics in genetics, it can't cover everything. Check out the bonus chapter on cloning.
* **Updates to the book, if any.**

Where to Go from Here

With *Genetics For Dummies*, 3rd Edition, you can start anywhere, in any chapter, and get a handle on what you're interested in right away. We make generous use of cross-references throughout the book to help you get background details that you may have skipped earlier. The table of contents and index can point you to specific topics in a hurry, or you can just start at the beginning and work your way straight through. If you read the book from front to back, you'll get a short course in genetics in the style and order that it's often taught in colleges and universities — Mendel first and DNA second.

1
The Lowdown on Genetics: Just the Basics

IN THIS PART . . .

Discover the basics of genetics and the various careers in the field.

Learn how cells divide and how chromosomes are divvyed up among those cells.

Learn about Mendelian genetics and how genes and traits are inherited.

Understand how the inheritance of genes and traits is not always straightforward.

IN THIS CHAPTER

- » Defining the subject of genetics and its various subdivisions
- » A brief introduction to what is covered in this book
- » A review of some of the possible career opportunities in genetics

Chapter 1 Welcome to Genetics: What's What and Who's Who

Welcome to the complex and fascinating world of genetics. Genetics is all about physical traits and the DNA code that supplies the building plans for any organism. This chapter defines the field of genetics and explains what geneticists do. You get an introduction to the big picture and a glimpse at some of the details found in other chapters of this book.

What Is Genetics?

Genetics is the field of science that examines how traits are passed from one generation to the next. Simply put, genetics affects *everything* about *every* living thing on earth. An organism's *genes* are segments of DNA (deoxyribonucleic acid) that are the fundamental units of heredity. Genes play an essential role in how the organism looks, behaves, and reproduces. Because all biology depends on genes, understanding genetics as a foundation for all other life sciences, including agriculture and medicine, is critical.

From a historical point of view, genetics is still a young science. The principles that govern inheritance of traits by one generation from another were described (and promptly lost) less than 150 years ago. Around the turn of the 20th century, the laws of inheritance were rediscovered, an event that transformed biology forever. It wasn't until the 1950s that the importance of the star of the genetics show, DNA, was really understood. Now technology is helping geneticists push the envelope of knowledge every day.

Genetics is generally divided into four major subdivisions:

- **Classical, or Mendelian, genetics:** A discipline that describes how physical characteristics (traits) are passed along from one generation to another.
- **Molecular genetics:** The study of the chemical and physical structures of DNA, its close cousin RNA (ribonucleic acid), and proteins. Molecular genetics also covers how genes do their jobs.
- **Population genetics:** A division of genetics that looks at the genetic makeup of larger groups.
- **Quantitative genetics:** A highly mathematical field that examines the statistical relationships between genes and the traits with which they are associated.

In the academic world, many genetics courses begin with classical genetics and proceed through molecular genetics, with a nod to population and quantitative genetics. In general, this book follows the same path, because each division of knowledge builds on the one before it. That said, it's perfectly okay, and very easy, to jump around among disciplines. No matter how you take on reading this book, it provides lots of cross references to help you stay on track.

Classical genetics: Transmitting traits from generation to generation

At its heart, *classical genetics* is the genetics of individuals and their families. It focuses mostly on studying physical traits, or *phenotypes,* as a stand-in for the genes that control appearance.

Gregor Mendel, a humble monk and part-time scientist, founded the entire discipline of genetics. Mendel was a gardener with an insatiable curiosity to go along with his green thumb. His observations may have been simple, but his conclusions were jaw-droppingly elegant. This man had no access to technology, computers, or a pocket calculator, yet he determined, with keen accuracy, exactly how inheritance works.

Classical genetics is sometimes referred to as:

- **Mendelian genetics:** You start a new scientific discipline, and it gets named after you. Seems fair.
- **Transmission genetics:** This term refers to the fact that classical genetics describes how traits are passed on, or *transmitted,* from parents to their offspring.

No matter what you call it, classical genetics includes the study of cells and chromosomes, which we cover in Chapters 2 and 6. Cell division is the machine that drives inheritance, but you don't have to understand combustion engines to drive a car, right? Likewise, you can dive straight into simple inheritance in Chapter 3 and work up to more complicated forms of inheritance in Chapter 4 without knowing anything whatsoever about cell division. (Mendel didn't know anything about chromosomes and cells when he figured this whole thing out, by the way.)

The genetics of sex and reproduction are also part of classical genetics. Various combinations of genes and *chromosomes* (strands of DNA) determine sex, as in maleness and femaleness. But the subject of sex gets even more complicated and interesting: The environment plays a role in determining the sex of some organisms (like crocodiles and turtles), and other organisms can even change sex with a change of address. If this has piqued your interest, you can find out all the astonishing details in Chapter 6. (Of note, we use the term *sex* throughout this book instead of the term *gender*. *Sex* is what defines males and females from a biological perspective. A person's *gender*, on the other hand, may also be influenced by social and cultural factors, and may differ from one's biological sex.)

Classical genetics provides the framework for many subdisciplines. The study of chromosome disorders such as Down syndrome, which we cover in Chapter 13, relies on cell biology and an understanding of what happens during cell division. Genetic counseling, which we cover in Chapter 15, also relies on understanding patterns of inheritance to interpret people's medical histories from a genetics perspective. In addition, forensics, covered in Chapter 18, uses Mendelian genetics to determine paternity and to work out who's who with DNA fingerprinting.

Molecular genetics: DNA and the chemistry of genes

Classical genetics concentrates on studying outward appearances, while the study of actual genes falls under the heady title of *molecular genetics.* The area of operations for molecular genetics includes all the machinery that runs cells and manufactures the structures called for by the plans found in genes. The focus of molecular genetics includes the physical and chemical structures of the double helix, DNA, which is broken down in all its glory in Chapter 5. The messages hidden in your DNA (your genes) constitute the building instructions for your appearance and everything else about you — from how your muscles function and how your eyes blink to your blood type, your susceptibility to particular diseases, and everything in between. How that DNA (and the immense amount of information it contains) is packaged in your cells is covered in Chapter 6, which reviews the structure and function of chromosomes.

Your genes are expressed through a complex system of interactions that begins with *transcription* — copying DNA's messages into a somewhat temporary form called RNA, which is short for *ribonucleic acid* and is covered in Chapter 9. RNA carries the DNA message through the process of *translation*, covered in Chapter 10, which in essence is like taking a blueprint to a factory to guide the manufacturing process. Where your genes are concerned, the factory makes the proteins (from the RNA blueprint) that get folded in complex ways to make the various components of the cells and tissues in the human body. The study of *gene expression* (how genes get turned on and off, which we review in Chapter 11) and how the genetic code works at the levels of DNA and RNA are considered parts of molecular genetics.

Research on the causes of cancer and the hunt for better treatments, which we address in Chapter 14, focuses on the molecular side of things because tumors result from changes in the DNA, called *mutations*. Chapter 12 covers mutations in detail. Gene therapy, covered in Chapter 16, and genetic engineering, covered in Chapter 19, are subdisciplines of molecular genetics.

Population genetics: Genetics of groups

Much to the chagrin of many undergrads, many aspects of genetics are surprisingly mathematical. One area in which calculations are used to describe what goes on genetically is population genetics.

REMEMBER

If you use Mendelian genetics and examine the inheritance patterns of many different individuals who have something in common, like geographic location, you can study population genetics. *Population genetics* is the study of the genetic diversity of a subset of a particular species (for details, you can flip ahead to Chapter 17). Basically, it's a search for patterns that help describe the genetic signature of a particular group, such as the consequences of migration, isolation from other populations, and mating choices.

Population genetics helps scientists understand how the collective genetic diversity of a population influences the health of individuals within the population. For example, cheetahs are lanky cats; they're the speed demons of Africa. Population genetics has revealed that all cheetahs are extremely genetically similar; in fact, they're so similar that a skin graft from one cheetah would be accepted by any other cheetah. Because the genetic diversity of cheetahs is so low, conservation biologists fear that a disease could sweep through the population and kill off all the individuals of the species. It's possible that no animals would be resistant to the disease, and therefore, none would survive, leading to the extinction of this amazing predator.

Evolutionary genetics is a type of population genetics that involves studying how traits change over time. We review evolutionary genetics in Chapter 17. Describing the genetics of populations from a mathematical standpoint is also critical to forensics, as explained in Chapter 18. To pinpoint the uniqueness of one DNA fingerprint, geneticists need to sample the genetic fingerprints of many individuals and decide how common or rare a particular pattern may be. Likewise, medicine uses population genetics to determine how common particular DNA changes are and to develop new medicines to treat disease (discussed in Chapter 22).

Quantitative genetics: Getting a handle on heredity

Quantitative genetics examines traits that vary in subtle ways and relates those traits to the underlying genetics of an organism. A combination of whole suites of genes and environmental factors controls characteristics like retrieving ability in dogs, egg size or number in birds, and running speed in humans. Mathematical in nature, quantitative genetics takes a rather complex statistical approach to estimate how much variation in a particular trait is due to the environment and how much is actually genetic.

One application of quantitative genetics is determining how heritable a particular trait is. This measure allows scientists to make predictions about how offspring will turn out based on characteristics of the parent organisms. Heritability gives some indication of how much a characteristic (like seed production) can change when selective breeding (or, in evolutionary time, natural selection) is applied.

Genetics as a Career

Studying genetics can lead to a variety of career opportunities, the most common of which we describe in the following sections. The daily life for someone working in genetics can include working in the lab, teaching in the classroom, or interacting with patients and their families. In this section, you'll first discover what a typical genetics lab is like, and then get a quick rundown of a variety of career paths in the ever-expanding field of genetics.

Exploring a genetics lab

A genetics lab is a busy place. It's full of equipment and supplies and researchers toiling away at their workstations (called *lab benches*, even though the bench is really just a raised, flat surface that's conducive to working while standing up). Depending on the lab, you may see people looking very official in white lab coats or researchers dressed more casually in jeans and T-shirts. Every lab contains some or all of the following:

- Disposable gloves to protect workers from chemical exposure and to protect DNA and other materials from contamination.
- Pipettes (for measuring even the tiniest droplets of liquids with extreme accuracy), glassware (for liquid measurement and storage), and vials and tubes (for chemical reactions).
- Electronic balances for making super-precise measurements of mass.
- Chemicals and ultrapure water.
- A refrigerator, a freezer, and an ultracold freezer for storing samples.

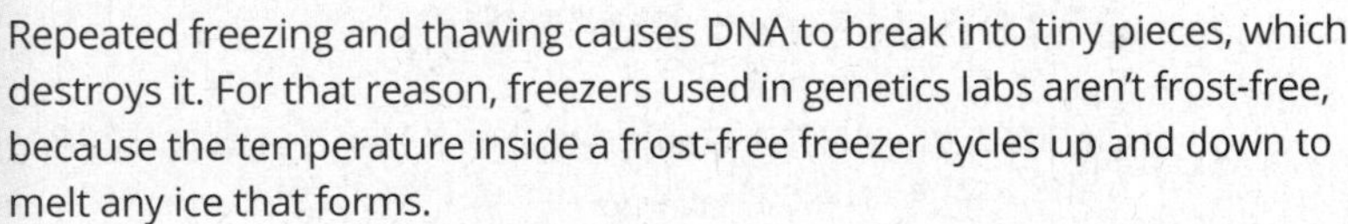

Repeated freezing and thawing causes DNA to break into tiny pieces, which destroys it. For that reason, freezers used in genetics labs aren't frost-free, because the temperature inside a frost-free freezer cycles up and down to melt any ice that forms.

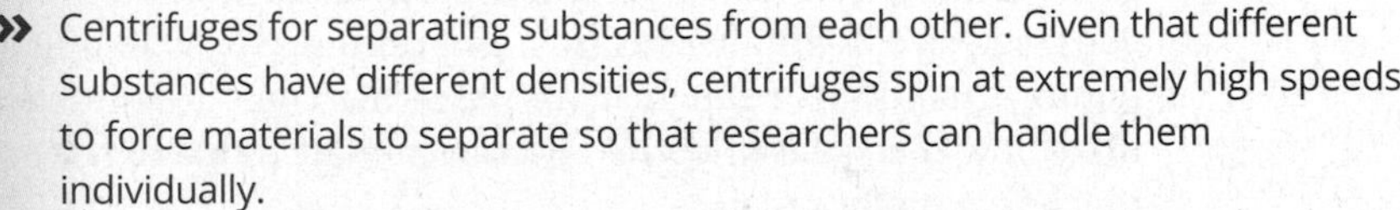

- Centrifuges for separating substances from each other. Given that different substances have different densities, centrifuges spin at extremely high speeds to force materials to separate so that researchers can handle them individually.
- Incubators for growing bacteria under controlled conditions. Researchers often use bacteria for experimental tests of how genes work.
- Autoclaves for sterilizing glassware and other equipment using extreme heat and pressure to kill bacteria and viruses.
- Complex pieces of equipment that are used to generate more copies of DNA fragments or to determine the sequence of segments of DNA.
- Lab notebooks for recording every step of every reaction or experiment in nauseating detail. Geneticists must fully replicate (run over and over) every experiment to make sure the results are valid. The lab notebook is also a legal document that can be used in court cases, so precision and completeness are musts.
- Computers packed with software for analyzing results and for connecting via the Internet to vast databases packed with genetic information. To get the addresses of some useful websites, see the sidebar, "Great genetics websites to explore," which is located toward the end of this chapter.

Researchers in the lab use the various pieces of equipment and supplies from the preceding list to conduct experiments and run chemical reactions. Some of the common activities that occur in the genetics lab include:

- Separating DNA from the rest of a cell's contents.
- Mixing chemicals that are used in reactions and experiments designed to analyze DNA samples.
- Growing special strains of bacteria and viruses to aid in examining short stretches of DNA.
- Using DNA sequencing to learn the order of bases that compose a DNA strand.
- Setting up polymerase chain reactions, or PCR, a powerful process that allows scientists to analyze even very tiny amounts of DNA.
- Analyzing the results of DNA sequencing by comparing sequences from many different organisms (you can find this information in a massive, publicly available database — `https://www.ncbi.nlm.nih.gov/genbank/`).
- Comparing DNA fingerprints from several individuals to identify perpetrators or to assign paternity.
- Holding weekly or daily meetings where everyone in the lab comes together to discuss results or plan new experiments.

Sorting through jobs in genetics

Whole teams of people contribute to the study of genetics. The following are just a few job descriptions for you to mull over if you're considering a career in genetics. For many of these jobs, you need a graduate degree or other training beyond college. So first, we discuss what it is like to be a graduate student studying genetics or someone who has just finished graduate school and is getting additional training before hitting the workforce. You should know, however, that even though these positions are considered part of your education, they really are like full-time jobs. (Or more than full-time in most cases!)

Graduate student and post-doc

At most universities, genetics labs are full of *graduate students* (often called, simply, *grad students*) working on either master's degrees or PhDs. In some labs, these students may be carrying out their own, independent research. On the other hand, many labs focus their work on a specific problem, like some specialized approach to studying cancer, and every student in that sort of lab works on some aspect of what his or her professor studies. Graduate students do a lot of the same things that lab techs do (see the following section), as well as design experiments, carry out those experiments, analyze the results, and then work to figure out what the results mean. Then, the graduate student writes a long document (called a *thesis* or *dissertation*) to describe what was done, what it means, and how it fits in with other people's research on the subject. While working in the lab, graduate students take classes and are subjected to grueling exams (trust us on the grueling part).

All graduate students must hold a bachelor's degree. Performance on the standardized GRE (Graduate Record Exam) determines eligibility for admission to graduate programs and may be used for selection for fellowships and awards.

TIP

If you're going to be staring down the GRE in the near future, you may want to get a leg up by checking out *GRE For Dummies with Online Practice*, 9th Edition, by Ron Woldoff (Wiley).

In general, it takes two or three years to earn a master's degree. A doctorate (denoted by *PhD*) usually requires anywhere from four to seven years of education beyond the bachelor's level.

After graduating with a PhD, a geneticist-in-training may need to get more experience before hitting the job market. Positions that provide such experience are collectively referred to as *post-docs* (post-doctoral fellowships). A person holding a post-doc position is usually much more independent than a grad student when it comes to research. The post-doc often works to learn new techniques or to acquire a specialty before moving on to a position as a professor or a research scientist.

Lab tech

Lab technicians, often called *lab techs*, handle most of the day-to-day work in the lab. However, the exact tasks a lab tech performs may depend on the type of laboratory and what daily activities are necessary for the lab to run smoothly. In general, lab techs mix chemicals for everyone else in the lab to use in experiments. Techs may also prepare other types of materials such as bacterial cultures, yeast cultures, or other biological samples. In addition, techs are usually responsible for keeping all the necessary supplies straight and washing the glassware — not a glamorous job but a necessary one, because labs use tons of glass beakers and flasks that need to be cleaned. When it comes to actual experiments, the responsibilities of a lab technician may vary, often with the experience of the lab tech.

The educational background needed to be a lab tech varies with the amount of responsibility a particular position demands. Most lab techs have a minimum of a bachelor's degree in biology or some related field and need some background in microbiology to understand and carry out the techniques of handling bacteria safely and without contaminating cultures. All lab techs must be good record-keepers, because every single activity in the lab must be documented in writing.

Research scientist

Research scientists often work in private industries, designing experiments and directing the activities of lab techs. All sorts of industries employ research scientists, including:

- Pharmaceutical companies, which use research scientists to conduct investigations on how drugs affect gene expression and to develop new treatments, such as gene therapy.
- Forensics labs, which use research scientists to analyze DNA found at crime scenes and to compare DNA fingerprints.
- Companies that analyze information generated by genome projects (human and others).
- Companies that support the work of other genetics labs by designing and marketing products used in research, such as kits used to extract DNA or run DNA fingerprints.

A research scientist usually holds a master's degree or a PhD. With only a bachelor's degree, several years of experience as a lab tech may suffice. Research scientists need to be able to design experiments and analyze results using statistics. Good record-keeping and strong communication skills (especially in writing) are musts. Most research scientists also need to be capable of managing and supervising people. In addition, financial responsibilities may include keeping up with expenditures, ordering equipment and supplies, and wrangling salaries of other personnel.

College or university professor

Professors do everything that research scientists do with the added responsibilities of teaching courses, writing proposals to get funds to support research, and writing papers on their research results for publication in reputable, peer-reviewed journals. Professors also supervise the lab techs, graduate students, and post-docs who work in their labs, which entails designing research projects and then ensuring that the projects are done correctly in the right amount of time (and under budget!).

Small schools may require a professor to teach as many as three courses every semester. Upper-tier institutions (think Big Ten or Ivy League) may require only one course of instruction per year (although they are likely expected to spend more time on research, publishing findings, and obtaining grant funding). Genetics professors teach the basics as well as advanced and specialty courses like population genetics or evolutionary genetics.

To qualify for a professorship, universities require a minimum of a PhD, and most require additional post-doctoral experience. Job candidates must have already published research results to demonstrate the ability to do relevant research. Most universities also look for evidence that the professor-to-be will be successful at getting grants, which means the candidate must usually land a grant before getting a job.

Clinical laboratory director

Clinical laboratory directors run the laboratories that perform genetic testing for patients. These laboratories must receive special certifications that ensure that the tests they perform are accurate and the results can be used for clinical decision-making (certification that is not needed in research laboratories). The responsibilities of a clinical laboratory director include analyzing and interpreting test results for patients, as well as supervising all laboratory operations. In addition, they must have strong communication skills and be able to interact with clinicians, laboratory staff, and patients.

Clinical laboratory directors generally have a doctoral degree (a PhD in genetics or a related field, or a medical degree) and have successfully completed a postdoctoral fellowship in a clinical genetics laboratory. They also need to take and pass a board examination in their area of expertise.

Clinical geneticist

Clinical geneticists work with patients who have or are at-risk for genetic conditions. Their role generally includes the diagnosis of genetic disorders, as well as the management and treatment. They can see patients of any age (from before birth until late in life) and in a variety of settings (such as in-patients in the hospital, out-patients in a specialty clinic, or in a private doctor's office). They evaluate patients using physical examinations, laboratory tests, and other medical evaluations (such as ultrasounds, X-rays, or magnetic resonance imaging [MRI]). They also evaluate the need for genetic testing and determine which test is the most appropriate for each patient and, potentially, any at-risk family members. Clinical geneticists are also trained to counsel patients and their families about their risks related to any heritable conditions running through the family. They typically work as part of a heathcare team, alongside genetic counselors, nurses, and clinic managers. In addition, they often participate in the training of medical students, medical genetics residents or fellows, and genetic counseling graduate students.

Clinical geneticists are physicians who have completed a residency in medical genetics, generally after having received training in pediatrics, obstetrics and gynecology, maternal and fetal medicine, or internal medicine. In addition, to become a board-certified clinical geneticist, a physician needs to take and pass an extensive examination covering everything they might need to know about genetics. Clinical geneticists can then work in a practice that sees patients referred for a wide range of reasons, or they can focus on specific areas like prenatal genetics, metabolic genetics, neurogenetics, or cancer genetics.

Genetic counselor

Genetic counselors are healthcare professionals with training in both genetics and psychology. Although they can work in a variety of settings, genetic counselors generally work as part of a healthcare team to help patients who have a personal or family history of a genetic condition, or individuals who may have a higher risk of having a child with a genetic condition based on genetic testing. Key parts of a genetic counselor's job are to collect and analyze medical histories, obtain and interpret family histories, and help individuals make decisions that are best for themselves and their families. The genetic counselor usually works directly with the patient to assemble all their personal and family medical histories into a family tree and then looks for patterns to determine which traits or conditions may be hereditary. Genetic counselors can also calculate the likelihood that any given family member may have for conditions running in the family. They are trained to conduct careful and thorough interviews to make sure that no information is missed or left out.

Genetic counselors usually hold a master's degree, although some also have a doctoral degree in a related field. Training includes courses in genetics (including clinical, molecular, and population) and psychology, as well as many hours working with patients to hone interview and analysis skills (under the close supervision of experienced professionals, of course). The position requires excellent record-keeping skills and strict attention to detail. Genetic counselors also need to be good at interacting with all kinds of people, including research scientists and physicians. And the ability to communicate very well, both in writing and verbally, is a must. Like medical geneticists, board certification is necessary and requires passing an extensive examination that includes all aspects of genetics and genetic counseling. A genetic counseling license is also required in order to practice as a genetic counselor in certain states.

REMEMBER

One of the most essential skills of a genetic counselor is the ability to be nonjudgmental and nondirective. The counselor must be able to analyze a family history without bias or prejudice, inform the patient of his or her options, and help the patient make decisions without directing him or her to a particular course of action. Furthermore, the counselor must keep all information about his or her patients confidential, sharing information only with authorized personnel such as the person's own physician in order to protect the patient's privacy.

Genetic counseling assistant

A relatively new position in the field of genetics is that of a genetic counseling assistant. Genetic counseling assistants work alongside genetic counselors and help them with a variety of tasks. They may interact with patients, helping to obtain all the information necessary for a clinical visit, such as previous medical records and a patient's family medical history. They also perform office-related tasks, such as scheduling patients, gathering and organizing medical forms, and filling out paperwork. Genetic counseling assistants often have a bachelor's degree in a science- or medical-related field, and many are pursuing (or plan to pursue) a master's degree in genetic counseling.

TIP

GREAT GENETICS WEBSITES TO EXPLORE

The Internet is an unparalleled source of information about genetics. With just a few mouse clicks, you can find the latest discoveries and attend the best courses ever offered on the subject. Here's a quick sample.

- To see a great video that explains genetics and gives it a human face, check out "Cracking the Code of Life": `https://www.pbs.org/wgbh/nova/genome/program.html`.
- New discoveries are unveiled every day. To stay current, log on to `www.sciencedaily.com/news/plants_animals/genetics/` and `https://www.sciencenews.org/topic/genetics`.
- For students, `http://learn.genetics.utah.edu/` can't be beat. From the basics of heredity to virtual labs to cloning, it's all there in easy-to-grasp animations and language.
- Want to get all the details about genes and diseases? The Genetics Home Reference provides straightforward explanations on numerous topics: `https://ghr.nlm.nih.gov/`. You could also start at `https://www.ncbi.nlm.nih.gov/books/NBK22183/` for a review of the basics. More advanced (and greatly detailed) information is available at Online Mendelian Inheritance in Man (OMIM): `www.ncbi.nlm.nih.gov/omim/`.
- If you're interested in a career in genetics, the American Society for Human Genetics is ready to help: `https://www.ashg.org/education/career_flowchart.shtml`.

IN THIS CHAPTER

» Getting to know the cell

» Understanding the basics of chromosomes, DNA, and genes

» Exploring simple cell division

» Appreciating the complexities of meiosis

Chapter 2
Basic Cell Biology

Genetics and the study of how cells work are closely related. The process of passing genetic material from one generation to the next depends completely on how cells grow and divide. To reproduce, a simple organism such as bacteria or yeast simply copies its DNA (through a process called *replication*, which is covered later in Chapter 7) and splits in two. But organisms that reproduce sexually go through a complicated dance that includes mixing and matching strands of DNA (a process called *recombination*) and then halving the amount of DNA for special sex cells, allowing completely new genetic combinations for their offspring. These amazing processes are part of what makes you unique.

In this chapter, we provide a brief introduction to cell structure, DNA, and chromosomes. In addition, you need to be familiar with the processes of *mitosis* (cell division) and *meiosis* (the production of sex cells) to appreciate how genetics works. So come inside your cell and let us introduce you to the basics. Later in this book, we will spend more time on the details of DNA and chromosomes, since these topics lay the groundwork for all things in genetics.

Looking Around Your Cell

There are two basic kinds of organisms, distinguished by whether or not they have a *nucleus* (a compartment filled with DNA surrounded by a membrane):

- **Prokaryotes:** Organisms whose cells lack a nucleus and therefore have DNA floating loosely in the liquid center of the cell.
- **Eukaryotes:** Organisms that have a well-defined nucleus to house and protect the DNA.

The basic qualities of prokaryotes and eukaryotes are similar but not identical. Because all living things fall into these two groups, understanding the differences and similarities between cell types is important. In this section, you will learn how to distinguish the two kinds of cells from each other, and you will get a quick tour of the insides of cells — both with and without nuclei (plural of nucleus). Figure 2-1 shows you the structure of each type of cell.

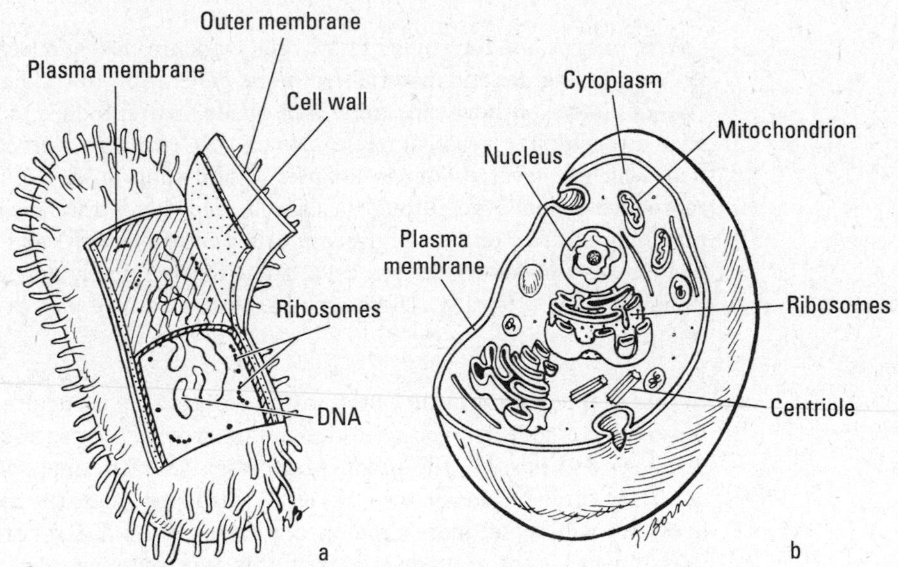

FIGURE 2-1: A prokaryotic cell (a) is very simple compared to a eukaryotic cell (b).

Cells without a nucleus

Scientists classify organisms composed of cells without nuclei as *prokaryotes*, which means "before nucleus." Prokaryotes are the most common forms of life on earth. You are, at this very moment, covered in and inhabited by trillions of prokaryotic cells: bacteria. Much of your life and your body's processes depend on these arrangements; for example, the digestion going on in your intestines is partially powered by bacteria that break down the food you eat. Most of the bacteria in your body are completely harmless, but some species of bacteria can be vicious and deadly, causing rapidly transmitted diseases such as cholera, a severe intestinal disease found most often in populations without a clean supply of water.

All bacteria, regardless of temperament, are simple, one-celled, prokaryotic organisms. None has cell nuclei, and all are small cells with relatively small amounts of DNA (you can flip to Chapter 8 for more on the amounts of DNA different organisms possess).

The exterior of a prokaryotic cell is encapsulated by a *cell wall* that serves as the bacteria's only protection from the outside world. A *plasma membrane* (*membranes* are thin sheets of molecules that act as a barrier) regulates the exchange of nutrients, water, and gases that nourish the bacterial cell. DNA, usually in the form of a single, hoop-shaped piece, floats around inside the cell; segments of DNA like this one are called *chromosomes* (see the section "Examining the basics of chromosomes" later in the chapter). The liquid interior of the cell is called the *cytoplasm.* The cytoplasm provides a cushiony, watery home for the DNA and other cell machinery that carry out the business of living.

Cells with a nucleus

Scientists classify organisms that have cells with nuclei as *eukaryotes,* which means "true nucleus." Eukaryotes range in complexity from simple, one-celled organisms to complex, multicellular organisms like you. Eukaryotic cells are fairly complicated and have numerous parts to keep track of (refer to Figure 2-1). Like prokaryotes, eukaryotic cells are held together by a *plasma membrane,* and sometimes a *cell wall* surrounds the membrane (plants, for example, have cell walls). But that's where the similarities end.

REMEMBER

The most important feature of the eukaryotic cell is the *nucleus* — the membrane-surrounded compartment that houses the DNA that's divided into one or more chromosomes. The nucleus protects the DNA from damage during day-to-day living. Eukaryotic chromosomes are usually long, string-like segments of DNA instead of the hoop-shaped ones found in prokaryotes. Another hallmark of eukaryotes is the way the DNA is packaged: Eukaryotes usually have much larger amounts of DNA than prokaryotes, and to fit all that DNA into the tiny cell nucleus, it must be tightly wound around special proteins. (For all the details about DNA packaging in eukaryotes, flip to Chapters 5 and 6.)

Unlike prokaryotes, eukaryotes have all sorts of cell parts, called *organelles*, that help each cell do what it's supposed to do. The organelles float around in the watery cytoplasm outside the nucleus. Two of the most important organelles are:

- **Mitochondria:** The powerhouses of the eukaryotic cell, mitochondria pump out energy by converting glucose to ATP (adenosine triphosphate). ATP acts like a battery of sorts, storing energy until it's needed for day-to-day living. All eukaryotes have mitochondria.
- **Chloroplasts:** These organelles are unique to plants and algae. They process the energy from sunlight into sugars needed to nourish the plant cells.

Eukaryotic cells are able to carry out behaviors that prokaryotes can't. For example, only eukaryotic cells are capable of ingesting fluids and particles for nutrition; prokaryotes must transport materials through their cell walls, a process that severely limits their dietary options.

In most multicellular eukaryotes, cells come in two basic varieties: body cells (called *somatic* cells) or sex cells. The two cell types have different functions and are produced in different ways.

Somatic cells

Somatic cells are produced by simple cell division called *mitosis* (see the section "Mitosis: Splitting Up" for details). Somatic cells of multicellular organisms like humans are differentiated into special cell types. Skin cells and muscle cells are both somatic cells, for instance, but if you were to examine your skin cells under a microscope and compare them with your muscle cells, you'd see that their structures are very different. The various cells that make up your body all have the same basic components (membrane, organelles, and so on), but the arrangements of the elements change from one cell type to the next so that they can carry out various jobs such as digestion (intestinal cells), energy storage (fat cells), or oxygen transport to your tissues (blood cells). Changes that occur only in the DNA of somatic cells (that is, changes that occur over the course of a person's lifetime) cannot be passed from parent to child.

Sex cells

Sex cells are specialized cells used for reproduction. Only eukaryotic organisms engage in sexual reproduction, which we cover in detail at the end of this chapter in the section "Mommy, where did I come from?" *Sexual reproduction* combines genetic material from two organisms and requires special preparation in the form of a reduction in the amount of genetic material allocated to sex cells — a process called *meiosis* (see "Meiosis: Making Cells for Reproduction" later in the chapter for an explanation). In humans, the two types of sex cells are eggs and sperm. Unlike somatic cells, changes in the DNA of sex cells can be passed from parent to child.

What's in a nucleus?

The nucleus of a eukaryotic organism is home to most of its DNA. (The mitochondria and chloroplasts also have DNA, which we discuss more in Chapter 5.) In humans, nearly all the body's cells have a nucleus filled with DNA (mature red blood cells do not). Each of your cells contains two copies of your *genome* – that is, all the DNA you inherited from your mother (minus the mitochondrial DNA) and all the DNA you inherited from your father. Your genome contains all the information necessary for the human body to grow and develop and for the various parts of the body to function every day.

The amount of information carried in your DNA is staggering. It is basically the "how-to guide" for your body. DNA is generally measured in base pairs (which we discuss in detail in Chapter 5). Your genome has approximately 3 billion base pairs. With two copies of your genome, the DNA from a *single cell* would measure just over 6 feet if stretched out and laid end-to-end!

In order to organize all that information and to fit it into each of your cells, your DNA is packaged into *chromosomes*. So, you can think of your genome as a multivolume how-to guide, with each chromosome being a volume that belongs to the set. Your genome is contained in 23 chromosomes, and each cell contains two sets of those 23 chromosomes (for a total of 46 chromosomes in every cell).

Examining the basics of chromosomes

Chromosomes are threadlike strands composed of DNA. To pass genetic traits from one generation to the next, the chromosomes must be copied (see Chapter 7), and then the copies must be divvied up. Most prokaryotes have only one circular chromosome that, when copied, is passed on to the *daughter cells* (new cells created by cell division). Eukaryotes have more complex problems to solve (like divvying up half the chromosomes to make sex cells), and their chromosomes behave differently during mitosis and meiosis. This section gets into the intricacies of chromosomes in eukaryotic cells because they're so complex.

Counting out chromosomes

Each eukaryotic organism has a specific number of chromosomes per cell — but it is different for each organism. For example, the human genome is contained into 23 chromosomes, while the fruit fly genome is contained in only 4 chromosomes. Chromosomes come in two varieties:

- **Sex chromosomes:** These chromosomes determine sex — male or female. Human cells contain two sex chromosomes. Females should have two X chromosomes, while males should have an X chromosome and a Y chromosome. (To find out more about how sex is determined by the X and Y chromosomes, flip to Chapter 6.)

- **Autosomal chromosomes:** *Autosomal* simply refers to non-sex chromosomes. Sticking with the human example, if you do the math, you can see that humans have 44 autosomal chromosomes.

Ah, but there's more. In humans, chromosomes come in pairs. That means you have 22 pairs of uniquely shaped autosomal chromosomes plus 1 pair of sex chromosomes for a total of 23 chromosome pairs. Your autosomal chromosomes are identified by numbers — 1 through 22 — and are generally arranged based on size, with chromosome 1 being the largest and chromosome 22 being the smallest. So you have two chromosome 1s, two 2s, and so on. Figure 2-2 shows you how all human chromosomes are divided into pairs and numbered. A *karyotype* like the one pictured in Figure 2-2 is one way chromosomes are examined; you discover more about karyotyping in Chapter 13.

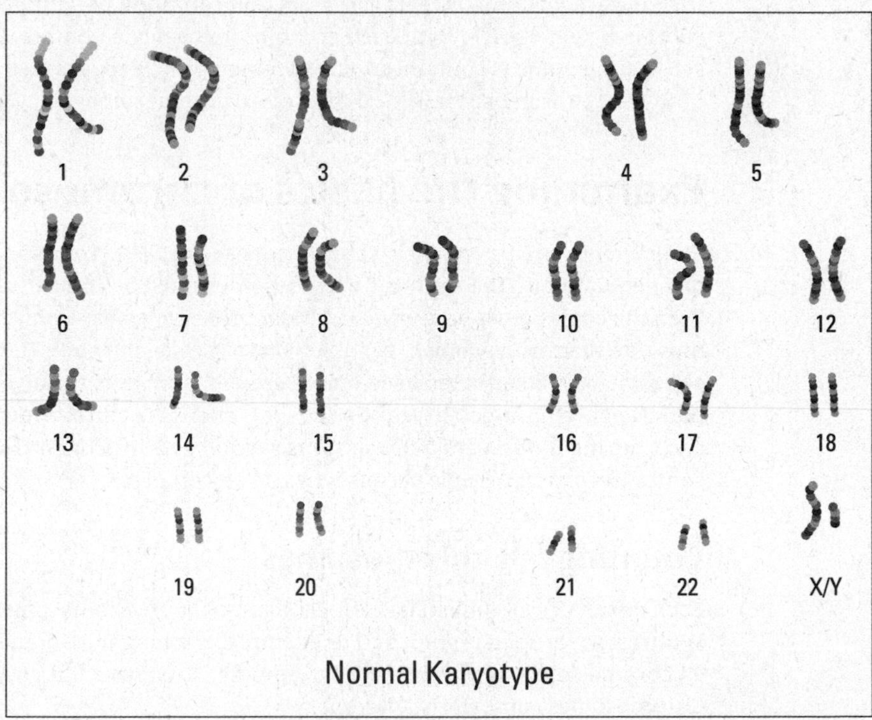

FIGURE 2-2: The 46 human chromosomes are divided into 23 pairs.

When chromosomes are sorted into pairs, the individual chromosomes in each pair are considered *homologous*, meaning that the paired chromosomes are identical to one another with respect to which genes they carry. In addition, your homologous chromosomes are identical in shape and size. These pairs of chromosomes are sometimes referred to as *homologs* for short.

Chromosome numbers can be a bit confusing. Humans are *diploid*, meaning we have two copies of each chromosome. Some organisms (like bees and wasps) have only one set of chromosomes (cells with one set of chromosomes are called *haploid*); other organisms have three, four, or as many as sixteen copies of each chromosome! The number of chromosome sets held by a particular organism is called the *ploidy*. For more on chromosome numbers, see Chapter 13.

The total number of chromosomes doesn't tell you what the ploidy of an organism is. For that reason, the number of chromosomes an organism has is often listed as some multiple of *n*. A single set of chromosomes referred to by the *n* is the haploid number. Humans are $2n = 46$, indicating that humans are diploid and their total number of chromosomes is 46. Human sex cells such as eggs or sperm are haploid (see "Mommy, where did I come from?" later in this chapter).

TECHNICAL STUFF

Geneticists believe that the homologous pairs of chromosomes in humans started as one set (that is, *haploid*), and the entire set was duplicated at some point in some distant ancestor, many millions of years ago.

Examining chromosome anatomy

Chromosomes are often depicted in stick-like forms, like those you see in Figure 2-3. Chromosomes don't look like sticks, though. In fact, most of the time they're loose and string-like. Chromosomes only take on this distinctive shape and form when cell division is about to take place (during metaphase of meiosis or mitosis). They're often drawn this way so that the special characteristics of eukaryotic chromosomes are easier to see. Figure 2-3 points out the important features of eukaryotic chromosomes. The part of the chromosome that appears pinched (located in the middle of the chromosomes) is called the *centromere*. The ends of the chromosomes are called *telomeres*. Telomeres are made of densely packed DNA and serve to protect the DNA messages that the chromosome carries.

So Where Are My Genes?

The differences in shapes and sizes of chromosomes are easy to see, but the most important differences between chromosomes are hidden deep inside the DNA. Chromosomes carry *genes* — sections of DNA that provide the building plans for specific proteins and that are associated with various traits. The genes tell the body how, when, and where to make all the structures that are necessary for the processes of living. In continuing the "how-to guide" analogy, each gene could be a chapter with a very specific set of instructions. Your genome contains approximately 22,000 genes, with each chromosome containing a varying number of genes, from fewer than 300 genes on the smallest chromosome (the Y chromosome) to more than 3000 genes on the largest (chromosome 1). The smallest human genes are only a few hundred base pairs in length. The largest is the gene that encodes the muscle protein dystrophin, which has 2.4 million base pairs.

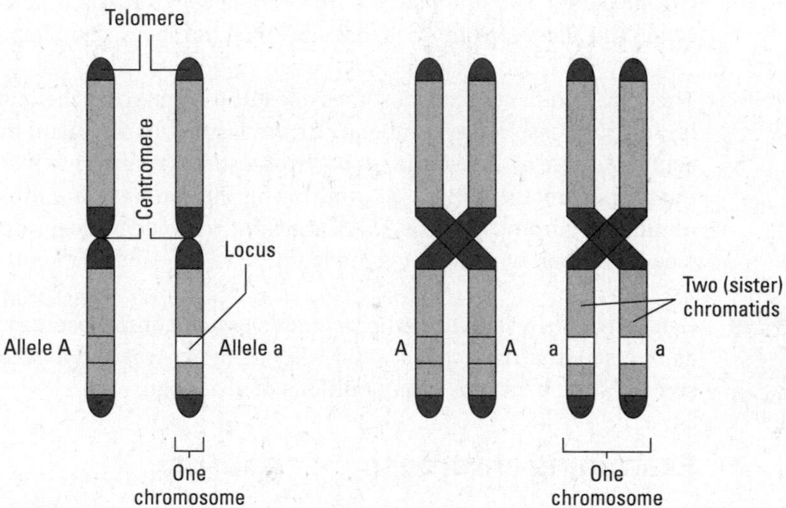

FIGURE 2-3: Basic structure of eukaryotic chromosomes.

You have two copies of each gene, with the exception of genes on the sex chromosomes in males (who have only one copy of most genes on the X and Y chromosomes). You inherit one copy of each gene from each parent. Each pair of homologous chromosomes carries the same — but not necessarily identical — genes. For example, both chromosomes of a particular homologous pair may contain a gene that controls hair color, but one can be a "brown hair" version of the gene and the other can be a "blond hair" version of the gene — alternative versions of genes are called *alleles* (refer to Figure 2-3). One of these alleles would be from your mother and the other from your father.

Any given gene can have one or more alleles. In Figure 2-3, one chromosome carries the allele *A* while its homolog carries the allele *a* (the relative size of a gene is very small relative to the whole chromosome; the alleles for a given gene are large here so you can see them). The alleles are associated with different physical traits *(phenotypes)* you see in organisms, like hair color or flower shape. You can find out more about how alleles affect phenotype in Chapter 3.

Each point along the chromosome is called a *locus* (Latin for "place"). The plural of locus is *loci* (pronounced *low*-sigh). Most of the phenotypes that you see are produced by multiple genes (that is, genes occurring at different loci and often on different chromosomes) acting together. For instance, human eye color is determined by at least three genes that reside on two different chromosomes. You can find out more about how genes are arranged along chromosomes in Chapter 11.

Mitosis: Splitting Up

Most cells have simple lifestyles: They grow, divide, and eventually die. Figure 2-4 illustrates the basic life cycle of a typical somatic, or body, cell.

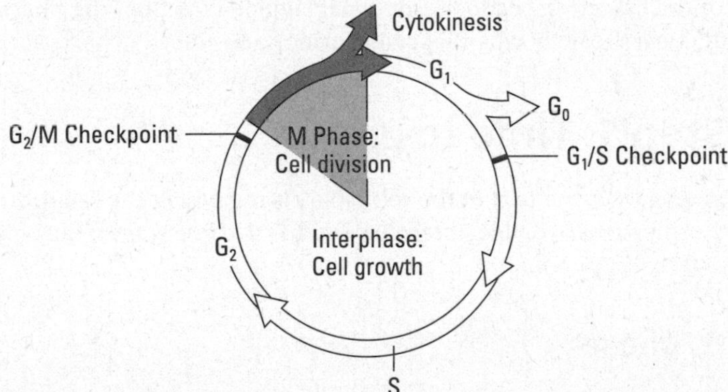

FIGURE 2-4: The cell cycle: mitosis, cell division, and all points in between.

The *cell cycle*, the stages a cell goes through from one division to another, is tightly regulated; some cells divide all the time, and others never divide at all. Your body uses mitosis to provide new cells when you grow and to replace cells that wear out or become damaged from injury. Talk about multitasking — you're going through mitosis right now, while you read this book! Some cells divide only part of the time, when new cells are needed to handle certain jobs like fighting infection. Cancer cells, on the other hand, get carried away and divide too often. In Chapter 14, you can find out how the cell cycle is regulated and what happens when it goes awry.

REMEMBER

The cell cycle includes *mitosis* (the process of dividing the cell nucleus) and cell division. The result of each round of the cell cycle is a simple cell division that creates two identical new cells from one original cell. During the cell cycle, all DNA present in the cell is copied (see Chapter 7), and when the original cell divides, a complete collection of all the chromosomes (in humans, 23 pairs) goes to each of the two resulting cells. Prokaryotes and some simple eukaryotic organisms use this process to reproduce themselves. (More complex eukaryotic organisms use *meiosis* for sexual reproduction, in which each of the two sex cells sends only one copy of each chromosome into the eggs or sperm. You can read all about that in the section "Meiosis: Making Cells for Reproduction," later in this chapter.)

REMEMBER

You should remember two important points about mitosis:

- **Mitosis produces two identical cells.** The new daughter cells are identical to each other *and* to the cell that divided to create them (the mother cell).
- **Cells created by mitosis have exactly the same number of chromosomes as the original cell did.** If the original cell had 46 chromosomes, the new cells each have 46 chromosomes.

Mitosis is only one of the major phases in the cell cycle; the other is *interphase.* In the following sections, you will be guided through the phases of the cell cycle and learn exactly what happens during each one.

Step 1: Time to grow

Interphase is the part of the cell cycle during which the cell grows, copies its DNA, and prepares to divide. Interphase occurs in three stages: the G_1 phase, the S phase, and the G_2 phase.

G_1 phase

When a cell begins life, such as the moment an egg is fertilized, the first thing that happens is the original cell starts to grow. This period of growth is called the *G_1 phase* of interphase, or the first gap phase. Lots of things happen during G_1, including physical growth of the cell and the creation of more organelles.

Some cells opt out of the cell cycle permanently, stop growing, and exit the process at G_0. Your brain cells, for example, have retired from the cell cycle. Mature red blood cells and muscle cells don't divide, either. In fact, human red blood cells have no nuclei and thus possess no DNA of their own.

If the cell in question plans to divide, though, it can't stay in G_1 forever. Actively dividing cells go through the whole cell cycle every 24 hours or so. After a predetermined period of growth that lasts from a few minutes to several hours, the cell arrives at the first checkpoint (refer to Figure 2-4). When the cell passes the first checkpoint, there's no turning back.

TECHNICAL STUFF

Various proteins control when the cell moves from one phase of the cycle to the next. At the first checkpoint, proteins called *cyclins* and enzymes called *kinases* control the border between G_1 and the next phase. Cyclins and kinases interact to cue up the various stages of the merry-go-round of cell division. Two proteins, G_1 cyclin and cyclin dependent kinase (CDK), hook up to escort the cell over the border from G_1 to S — the next phase.

S phase

S phase is the point at which the cell's DNA is replicated (here, *S* refers to *synthesis*, or copying, of the DNA). When the cell enters the S phase, activity around the chromosomes really steps up. All the chromosomes must be copied to make exact replicas that later are passed on to the newly formed daughter cells produced by cell division. DNA replication is a very complex process that gets full coverage in Chapter 7.

REMEMBER

For now, all you need to know is that all the cell's chromosomes are copied during the S phase, and the copies stay together as a unit (joined at the centromere) when the cell moves from the S phase into G_2 — the final step in interphase. The replicated chromosomes are called *sister chromatids* (refer to Figure 2-3), which are alike in every way. They carry the exact same copies of the exact same genes. During mitosis (or meiosis), the sister chromatids are divided up and sent to the daughter cells as part of the cell cycle.

G_2 phase

The *G_2 phase* leads up to cell division. It's the last phase before actual mitosis gets underway. G_2, sometimes called *Gap 2*, gives the cell time to get bigger before splitting into two smaller cells. Another set of cyclins and CDK work together to push the cell through the second checkpoint located at the border between G_2 and mitosis. (For details on the first checkpoint, see the previous section, "G_1 phase.") As the cell grows, the chromosomes, now copied and hooked together as sister chromatids, stay together inside the cell nucleus. (The DNA is still "relaxed" at this point and hasn't yet taken on the fat, sausage-shaped appearance it assumes during mitosis.) After the cell crosses the G_2/M checkpoint (refer to Figure 2-4), the business of mitosis formally gets underway.

Step 2: Divvying up the chromosomes

In the cell cycle, *mitosis* is the process of dividing up the newly copied chromosomes (that were created in interphase; see the preceding section) to make certain that the new cells each get a full set. Generally, mitosis is divided into four phases, which you can see in Figure 2-5 and read about in the following sections.

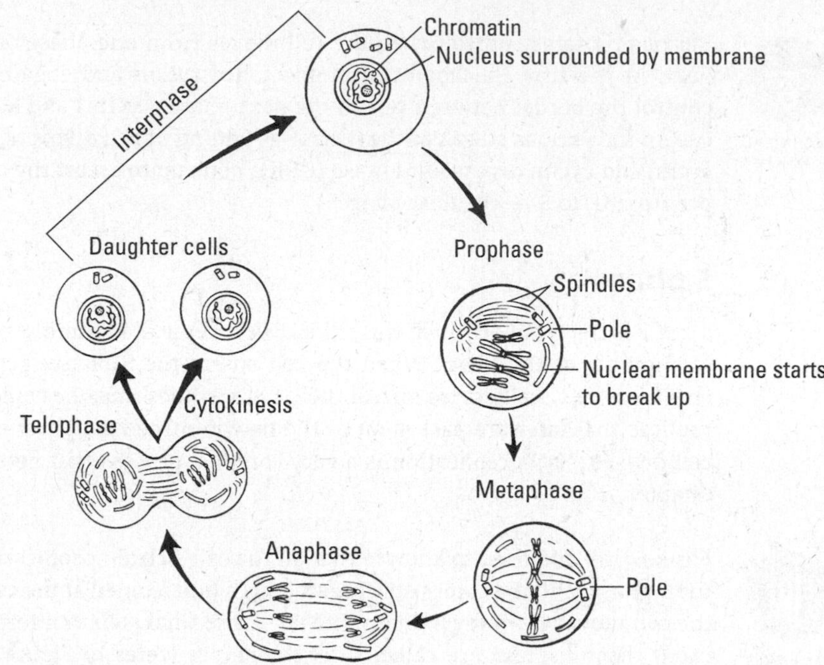

FIGURE 2-5: The process of mitosis, broken into four stages: prophase, metaphase, anaphase, and telophase.

TIP

The phases of mitosis are a bit artificial, because the movement doesn't stop at each point; instead, the chromosomes cruise right from one phase to the next. But dividing the process into phases is useful for understanding how the chromosomes go from being all mixed together to neatly parting ways and getting into the proper, newly formed cells.

Prophase

During *prophase*, the chromosomes get very compact and condensed, taking on the familiar sausage shape. During interphase (see the section "Step 1: Time to grow," earlier in this chapter), the DNA that makes up the chromosomes is tightly wound around special proteins, sort of like string wrapped around beads. The whole "necklace" is wound tightly on itself to compress the enormous DNA molecules to sizes small enough to fit inside the cell nucleus. But even when coiled during interphase, the chromosomes are still so threadlike and tiny that they're essentially invisible. That changes during prophase, when the chromosomes become so densely packed that you can easily see them with an ordinary light microscope.

TIP

By the time they reach prophase, chromosomes have duplicated to form sister chromatids (refer to Figure 2-3 earlier in this chapter). Sister chromatids of each chromosome are exact twin copies of each other. While the sister chromatids are attached during the early stages of mitosis, they are considered to be one chromosome. Once they are separated from each other in the later stages of mitosis, they will count as two separate chromosomes.

As the chromosomes or chromatids condense, the cell nucleus starts breaking up, allowing the chromosomes to move freely across the cell as the process of mitosis progresses.

Metaphase

Metaphase is the point when the chromosomes all line up in the center of the cell. After the nuclear membrane dissolves and prophase is complete, the chromosomes go from being a tangled mass to lining up in a more or less neat row in the center of the cell (refer to Figure 2-5). Threadlike strands called *spindles* grab each chromosome around its waist-like centromere. The spindles are attached to points on either side of the cell called *poles.*

TIP

Sometimes, scientists use geographic terms to describe the positions of chromosomes during metaphase: The chromosomes line up at the equator and are attached to the poles. This trick may help you better visualize the events of metaphase.

Anaphase

During *anaphase,* the sister chromatids are pulled apart, and the resulting halves migrate to opposite poles (refer to Figure 2-5). At this point, it's easy to see that the chromatids are actually chromosomes. Every sister chromatid gets split apart so that the cell that's about to be formed ends up with a full set of all the original cell's chromosomes.

Telophase

Finally, during *telophase,* nuclear membranes begin to form around the two sets of separated chromosomes (refer to Figure 2-5). The chromosomes begin to relax and take on their usual interphase form. The cell itself begins to divide as telophase comes to an end.

Step 3: The big divide

When mitosis is complete and new nuclei have formed, the cell divides into two smaller, identical cells. The division of one cell into two is called *cytokinesis* (*cyto* meaning "cell" and *kinesis* meaning "movement"). Technically, cytokinesis happens after metaphase is over and before interphase begins. Each new cell has a full set of chromosomes, just as the original cell did. All the organelles and cytoplasm present in the original cell are divided up to provide the new cell with all the machinery it needs for metabolism and growth. The new cells are now at interphase (specifically, the G_1 stage) and are ready to begin the cell cycle again.

Meiosis: Making Cells for Reproduction

Meiosis is a cell division that includes reducing the chromosome number as preparation for sexual reproduction. Meiosis reduces the amount of DNA by half so that when fertilization occurs, each offspring gets a full set of chromosomes. As a result of meiosis, the cell goes from being diploid to being haploid. Or, to put it another way, the cell goes from being $2n$ to being n. In humans, this means that the cells produced by meiosis (either eggs or sperm) have 23 chromosomes each — one copy of each of the homologous chromosomes. (See the section "Counting out chromosome numbers," earlier in this chapter, for more information.)

REMEMBER

Meiosis has many characteristics in common with mitosis. The stages go by similar names, and the chromosomes move around similarly, but the products of meiosis are completely different from those of mitosis. Whereas mitosis ends with two identical cells, meiosis produces *four* cells each with *half* the amount of DNA that the original cell contained. Furthermore, with meiosis, the homologous chromosomes go through a complex exchange of segments of DNA called *recombination.* Recombination is one of the most important aspects of meiosis and leads to genetic variation that allows each individual produced by sexual reproduction to be truly unique.

Meiosis goes through two rounds of division: meiosis I and the sequel, meiosis II. Figure 2-6 shows the progressing stages of both meiosis I and meiosis II. Unlike lots of movie sequels, the sequel in meiosis is *really necessary.* In both rounds of division, the chromosomes go through stages that resemble those in mitosis. However, the chromosomes undergo different actions in meiotic prophase, metaphase, anaphase, and telophase.

REMEMBER

Students often get stuck on the phases of meiosis and miss its most important aspects: recombination and the division of the chromosomes. To prevent that sort of confusion, we don't break down meiosis by phases. Instead, we focus on the activities of the chromosomes themselves.

In meiosis I:

- The homologous pairs of chromosomes line up side by side and exchange parts. This is called *crossing-over* or *recombination,* and it occurs during prophase I.
- During metaphase I, the paired homologous chromosomes line up at the equator of the cell (called the *metaphase plate*), and homologs go to opposite poles during the first round of anaphase.
- The cell divides after telophase I, reducing the amount of genetic material by half, and enters a second round of division — meiosis II.

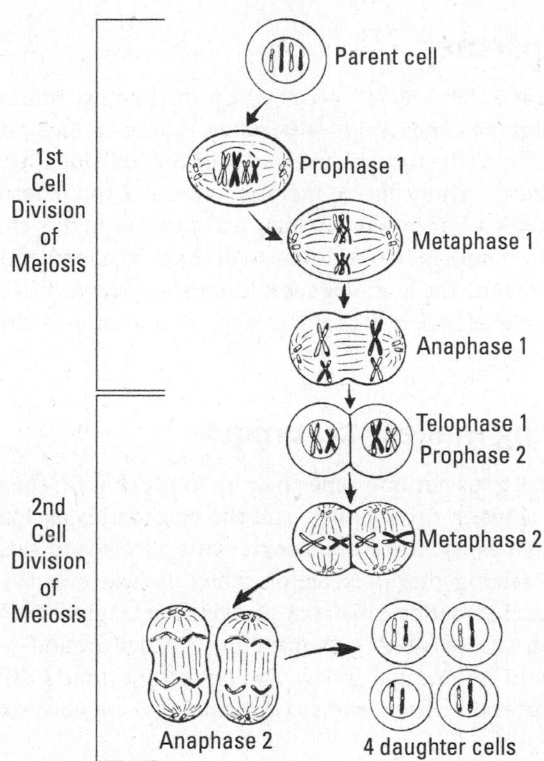

FIGURE 2-6: The phases of meiosis.

During meiosis II:

- The individual chromosomes (still present as attached sister chromatids) condense during prophase II and line up at the metaphase plates of both cells (metaphase II).
- The chromatids separate and go to opposite poles (anaphase II).
- The cells divide, resulting in a total of *four* daughter cells, each possessing *one* copy of each chromosome.

Meiosis I

Cells that undergo meiosis start in a phase similar to the interphase that precedes mitosis. The cells grow in a G_1 phase, undergo DNA replication during S, and prepare for division during G_2. (To review what happens in each of these phases, flip back to the section "Step 1: Time to grow.") When meiosis is about to begin, the chromosomes condense. By the time meiotic interphase is complete, the chromosomes have been copied and are hitched up as sister chromatids, just as they would be in mitosis. Next up are the phases of meiosis I, which we profile in the sections that follow.

Find your partner

During prophase I (labeled "I" because it's in the first round of meiosis), the homologous chromosomes find each other. These homologous chromosomes originally came from the mother and father of the individual whose cells are now undergoing meiosis. Thus, during meiosis, maternal and paternal chromosomes, as homologs, line up side by side. In Figure 2-2 earlier in this chapter, you can see an entire set of 46 human chromosomes. Although the members of the pair seem identical, they're not. The homologous chromosomes have different combinations of alleles at the thousands of loci along each chromosome (a single allele is illustrated in Figure 2-3).

Recombining makes you unique

When the homologous chromosomes pair up in prophase I, the chromatids of the two homologs actually zip together, and the chromatids exchange parts of their sequences. Enzymes cut the chromosomes into pieces and seal the newly combined strands back together in an action called *crossing-over*. When crossing-over is complete, the chromatids consist of part of their original DNA and part of their homolog's DNA. The loci don't get mixed up or turned around — the chromosome sequence stays in its original order. The only thing that's different is that the maternal and paternal chromosomes (as homologs) are now mixed together.

Figure 2-7 illustrates crossing-over in action. The figure shows one pair of homologous chromosomes and two loci. At both loci, the chromosomes have alternative forms of the genes. In other words, the alleles are different: Homolog one has *A* and *b*, and homolog two has *a* and *B*. When replication takes place, the sister chromatids are identical (because they're exact copies of each other). After crossing-over, the two sister chromatids have exchanged arms. Thus, each homolog has a sister chromatid that's different.

Partners divide

The recombined homologs line up at the metaphase equator of the cell (refer to Figure 2-6). The nuclear membrane begins to break down, and in a process similar to mitotic anaphase, spindle fibers grasp the homologous chromosomes by their centromeres and pull them to opposite sides of the cell.

At the end of the first phase of meiosis, the cell undergoes its first round of division (telophase I, followed by cytokinesis I). The newly divided cells each contain one set of chromosomes, the now partnerless homologs, still in the form of replicated sister chromatids.

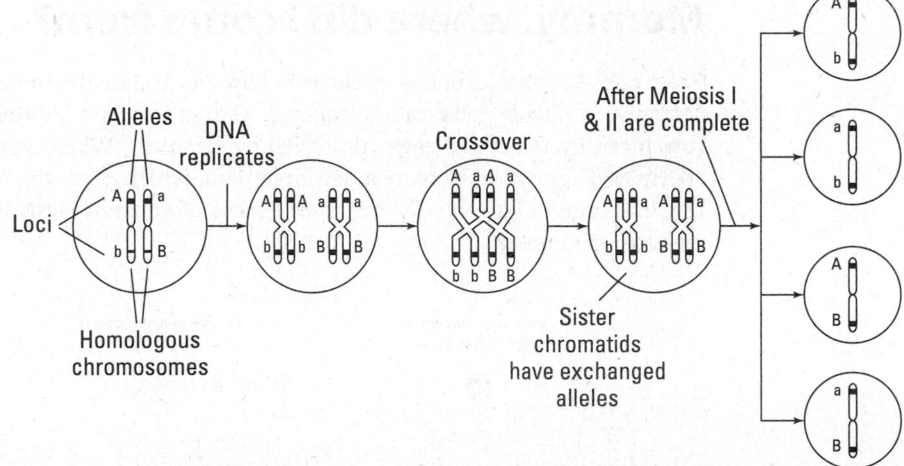

FIGURE 2-7: Crossing-over creates unique combinations of alleles during meiosis.

REMEMBER

When the homologs line up, maternal and paternal chromosomes pair up, but it's random as to which side of the equator each one ends up on. Therefore, each pair of homologs divides independently of every other homologous pair. This is the basis of the principle of independent assortment, which we cover in Chapters 3 and 4.

Following telophase I, the cells enter an in-between round called *interkinesis* (which means "between movements"). The chromosomes relax and lose their fat, ready-for-metaphase appearance. Interkinesis is just a "resting" phase in preparation for the second round of meiosis.

Meiosis II

Meiosis II is the second phase of cell division that produces the final product of meiosis: cells that contain only one copy of each chromosome. The chromosomes condense once more to their now-familiar fat sausage shapes. Keep in mind that each cell has only a single set of chromosomes, which are still in the form of sister chromatids.

During metaphase II, the chromosomes line up along the equator of the cells, and spindle fibers attach at the centromeres. In anaphase II, the sister chromatids are pulled apart and move to opposite poles of their respective cell. The nuclear membranes form around the now single chromosomes (telophase II). Finally, cell division takes place. At the end of the process, each of the four cells contains one single set of chromosomes (refer to Figure 2-6).

Mommy, where did I come from?

From gametogenesis, honey. Meiosis in humans and in all animals that reproduce sexually produces cells called *gametes*. Gametes come in the form of sperm (produced by males) or eggs (produced by females). When conditions are right, sperm and egg unite to create a new organism, which takes the form of a *zygote* (a fertilized egg). Figure 2-8 shows the process of *gametogenesis* (the production of gametes) in humans.

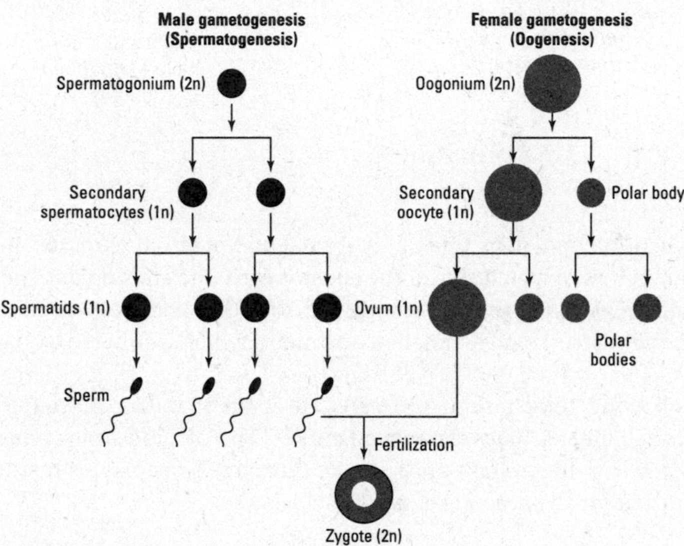

FIGURE 2-8: Gametogenesis in humans.

For human males, special cells in the male's sexual organs (testes) produce *spermatogonia*. Spermatogonia are $2n$ — they contain a full diploid set of 46 chromosomes (see the earlier section "Counting out chromosomes"). After meiosis I, each single spermatogonium has divided into two cells called *secondary spermatocytes*. These spermatocytes contain only one copy of each homolog (as sister chromatids). In meiosis II, the chromatids in the secondary spermatocytes separate and the cells divide to become *spermatids*. Spermatids then mature to become sperm cells with one copy of each chromosome. Thus, sperm cells are haploid and contain 23 chromosomes. Because males have X and Y sex chromosomes, half their sperm (men produce literally millions) contain Xs and half contain Ys.

Human females produce eggs in much the same way that men produce sperm. Egg cells, which are produced by the ovaries, start as diploid *oogonia* (that is, $2n = 46$). The big difference between egg and sperm production is that at the end of meiosis II, only one mature, haploid (23 chromosomes) sex cell (as an egg) is produced instead of four (refer to Figure 2-8). The other three cells produced are called *polar bodies;* the polar bodies aren't actual egg cells and can't be fertilized to produce offspring.

Why does the female body produce one egg cell and three polar bodies? Egg cells need large amounts of cytoplasm to nourish the zygote in the period between fertilization and when the mother starts providing the growing embryo with nutrients and energy through the placenta. The easiest way to get enough cytoplasm into the egg when it needs it most is to put less cytoplasm into the other three cells produced in meiosis II.

IN THIS CHAPTER

» Appreciating the work of Gregor Mendel

» Understanding inheritance, dominance, and segregation of alleles

» Solving basic genetics problems using probability

Chapter 3

Visualize Peas: Discovering the Laws of Inheritance

All the physical traits of any living thing originate in that organism's genes. Look at the leaves of a tree or the color of your own eyes. How tall are you? What color is your dog's or cat's fur? Can you curl or fold your tongue? Got hair on the backs of your fingers? All that and much more came from genes passed down from parent to offspring. Even if you don't know much about how genes work or even what genes actually are, you've probably already thought about how physical traits can be inherited. Just think of the first thing most people say when they see a newborn baby: Who does he or she look most like, mommy or daddy?

The *laws of inheritance* — how traits are transmitted from one generation to the next — were discovered less than 200 years ago. In the early 1850s, Gregor Mendel, an Austrian monk with a love of gardening, looked at the physical world around him and, by simply growing peas, categorized the patterns of genetic inheritance that are still recognized today. In this chapter, you discover how Mendel's peas changed the way scientists view the world. If you skipped Chapter 2, don't worry — Mendel didn't know anything about mitosis or meiosis when he formulated the laws of inheritance.

Mendel's discoveries have had an enormous impact on our understanding of the role genetics plays in the field of medicine. If you're interested in how genetics affects your health (Part 3), reading this chapter and getting a handle on the laws of inheritance will help you.

Gardening with Gregor Mendel

For centuries before Mendel planted his first pea plant, scholars and scientists argued about how inheritance of physical traits worked. It was obvious that *something* was passed from parent to offspring, because diseases and physical traits seemed to run in families. And farmers knew that by breeding plants and animals with certain physical features that they valued, they could create varieties that produced desirable products, like higher yielding maize, stronger horses, or hardier dogs. But just how inheritance worked and exactly what was passed from parent to child remained a mystery.

Enter the star of our gardening show, Gregor Mendel, an Austrian monk and botanist. Mendel was, by nature, a curious person. As he wandered around the gardens of the monastery where he lived in the mid-19th century, he noticed that his pea plants looked different from one another in a number of ways. Some were tall and others short. Some had green seeds, and others had yellow seeds. Mendel wondered what caused the differences he observed and decided to conduct a series of simple experiments. He chose seven characteristics of pea plants for his experiments, as you can see in Table 3-1:

TABLE 3-1 **Seven Traits of Pea Plants Studied by Gregor Mendel**

Trait	Common Form	Uncommon Form
Seed color	Yellow	Green
Seed shape	Round	Wrinkled
Seed coat color	Gray	White
Pod color	Green	Yellow
Pod shape	Inflated	Constricted
Plant height	Tall	Short
Flower position	Along the stem	At the tip of the stem

For ten years, Mendel patiently grew many varieties of peas with various seed colors, seed shapes, plant heights, and so on. In a process called *crossing*, he mated parent plants to see what their offspring would look like. When he passed away in 1884, Mendel was unaware of the magnitude of his contribution to science. A full 34 years passed after publication of his work (in 1868) before anyone realized what this simple gardener had discovered. (For the full story on how Mendel's research was lost and found again, flip to Chapter 21.)

If you don't know much about plants, understanding how plants reproduce may help you appreciate what Mendel did. To mate plants, you need flowers and the dusty substance they produce called *pollen* (the plant equivalent of sperm). Flowers have structures called *ovaries* (see Figure 3-1); the ovaries are hidden inside the *pistil* and are connected to the outside world by the *stigma.* Pollen is produced by structures called *stamen.* Like those of animals, the ovaries of plants produce eggs that, when exposed to pollen (in a process called *pollination*), are fertilized to produce seeds. Under the right conditions, the seeds sprout to become plants in their own right. The plants growing from seeds are the offspring of the plant(s) that produced the eggs and the pollen. Fertilization can happen in one of two ways:

- **Out-crossing:** Two plants are crossed, and the pollen from one can be used to fertilize the eggs of another.
- **Self-pollination (or selfing):** Some flowers produce both flowers and pollen, in which case the flower may fertilize its own eggs. Not all plants can self-fertilize, but Mendel's peas could.

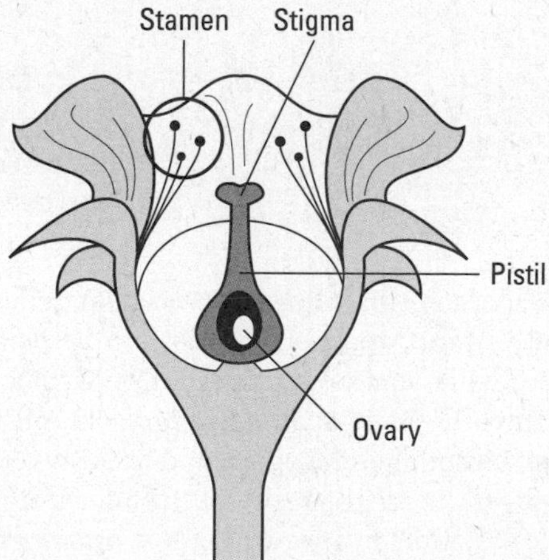

FIGURE 3-1: Reproductive parts of a flower.

Speaking the Language of Inheritance

You probably already know that genes are passed from parent to offspring and that somehow, genes are responsible for the physical traits (*phenotype*, such as hair color) you observe in yourself and the people and organisms around you. The simplest possible definition of a *gene* is an inherited factor that determines some trait; however, technically speaking, a gene is a segment of DNA that provides the instructions for making a specific protein.

REMEMBER

Genes come in different forms, or versions, called *alleles*. An individual's alleles determine the phenotype (physical trait) associated with that particular gene. The combination of alleles at a specific place in your DNA is known as a *genotype*. Genes occupy *loci* — specific locations along the strands of your DNA (*locus* is the singular form). Different traits (like hair texture and hair color) are determined by genes that occupy different loci, often on different chromosomes (see Chapter 2 for the basics of chromosomes). Take a look at Figure 3-2 to see how alleles are arranged in various loci along two pairs of generic chromosomes.

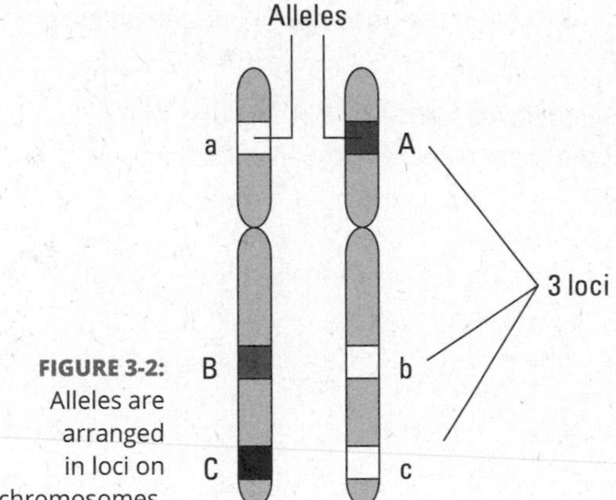

FIGURE 3-2: Alleles are arranged in loci on chromosomes.

In humans and many other organisms, alleles of particular genes come in pairs. If both alleles are identical in form, that locus is said to be *homozygous*, and the whole organism can be called a *homozygote* for that particular locus. If the two alleles aren't identical, then the individual is *heterozygous*, or a *heterozygote*, for that locus. Individuals can be both heterozygous and homozygous at different loci at the same time, and most traits are the result of the alleles at multiple loci. This is how all the phenotypic variation you see in a single organism is produced. For example, your eye color is controlled by several different genes (or loci), your hair color is controlled by a different set of genes, and your skin color by yet other set of genes. You can see how figuring out how complex sets of traits are inherited would be pretty difficult.

Simplifying Inheritance

When it comes to sorting out inheritance, it's easiest to start out with how one trait is transmitted from one generation to the next. This is the kind of inheritance, sometimes called *simple inheritance*, that Mendel started with when first studying his pea plants.

Mendel's choice of pea plants and the traits he chose to focus on had positive effects on his ability to uncover the laws of inheritance.

- **The original parent plants Mendel used in his experiments were true breeding.** When true breeders are allowed to self-fertilize, the exact same physical traits show up, unchanged, generation after generation. True-breeding tall plants always produce tall plants, true-breeding short plants always produce short plants, and so on. He knew this when he began his experiments.
- **Mendel studied traits that had only two forms, or *phenotypes,* for each characteristic (like short or tall).** He deliberately chose traits that were either one type or another, like tall or short, or green-seeded or yellow-seeded. Studying traits that come in only two forms made the inheritance of traits much easier to sort out. (Chapter 4 covers traits that have more than two phenotypes.)
- **Mendel worked only on traits whose genes were *not* located on the chromosomes that determine sex (explained in Chapter 6), and traits where one version of the associated gene is dominant over the other.** Of course, he did not know about chromosomes or genes or inheritance patterns when he devised his experiments. (Chapter 4 discusses more complicated forms of inheritance.)

Before his pea plants began producing pollen, Mendel opened the flower buds. He cut off either the pollen-producing part (the *stamen*) or the pollen-receiving part (the *stigma*) to prevent the plant from self-fertilizing. After the flower matured, he transferred pollen by hand — okay, not technically his hand; he used a tiny brush — from one plant (the "father") to another (the "mother"). Mendel then planted the seeds (the offspring) that resulted from this "mating" to see which physical traits each cross produced. The following sections explain the three laws of inheritance that Mendel discovered from his experiments.

Establishing dominance

For his experiments, Mendel crossed true-breeding plants that produced round seeds with true breeders that produced wrinkled seeds, crossed short true breeders with tall true breeders, and so on. Crosses of parent organisms that differ by only one trait, like seed shape or plant height, are called *monohybrid crosses.* Mendel patiently moved pollen from plant to plant, harvested and planted seeds, and observed the results after the offspring plants matured. His plants produced literally thousands of seeds, so his garden must have been quite a sight.

To describe Mendel's experiments and results, the parental generation is referred to by the letter *P*. The first generation of offspring is referred to as *F1*. If F1 offspring are mated to each other (or allowed to self-fertilize), the next generation is referred to as *F2* (see Figure 3-3 for the generation breakdown).

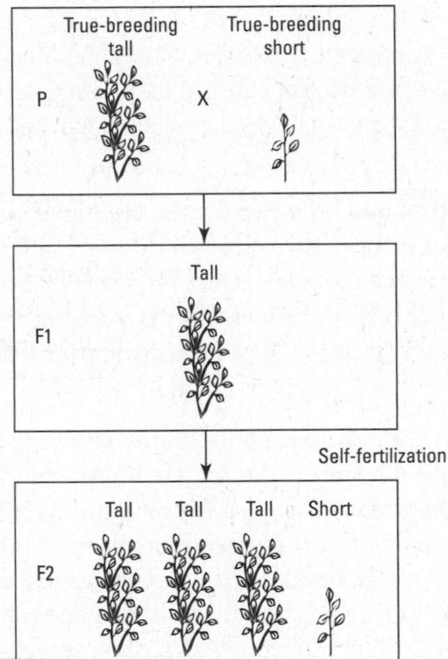

FIGURE 3-3: Monohydrid crosses illustrate how simple inheritance works.

REMEMBER

The results of Mendel's experiments were amazingly consistent. In every case when he mated true breeders of different phenotypes, all the F1 offspring had the same phenotype as one or the other parent plant. For example, when Mendel crossed a true-breeding tall parent with a true-breeding short parent, *all* the F1 offspring were tall. This result was surprising because until then, many people thought inheritance was a blending of the characteristics of the two parents — Mendel had expected his first-generation offspring to be medium height.

If Mendel had just scratched his head and stopped there, he wouldn't have learned much. But he allowed the F1 offspring to self-fertilize, and something interesting happened: About 25 percent of the F2 offspring were short, and the rest, about 75 percent, were tall (refer to Figure 3-3).

From that F2 generation, when allowed to self-fertilize, his short plants were true breeders — all produced short progeny. His F2 tall plants produced both tall and short offspring. About one-third of his tall F2s bred true as tall. The rest produced tall and short offspring in a 3:1 ratio (that is, ¾ tall and ¼ short; refer to Figure 3-3).

After thousands of crosses, Mendel came to the accurate conclusion that the factors that determine seed shape, seed color, pod color, plant height, and so on are acting sets of two. He reached this understanding because *one* phenotype showed up in the F1 offspring, but *both* phenotypes were present among the F2 plants. The result in the F2 generation told him that whatever it was that controlled a particular trait (such as plant height) had been present but somehow hidden in the F1 offspring.

REMEMBER

Mendel quickly figured out that certain traits seem to act like rulers, or dominate, other traits. *Dominance* means that one factor masks the presence of another. Round seed shape dominated wrinkled. Tall height dominated short. Yellow seed color dominated green. Mendel rightly determined the genetic principle of *dominance* by strictly observing phenotype in generation after generation and cross after cross. When true tall and short plants were crossed, each F1 offspring got one height-determining factor from each parent. Because tall is *dominant* over short, all the F1 plants were tall. Mendel found that the only time *recessive* characters (traits that are masked by dominant traits) were expressed was when the two factors were alike, as when short plants self-fertilized.

Segregating alleles

Segregation is when things get separated from each other. In the genetic sense, what's separated are the two factors — the alleles of the gene (one from each parent) — that determine phenotype. Figure 3-4 traces the segregation of the alleles for seed color through three generations. The shorthand for describing alleles is typically a capital letter for a dominant trait and the same letter in lowercase for a recessive trait. In this example, the capital letter *Y* is used for the dominant allele that makes yellow seeds; the lowercase *y* is used for the recessive allele that, when homozygous, makes seeds green. This shorthand is generally used for situations in where there are only two alleles. To see an example in which there are multiple alleles and phenotypes, see Chapter 4.

REMEMBER

The letters or symbols you use for various alleles and traits are arbitrary. Just make sure you're consistent in how you use letters and symbols, and don't get them mixed up.

In the segregation example featured in Figure 3-4, the parents (in the P generation) are homozygous (which is what makes them true breeding for that trait). Each individual parent plant has a certain genotype — a combination of alleles — that determines its phenotype. Because pea plants are *diploid* (meaning they have two copies of each gene; see Chapter 2), the genotype of each plant is described using two letters. For example, a true-breeding yellow-seeded plant would have the homozygous genotype YY, and green-seeded plants are homozygous yy. The *gametes* (sex cells, as in pollen or eggs) produced by each plant bear only one allele. (Sex cells are *haploid*; see Chapter 2 for the details on how meiosis produces haploid gametes.) Therefore, the true breeders can produce gametes of only one type — YY plants can only make Y gametes and yy plants can only produce y gametes. When a Y pollen and a y egg (or vice versa, y pollen and Y egg) get together, they make a Yy offspring — this is the heterozygous F1 generation.

REMEMBER

The bottom line of the principle of segregation is this parsing out of the pairs of alleles into gametes. Each gamete gets one and only one allele for each locus; this is the result of homologous chromosomes parting company during the first round of meiosis (see Chapter 2 for more on how chromosomes split during meiosis). When the F1 generation self-fertilizes (to create the F2 generation), each plant produces two kinds of gametes: Half are Y, and the other half are y. Segregation makes four combinations of zygotes possible: YY, Yy, yY, or yy. (Yy and yY look redundant, but they're genetically significant because they represent different contributions [y or Y] from each parent.) Phenotypically, Yy, yY, and YY all look alike: yellow seeds. Only yy makes green seeds. The ratio of genotypes is 1:2:1 (¼ homozygous dominant: ½ heterozygous: ¼ homozygous recessive), and the ratio of phenotypes is 3 to 1 (dominant phenotype to recessive phenotype).

If allowed to self-fertilize in the F3 generation, yy parents make yy offspring, and YY parents produce only YY offspring. The Yy parents again make YY, Yy, and yy offspring in the same ratios observed in the F2: ¼ YY, ½ Yy, and ¼ yy.

Scientists now know that what Mendel saw acting in sets of two were genes. Single pairs of genes (that is, one locus) control each trait. That means that plant height is at one locus (controlled by one gene), seed color at a different locus, seed shape at a third locus, and so on.

Declaring independence

As Mendel learned more about how traits were passed from one generation to the next, he carried out experiments with plants that differed in two or more traits. He discovered that the traits behaved independently — that is, that the inheritance of plant height had no effect on the inheritance of seed color, for example.

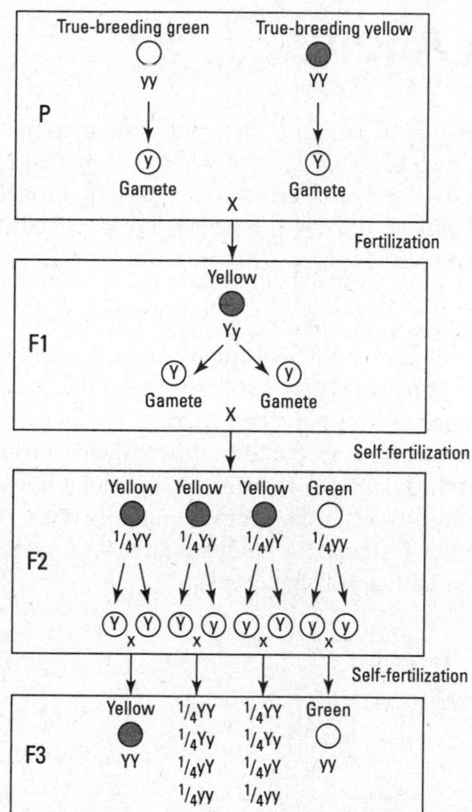

FIGURE 3-4: The principles of segregation and dominance as illustrated by three generations of pea plants with green and yellow seeds.

REMEMBER

The independent inheritance of traits is called the *law of independent assortment* and is a consequence of meiosis. When homologous pairs of chromosomes separate, they do so randomly with respect to each other. The movement of each individual chromosome is independent with respect to every other chromosome. It's just like flipping a coin: As long as the coin isn't rigged, one coin flip has no effect on another — each flip is an independent event. Genetically, what this random separation amounts to is that alleles on different chromosomes are inherited independently.

Segregation and independent assortment are closely related principles. *Segregation* tells you that alleles at the same locus on pairs of chromosomes separate and that each offspring has the same chance of inheriting a particular allele from a parent. *Independent assortment* means that every offspring also has the same opportunity to inherit any allele at any other locus (although this rule does have some exceptions, which are described in Chapter 4).

Finding Unknown Alleles

Mendel crossed parent plants in many different combinations to work out the identity of the hidden factors (which we now know as genes) that produced the phenotypes he observed. One type of cross was especially informative. A *testcross* is when any individual with an unknown genotype is crossed with a true-breeding individual with the recessive phenotype (in other words, a homozygote).

Each cross provides different information about the genotypes of the individuals involved. For example, Mendel could take any plant with any phenotype and test-cross it with a true-breeding recessive plant to find out which alleles the plant of unknown genotype carried. Here's how the testcross would work: A plant with the dominant phenotype, violet flowers, could be crossed with a true-breeding white flowered plant (ww). If the resulting offspring all had violet flowers, Mendel knew that the unknown genotype was homozygous dominant (WW). In Figure 3-5, you see the results of another testcross: A heterozygote (Ww) testcross yielded offspring of half white and half violet phenotypes.

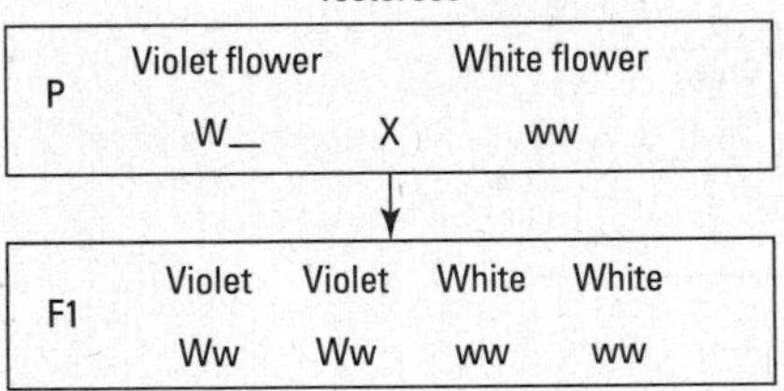

FIGURE 3-5: The results of testcrosses reveal unknown genotypes.

Applying Basic Probability to the Likelihood of Inheritance

Predicting the results of crosses is easy, because the rules of probability govern the likelihood of getting particular outcomes. The following are two important rules of probability that you should know:

- **Multiplication rule:** Used when the probabilities of events are independent of each other — that is, the result of one event doesn't influence the result of another. In other words, this rule is used with you want to know that probability of X happening *AND* Y happening at the same time, when X and Y are independent events. The combined probability of both events occurring is the product of the events, so you multiply the probabilities.

» **Addition rule:** Used when you want to know the probability of one event occurring as opposed to a second event that cannot occur at the same time (that is, a second mutually exclusive event). Put another way, you use this rule when you want to know the probability of one *OR* another event happening, but not both.

BEATING THE ODDS WITH GENETICS

When you try to predict the outcome of a certain event, like a coin flip or the sex of an unborn child, you're using probability. For many events, the probability is either-or. For instance, a baby can be either male or female, and a coin can land either heads or tails. Both outcomes are considered equally likely; as long as the coin isn't rigged somehow, you have a 50:50 chance of either. For many events, however, determining the likelihood of a certain outcome is more complicated. Deciding how to calculate the odds depends on what you want to know.

Take, for example, predicting the sex of several children born to a given couple. The probability of any baby being a boy is ½, or 50 percent. If the first baby is a boy, the probability of the second child being a boy is still 50 percent, because the events that determine sex are independent from one child to the next (see Chapter 2 for a rundown of how meiosis works to produce gametes for sex cells). That means the sex of one child has no effect on the sex of the next child. But if you want to know the probability of having two boys in a row, you multiply the probability of each independent event together: ½ × ½ = ¼, or 25 percent. If you want to know the probability of having two boys *or* two girls, you add the probabilities of the events together: ¼ (the probability of having two boys) + ¼ (the probability of having two girls) = ½, or 50 percent.

Genetic counselors use probability to determine the likelihood that someone has inherited a given condition and the likelihood that a person will pass on a condition if he or she has it or carries a gene mutation that can cause it. For example, a man and woman are each a carrier for a recessive disorder (each carries a gene mutation that can cause it, but both are unaffected since they also carry a normal copy of the gene as well), such as cystic fibrosis. For recessive conditions, both copies of the associated gene need to have a mutation in them for the person to be affected. The counselor can predict the likelihood that the couple will have an affected child. Just as in Mendel's flower crosses, each parent can produce two kinds of gametes, affected (has the gene mutation) or unaffected (does not have the gene mutation). The man produces half-affected and half-unaffected gametes, as does the woman. The probability that any child inherits an affected allele from the mom *and* an affected allele from the dad is ¼ (that's ½ × ½). The probability that a child will be affected *and* female is ⅛ (that's ¼ × ½). The probability that a child will be affected *or* a boy is ¾ (that's ¼ + ½).

For more details about the laws of probability, check out the sidebar "Beating the odds with genetics."

Here's how you apply the addition and multiplication rules for monohybrid crosses (crosses of parent organisms that differ only by one trait). Suppose you have two pea plants. Both plants have violet flowers, and both are heterozygous (Ww). Each plant will produce two sorts of gametes, W and w, with equal probability — that is, half of the gametes will be W and half will be w for each plant. To determine the probability of a certain genotype resulting from the cross of these two plants, you use the multiplication rule and multiply probabilities. For example, what's the probability of getting a heterozygote (Ww) from this cross?

Because both plants are heterozygous (Ww), the probability of getting a W from plant one is ½, and the probability of getting a w from plant two is also ½. The word *and* tells you that you need to multiply the two probabilities to determine the probability of the two events happening together. So, $\frac{1}{2} \times \frac{1}{2} = \frac{1}{4}$. But there's another way to get a heterozygote from this cross: Plant one could contribute the w, and plant two could contribute the W. The probability of this turn of events is exactly equal to the first scenario: $\frac{1}{2} \times \frac{1}{2} = \frac{1}{4}$. Thus, you have two equally probable ways of getting a heterozygote: wW or Ww. The word *or* tells you that you must add the two probabilities together to get the total probability of getting a heterozygote: $\frac{1}{4} + \frac{1}{4} = \frac{1}{2}$. Put another way, there's a 50 percent probability of getting heterozygote offspring when two heterozygotes are crossed.

Solving Simple Genetics Problems

Every genetics problem, from those on an exam to one that determines what coat color your dog's puppies may have, can be solved in the same manner. Here's a simple approach to any genetics problem:

1. **Determine how many traits you're dealing with.**
2. **Count the number of phenotypes.**
3. **Carefully read the problem to identify the question.** Do you need to calculate the ratios of genotype (for example, YY, Yy, or yy) or phenotype (yellow or green)? Are you trying to determine something about the parents or the offspring?
4. **Look for words that mean *and* and *or* to help determine which probabilities to multiply (*and*) and which to add (*or*).**

Deciphering a monohybrid cross

Imagine that you have your own garden full of the same variety of peas that Mendel studied. After reading this book, filled with enthusiasm for genetics, you rush out to examine your pea plants, having noticed that some plants are tall and others short. You know that last year you had one tall plant (which self-fertilized) and that this year's crop consists of the offspring of last year's one tall parent plant. After counting plants, you discover that 77 of your plants are tall, and 26 are short. What was the genotype of your original plant? What is the dominant allele?

You have two distinct phenotypes (tall and short) of one trait — plant height. You can choose any symbol or letter you please, but often, geneticists use a letter like *t* for short and then capitalize that letter for the other allele (here, *T* for tall).

One way to start solving the problem of short versus tall plants is to determine the ratio of one phenotype to the other. To calculate the ratios, add the number of offspring together: 77 + 26 = 103, and divide to determine the proportion of each phenotype: 77 ÷ 103 = 0.75, or 75 percent are tall. To verify your result, you can divide 26 by 103 to see that 25 percent of the offspring are short, and 75 percent plus 25 percent gives you 100 percent of your plants.

From this information alone, you've probably already realized (thanks to simple probability) that your original plant must have been heterozygous and that tall is dominant over short. As explained in the "Segregating alleles" section earlier in this chapter, a heterozygous plant (Tt) produces two kinds of gametes (T or t) with equal probability (that is, half the time the gametes have the dominant T allele and the other half they have the recessive t allele). The probability of getting a homozygous dominant (TT) genotype is $\frac{1}{2} \times \frac{1}{2} = \frac{1}{4}$ (that's the probability of getting T twice: T once *and* T a second time, like two coin flips in a row landing on heads). The probability of getting a heterozygous dominant (T and t, *or* t and T) is $\frac{1}{2} \times \frac{1}{2} = \frac{1}{4}$ (to get Tt) plus $\frac{1}{2} \times \frac{1}{2} = \frac{1}{4}$ (tT). The total probability of a plant with the dominant phenotype (TT *or* Tt *or* tT) is $\frac{1}{4} + \frac{1}{4} + \frac{1}{4} = \frac{3}{4}$. With 103 plants, you'd expect 77.25 (on average) of them to show the dominant phenotype — which is essentially what you observed.

Tackling a dihybrid cross

To become more comfortable with the process of solving simple genetics problems, you can tackle a problem that involves more than one trait: a *dihybrid cross*.

Here's the problem scenario. (If you're a rabbit breeder, please forgive this oversimplification.) Your roommate moves out and leaves behind two bunnies (you were feeding them anyway, and they're cute, so you don't mind). One morning you wake to find that your bunnies are now parents to a litter of babies.

- One is gray and has long fur.
- Two are black and have long fur.
- Two are gray and have short fur.
- Seven look just like the parents: black with short fur.

Besides the meaningful lesson about spaying and neutering pets, what can you discover about the genetics of coat color and hair length of your rabbits?

First, how many traits are you dealing with? You're dealing with two traits: color of fur and length of fur. Each trait has two phenotypes: Fur can be black or gray, and length of fur can be long or short.

The simplest method is to examine one trait at a time — in other words, look at the monohybrid crosses. (Jump back to the section "Deciphering a monohybrid cross" for a refresher.)

Both parents have short fur. How many of their offspring have short fur? Nine of twelve, and 9 ÷ 12 = ¾, or 75 percent. That means there are three short-haired bunnies to every one long-haired bunny.

Being identical in phenotype, the parents both have black coats. How many babies have black coats? Nine of the twelve. There's that comfortingly familiar ratio again! The ratio of black to gray is 9 to 3 (or 3 to 1).

From your knowledge of monohybrid crosses, you've probably guessed that the parent rabbits are heterozygous for coat color and, at the same time, are heterozygous for fur length. To be sure, you can calculate the probability of certain genotypes and corresponding phenotypes of offspring for two rabbits that are heterozygous at two loci (see Figure 3-6).

The phenotypic ratio observed in the rabbits' offspring (7:2:2:1) is consistent with what we would expect for a typical F2 generation in a dihybrid cross (9:3:3:1; refer to Figure 3-6), given that you don't have enough animals for it to be 9:3:3:1. The rarest phenotype is the one that's recessive for both traits; in this case, long hair and gray color are both recessive. The most common phenotype is the one that's dominant for both traits. The fact that seven of your twelve baby rabbits are black with short fur tells you that the probability of getting a particular allele for color and a particular allele for coat length is the product of two independent events. Coat color and hair length are coded by genes that are inherited independently — as you would expect under the principle of independent assortment.

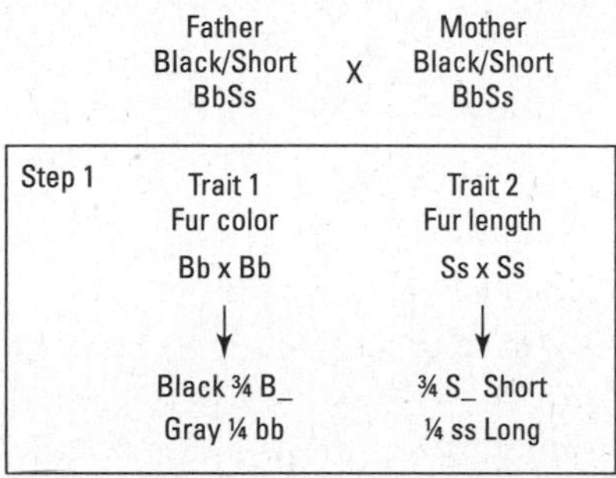

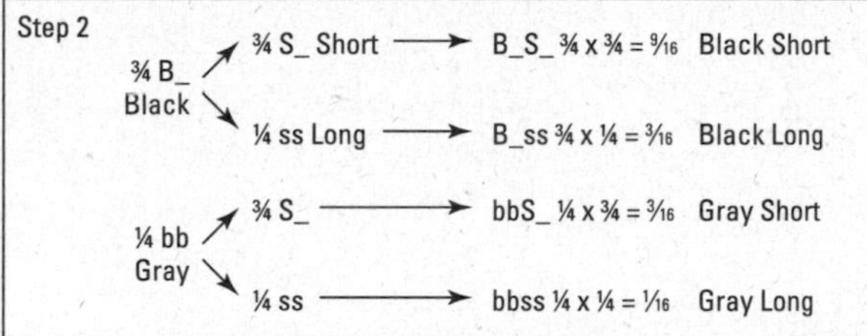

FIGURE 3-6: Genotypes and phenotypes resulting from a simple dihybrid cross.

IN THIS CHAPTER

- » Examining the variations of dominant alleles
- » Reviewing how simple inheritance becomes more complicated
- » Looking at some exceptions to Mendel's laws

Chapter 4
Law Enforcement: Mendel's Laws Applied to Complex Traits

Although nearly 150 years have elapsed since Gregor Mendel cultivated his pea plants (introduced in Chapter 3), the observations he made and the conclusions he drew still accurately describe how genes are passed from parent to offspring. The basic laws of inheritance — dominance, segregation, and independent assortment — continue to stand the test of time.

However, inheritance isn't nearly as simple as Mendel's experiments suggest. Dominant alleles don't always dominate, and genes aren't always inherited independently. Some genes mask their appearances, and some alleles can kill. This chapter explains exactly how Mendel was right, and wrong, about the laws of inheritance and how they're enforced.

Dominant Alleles Rule . . . Sometimes

If Mendel had chosen a plant other than the pea plant for his experiments, he may have come to some very different conclusions. The traits that Mendel studied show *simple dominance* — when the dominant allele's *phenotype,* or physical trait (a yellow seed, for example), masks the presence of the recessive allele. The recessive phenotype (a green seed in this example) is only expressed when both alleles are recessive, which is written as *yy.* (See Chapter 3 for the definitions of commonly used genetics terms such as *allele, recessive,* and *homozygote.*) But not all alleles behave neatly as dominant-recessive. Some alleles show incomplete dominance and, therefore, seem to display a blend of phenotypes from the parents. This section tells you how dominant alleles rule the roost — but only part of the time.

Wimping out with incomplete dominance

A trip to the grocery store can be a nice genetics lesson. Take eggplant, for example. Eggplant comes in various shades of (mostly) purple skin that are courtesy of a pair of alleles at a single locus interacting in different ways to express the phenotype — purple fruit color. Dark purple and white colors are both the result of homozygous alleles. Dark purple is homozygous for the dominant allele (PP), and white is homozygous for the recessive allele (pp). When crossed, dark purple and white eggplants yield light purple offspring — the intermediate phenotype. This intermediate color is the result of the allele for purple being incomplete in its dominance of the allele for white (which is actually the allele for no color).

REMEMBER

With *incomplete dominance,* the alleles are inherited in exactly the same way they always are: One allele comes from each parent. The alleles still conform to the principles of segregation and independent assortment, but the way those alleles are expressed (the phenotype) is different. (You can find out about exceptions to the independent assortment rules in the section "Genes linked together," later in this chapter.)

Here's how the eggplant cross works: The parent plants are PP (for purple) and pp (for white). The F1 generation is all heterozygous (Pp), just as you'd expect from Mendel's experiments (see Chapter 3). If this were a case of simple dominance, all the Pp F1 generation would be dark purple. But in this case of incomplete dominance, the F1 generation comes out light purple (sometimes called violet). The heterozygotes produce a less purple pigment, making the offspring lighter in color than homozygous purple plants.

In the F2 generation (the result of crossing Pp with Pp), half the offspring have violet fruits (corresponding with the Pp genotype). One-quarter of the offspring are dark purple (PP) and one-quarter are white (pp) — these are the homozygous offspring. Rather than the 3:1 phenotypic ratio (three dark purple eggplants and one white eggplant) you'd expect to see with simple dominance, with incomplete dominance, you see a 1:2:1 ratio (one dark purple eggplant, two light purple eggplants, and one white eggplant) — the exact ratio of the underlying genotype (PP, Pp, Pp, pp).

Keeping it fair with codominance

When alleles share equally in the expression of their phenotypes, the inheritance pattern is considered *codominant.* Both alleles are expressed fully as phenotypes instead of experiencing some intermediate expression (like what's observed in incomplete dominance).

You can see a good example of codominance in human blood types. If you've ever donated blood (or received a transfusion), you know that your blood type is extremely important. If you receive the wrong blood type during a transfusion, you can have a fatal allergic reaction. Blood types are the result of proteins that add various types of sugar tags (*antigens*) to your blood cells that allow your body to recognize the cells as a normal part of you.

Your antigens determine your blood type (see Figure 4-1). Two alleles code for antigens (type A antigen and type B antigen). Both act in a dominant manner and can lead to Type A or Type B blood, respectively. When a person inherits one allele of each (heterozygous), they will make both antigens simultaneously and in equal amounts (codominance), and therefore have Type AB blood. The situation with blood types gets more complicated by the presence of a third allele. The third allele that is recessive (O) is a broken version of the gene and does not attach any antigen to the red blood cells. If you have two copies of this broken allele (homozygous recessive) then you would have Type O blood. So, blood types in humans show two sorts of inheritance:

- Codominance (for A and B);
- Dominant-recessive (A or B paired with the O allele).

Type O is only expressed in the homozygous state. For more information on multiple alleles, check out the section "More than two alleles," later in this chapter.

	Group A	Group B	Group AB	Group O
Phenotype (Blood type)	A	B	AB	O
Antigens present	A antigen	B antigen	A and B antigens	None
Genotype	AA or AO	BB or BO	AB	OO

FIGURE 4-1: Human blood type.

Dawdling with incomplete penetrance

Some dominant alleles don't express their influence consistently. When dominant alleles are present but fail to show up as a phenotype, the condition is termed *incompletely penetrant. Penetrance* is the probability that an individual having a dominant allele will show the associated phenotype. *Complete penetrance* means every person having the allele shows the phenotype. Most dominant alleles have 100 percent penetrance — that is, the phenotype is expressed in every individual possessing the allele. However, other alleles may show reduced, or incomplete, penetrance, meaning that individuals carrying the allele have a reduced probability of having the trait.

Penetrance of disease-causing alleles like those responsible for certain cancers or other hereditary disorders complicate matters in genetic testing (see Chapter 15 to find out more about genetic testing for disease). For example, changes in the *BRCA1* gene, which is associated with breast and ovarian cancer, are incompletely penetrant. Studies estimate that approximately 70 percent of women carrying a disease-causing allele will develop cancer by the age of 70. Therefore, genetic tests indicating that someone carries the allele only point to increased risk, not a certainty of getting the disease. Women carrying a disease-causing allele need to be screened regularly for early signs of the disease, when treatment can be most effective, and may be offered prophylactic treatment before any signs of cancer appear.

TECHNICAL STUFF

Geneticists usually talk about penetrance in terms of a percentage. In this example, the breast cancer gene is 70 percent penetrant.

Regardless of penetrance, the degree to which an allele expresses the phenotype may differ from individual to individual; this variable strength is called *expressivity*. One trait with variable expressivity that shows up in humans is *polydactyly*, the condition of having more than ten fingers or toes. In persons with polydactyly, the expressivity of the trait is measured by the completeness of the extra digits — some people have tiny skin tags, and others have fully functional extra fingers or toes.

Alleles Causing Complications

The variety of forms that genes take (as alleles) accounts for the enormous diversity of physical traits you see in the world around you. For example, many alleles exist for eye color and hair color. In fact, several genes contribute to most phenotypes. Dealing with multiple loci and many alleles at each locus complicates inheritance patterns and makes them harder to understand. For many disorders, scientists don't fully understand the form of inheritance because variable expressivity and incomplete penetrance mask the patterns. Additionally, multiple alleles can interact as incompletely dominant, codominant, or dominant-recessive (see "Dominant Alleles Rule . . . Sometimes," earlier in this chapter, for the whole story). This section explains how various alleles of a single gene can complicate inheritance patterns.

More than two alleles

When it came to his pea plant research, Mendel deliberately chose to study traits that came in only two flavors. For instance, his peas had only two flower color possibilities: white and purple. The allele for purple in the common pea plant is fully dominant, so it shows up as the same shade of purple in both heterozygous and homozygous plants. In addition to being fully dominant, purple is completely penetrant, so every single plant that inherits the gene for purple flowers has purple flowers.

If Mendel had been a rabbit breeder instead of a gardener, his would likely be a different story. He may not have earned the title *Father of Genetics* because the broad spectrum of rabbit coat colors would make most anyone simply throw up their hands.

To simplify matters, consider one gene for coat color in bunnies. The *C* gene has four alleles that control the amount of pigment produced in the hair shaft. These four alleles give you four rabbit color patterns to work with. The various rabbit color alleles are designated by the letter *c* with superscripts:

- **Brown (c^+):** Brown rabbits are considered *wild-type,* which is generally considered the "normal" phenotype. Brown rabbits are brown all over.
- **Albino (c):** Rabbits homozygous for this color allele don't produce any pigment at all. Therefore, these white rabbits are considered *albino.* They have all-white coats, pink eyes, and pink skin.
- **Chinchilla (c^{ch}):** Chinchilla rabbits are solid gray (specifically, they have white hair with black tips).
- **Himalayan (c^h):** Himalayan rabbits are white but have dark hair on their feet, ears, and noses.

TECHNICAL STUFF

Wild-type is a bit of a problematic term in genetics. Generally, wild-type is considered the "normal" phenotype, and everything else is "mutant." *Mutant* is simply different, an alternative form that's not necessarily harmful. Wild-type tends to be the most common phenotype and is usually dominant over other alleles. You're bound to see wild-type used in genetics books to describe phenotypes such as eye color in fruit flies. Though rare, the mutant color forms occur in natural populations of animals. In the case of domestic rabbits, color forms other than brown are the product of breeding programs specifically designed to obtain certain coat colors.

REMEMBER

Although a particular trait can be determined by a number of different alleles (as in the four allele possibilities for rabbit coat color), any particular animal carries only two alleles at a particular locus at one time.

The C gene in rabbits exhibits a dominance hierarchy common among genes with multiple alleles. Wild-type is completely dominant over the other three alleles, so any rabbit having the c^+ allele will be brown. Chinchilla is incompletely dominant over Himalayan and albino. That means heterozygous chinchilla/Himalayan rabbits are gray with dark ears, noses, and tails. Heterozygous chinchilla/albinos are lighter than homozygous chinchillas. Albino is only expressed in animals that are homozygous (cc).

The color alleles in monohybrid crosses for rabbit color follow the same rules of segregation and independent assortment that apply to the pea plants that Mendel studied (see Chapter 3). The phenotypes for rabbit color are just more complex. For example, if you were to cross an albino rabbit (cc) with a homozygous chinchilla ($c^{ch}c^{ch}$), the F1 generation would all be heterozygous cc^{ch} and all have the same phenotype (light chinchilla). In the F2 generation (cc^{ch} mated with cc^{ch}) you'd get the expected 1:2:1 genotypic ratio (1 cc to 2 cc^{ch} to 1 $c^{ch}c^{ch}$), and the phenotypes would show a corresponding 1:2:1 ratio (one albino, two light chinchilla, one full chinchilla).

TECHNICAL STUFF

A total of five genes control coat color in rabbits. The section "Genes in hiding," later in this chapter, delves into how multiple genes interact to create fur color.

Lethal alleles

REMEMBER

Many alleles express unwanted traits (phenotypes) that indirectly cause suffering and death (such as the excessive production of mucus in the lungs of cystic fibrosis patients). Rarely, alleles may express a *lethal phenotype* — that is, death — immediately and thus are never expressed beyond the zygote. These alleles produce a 1:2 phenotypic ratio, because only heterozygotes and homozygous nonlethals survive to be counted.

The first lethal allele that scientists described was associated with yellow coat color in mice. Yellow mice are *always* heterozygous. When yellow mice are bred to other yellow mice, they produce yellow and non-yellow offspring in a 2:1 ratio, because all homozygous yellow mice die as embryos. Homozygous yellow has no real phenotype (beyond dead), because these animals never survive.

Lethal alleles are almost always recessive, and thus are expressed only in homozygotes. One notable exception is the gene that causes Huntington disease. Huntington disease (also known as Huntington chorea) is inherited as an autosomal dominant lethal disorder, meaning that persons with Huntington disease develop a progressive nerve disorder that causes involuntary muscle movement and loss of mental function. Huntington disease is expressed in adulthood and is always fatal. It has no cure and treatment is aimed at alleviating symptoms of the disease.

Making Life More Complicated

Many phenotypes are determined by the action of more than one gene at a time. Genes can hide the effects of each other, and sometimes one gene can control several phenotypes at once. This section looks at how genes make life more complicated (and more interesting).

When genes interact

If you don't mind returning to the produce section of your local grocery store (no more eggplants, we promise), you can observe the interaction of multiple genes to produce various colors of bell peppers. Two genes (R and C) interact to make these mild, sweet peppers appear red, brown, yellow, or green. You see four phenotypes as the result of two alleles at each locus.

Figure 4-2 shows the genetic breakdown of bell peppers. In the parental generation (P), you start with a homozygous dominant pepper (RRCC), which is red, crossed with a homozygous recessive (rrcc) green pepper. (This is a dihybrid cross — that is, one involving two genes — like the one described at the end of Chapter 3.) You can easily determine the expected genotypic ratios by considering each locus separately. For the F1 generation, that's really easy to do, because both loci are heterozygous (RrCc). Just like homozygous dominant peppers, fully heterozygous peppers are red. When the F1 peppers self-fertilize, the phenotypes of brown and yellow show up.

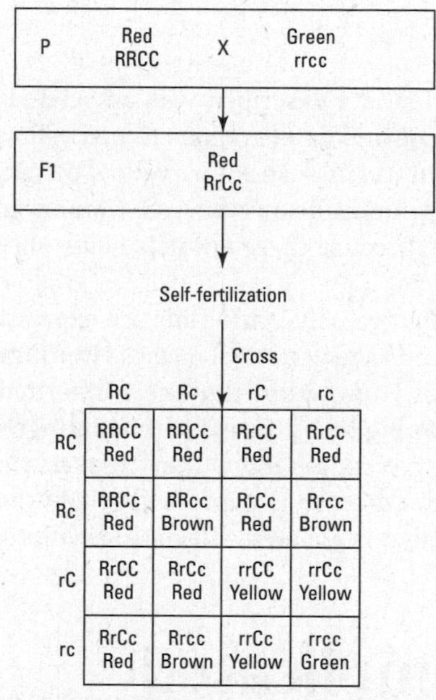

Conclusion: 9⁄16 Red 3⁄16 Brown 3⁄16 Yellow 1⁄16 Green

FIGURE 4-2: Genes interact to produce pigment in this dihybrid cross for pepper color.

Brown pepper color is produced by the genotype R_cc. The blank means that the R locus must have at least one dominant allele present to produce color, but the other allele can be either dominant or recessive. Yellow is produced by the combination rrC_. To make yellow pigment, the C allele must be either heterozygous dominant or homozygous dominant with a recessive homozygous R allele. The F2 generation shows the familiar 9:3:3:1 dihybrid phenotypic ratio (just like the rabbits do in Chapter 3). The loci assort independently, just as you'd expect them to.

Genes in hiding

As the preceding section explains, in pepper color, the alleles of two genes interact to produce color. But sometimes, genes hide or mask the action of other genes altogether. This occurrence is called *epistasis.*

A good example of epistasis is the way in which color is determined in horses. Like that of dogs, cats, rabbits, and humans, hair color in horses is determined by numerous genes. At least seven loci determine color in horses. To simplify mastering epistasis, you tackle the actions of only three genes: W, E, and A (see Table 4-1 for a rundown of the genes and their effects). One locus (W) determines the presence or absence of color. Two loci (E and A) interact to determine the distribution of red and black hair — the most common hair colors in horses.

A horse that carries one dominant allele for W will be albino — no color pigments are produced, and the animal has white skin, white hair, and pink eyes. (Homozygous dominant for the white allele is lethal; therefore, no living horse possesses the WW genotype.) All horses that are some color other than white are homozygous recessive (ww). (If you're a horse breeder, you know that this is really oversimplified here. Please forgive us.) Therefore, the dominant allele W shows *dominant epistasis* because it masks the presence of other alleles that determine color.

If a horse isn't white (that is, not albino), then two main genes are likely determining its hair color: E and A. When the dominant allele E is present, the horse has black hair (it may not be black all over, but it's black somewhere). Black hair is expressed because the E locus controls the production of two pigments, red and black. EE and Ee horses produce both black and red pigments. Homozygous recessive (ee) horses are red; in fact, they're always red regardless of what's happening at the A locus. Thus, ee is *recessive epistatic,* which means that in the homozygous recessive individual, the locus masks the action of other loci. In this case, the production of black pigment is completely blocked.

When a horse has at least one dominant allele at the E locus, the A locus controls the amount of black produced. The A locus (also called *agouti,* which is a dark brown color) controls the production of black pigments. A horse with the dominant A allele produces black only on certain parts of its body (often on its mane, tail, and legs — a pattern referred to as *bay*). Horses that are aa are simply black. However, the homozygous recessive E locus (*ee*) masks the A locus entirely (regardless of genotype), blocking black color completely.

This example of the genetics of horsehair color proves that the actions of genes can be complex. In this one example, you see a lethal allele (W) along with two other loci that can each mask the other under the right combination of alleles.

This potential explains why it can be so difficult to determine how certain conditions are inherited. Epistasis can act along with reduced penetrance to create extremely elusive patterns of inheritance — patterns that often can only be worked out by examining the DNA itself.

TABLE 4-1 **Genetics of Hair Color in Horses**

Genotype	Phenotype	Type of Epistasis	Effect
WW__	Lethal	No epistasis	Death
Ww__	Albino	Dominant	Blocks all pigments
wwE_aa	Black	Recessive	Blocks red
wwE_A_	Bay or brown	No epistasis	Both red and black expressed
wwee__	Red	Recessive	Blocks black

Genes linked together

Roughly 30 years after Mendel's work was rediscovered in 1900 and verified by the scientific community (see Chapter 21 for the whole story), the British geneticist Ronald A. Fisher realized that Mendel had been exceptionally lucky — either that or he'd cheated. Of the many, many traits Mendel could have studied, he published his results on seven traits that conform to the laws of segregation and independent assortment, have two alleles, and show dominant-recessive inheritance patterns. Fisher asserted that Mendel must have published the part of his data he understood and left out the rest. (After Mendel died, all his papers were burned, so we'll never know the truth.) The rest would include all the parts that make inheritance messy, like epistasis and *linkage*.

REMEMBER

Because of the way genes are situated along chromosomes, genes that are very close together spatially (that is, fewer than 50 million base pairs apart; see Chapter 5 to see how DNA is measured in base pairs) are more likely to be inherited together. When genes are so close together that they're inherited together (either all or part of the time), the genes are said to be *linked* (see Figure 4-3). The occurrence of linked genes means that not all genes are subject to independent assortment. To determine if genes are linked, geneticists carry out a process called *linkage analysis.*

The process of linkage analysis is really a determination of how often *recombination* (the mixing of information, also called *crossing-over,* contained on the two homologous chromosomes; see Chapter 2) occurs between two or more genes.

If genes are close enough together on the chromosome, they end up being linked more than 50 percent of the time. However, genes on the same chromosome can behave as if they were on different chromosomes, because during the first stage of meiosis (see Chapter 2), crossing-over occurs at many points along the two homologous chromosomes. If crossing-over splits two loci up more than 50 percent of the time, the genes on the same chromosome appear to assort independently, as if the genes were on different chromosomes altogether.

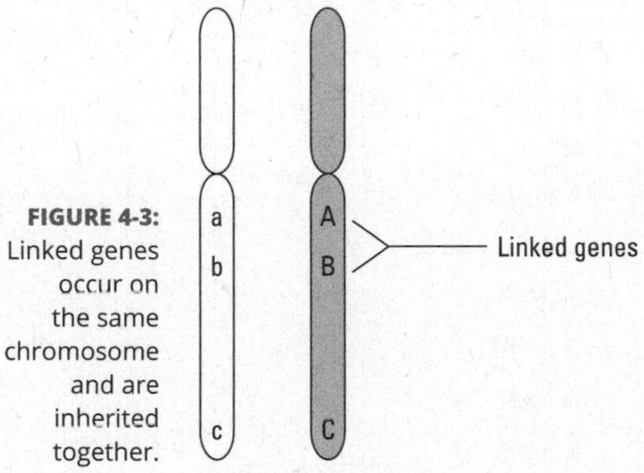

FIGURE 4-3: Linked genes occur on the same chromosome and are inherited together.

Generally, geneticists perform linkage analysis by examining dihybrid crosses (dihybrid means two loci; see Chapter 3) between a heterozygote and a homozygote. If you want to determine the linkage between two traits in fruit flies, for example, you choose an individual that's AaBb and cross it with one that's aabb. If the two loci, A and B, are assorting independently, you can expect to see the results shown in Figure 4-4. The heterozygous parent produces four types of gametes — AB, aB, Ab, and ab — with equal frequency. The homozygous parent can only make one sort of gamete — ab. Thus, in the F1 offspring, you see a 1:1:1:1 ratio.

But what if you see a completely unexpected ratio, like the one shown in Table 4-2? What does that mean? These results indicate that the traits are linked.

As you can see in Figure 4-5, the dihybrid parent makes four sorts of gametes. Even though the loci are on the same chromosome, the gametes don't occur in equal frequency. Most of the gametes show up just as they do on the chromosome, but crossover occurs between the two loci roughly 20 percent of the time, producing the two rarer sorts of gametes (each is produced about 10 percent of the time). Crossover occurs with roughly the same frequency in the homozygous parent, too, but because the alleles are the same, the results of those crossover events are invisible. Therefore, you can safely ignore that part of the problem.

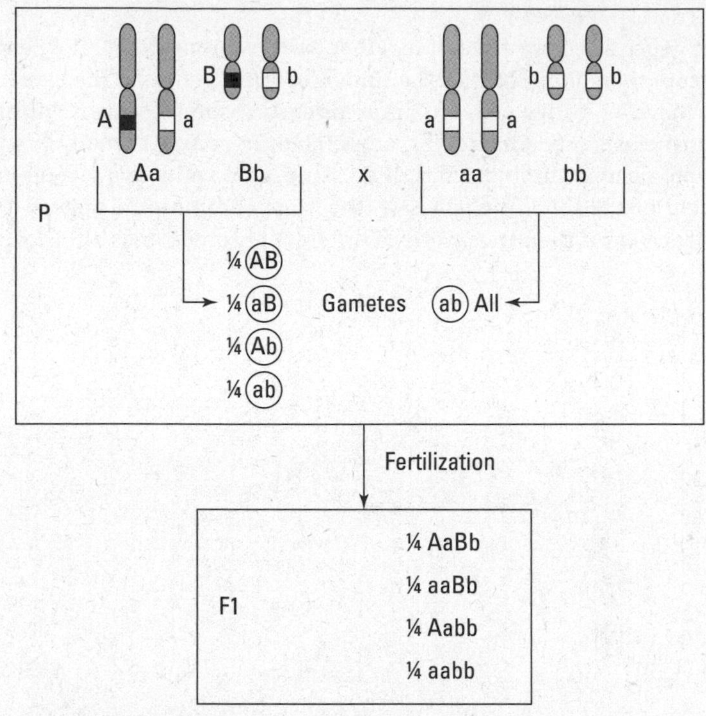

FIGURE 4-4: Typical results of a dihybrid testcross when traits assort independently.

TABLE 4-2

Linked Traits in a Dihybrid Testcross

Genotype	Number of Offspring	Proportion
Aabb	320	40%
aaBb	318	40%
AaBb	80	10%
Aabb	76	10%

TECHNICAL STUFF

To calculate *map distance,* or the amount of crossover, between two loci, you divide the total number of recombinant offspring by the total number of offspring observed. The *recombinant offspring* are the ones that have a genotype different from the parental genotype. This calculation gives you a proportion: percent recombination. One map unit distance on a chromosome is equal to 1 percent recombination. Generally, one map unit is considered to be 1 million base pairs long.

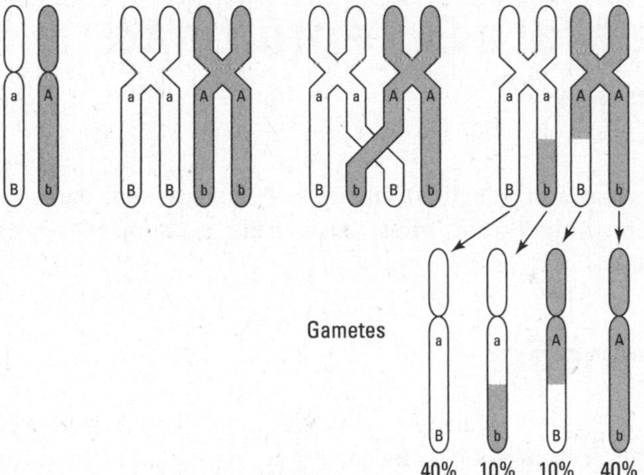

FIGURE 4-5: A dihybrid cross with linked genes.

TECHNICAL STUFF

As it turns out, genes for four of the traits Mendel studied were situated together on chromosomes. Two genes were on chromosome 1, and two were on chromosome 4; however, the genes were far enough apart that recombination was greater than 50 percent. Thus, all four traits appeared to assort independently, just as they would have if they'd been on four different chromosomes.

One gene with many phenotypes

Certain genes can control more than one phenotype that would appear to be completely unrelated. Genes that control multiple phenotypes are *pleiotropic.* Pleiotropy is very common; almost any major single gene disorder listed in the Online Mendelian Inheritance in Man database (`https://ncbi.nlm.nih.gov/omim`) shows pleiotropic effects.

Take, for example, phenylketonuria (PKU). This disease results from changes in a single gene and is inherited in an autosomal recessive manner (the gene is located on an autosome [non-sex chromosome] and disease-causing changes in both copies of the gene need to be present in order to have the condition). When persons with the homozygous recessive phenotype of PKU consume substances containing phenylalanine, their bodies lack the proper biochemical pathway to break down the phenylalanine into tyrosine. As a result, phenylalanine accumulates in the body, preventing normal brain development. The primary phenotype of persons with untreated PKU is intellectual disability, but the impaired biochemical pathway affects other phenotypic traits as well. Thus, PKU patients may also exhibit light hair color, skin problems, and seizures. All the phenotypic traits associated with PKU are associated with the single gene defect rather than the actions of more than one gene.

Uncovering More Exceptions to Mendel's Laws

As inheritance of genetic disorders is better studied, many exceptions to strict Mendelian inheritance rules arise. This section addresses four important exceptions.

Epigenetics

One of the biggest challenges to Mendel's laws comes from a phenomenon called *epigenetics.* The prefix *epi-* means "over" or "above." In epigenetics, organisms with identical alleles (including identical twins) may exhibit different phenotypes.

The difference in phenotypes doesn't come from changes in the genes themselves but from elsewhere in the chemical structure of the DNA molecule (you can find out all about DNA's chemical and physical structure in Chapter 5). What happens is that tiny chemical tags, called *methyl groups,* are attached to the DNA. In essence, the tags act like the operating system in your computer that tells the programs how often to work, where, and when. In the case of epigenetics, the tags can shut genes down or turn genes on. Not only that, but the tags can be passed from parent cell to daughter cell, as well as inherited from one generation to the next.

Some epigenetic effects are normal and useful: They control how your various cells look and behave, like the differences between a heart muscle cell and a skin cell. However, other tags act like mutations and cause diseases like cancer. (You can discover more about the role DNA plays in cancer in Chapter 14.) Epigenetics is an exciting area of genetics research that will yield answers to how the genetic code in your DNA is affected by aging, your environment, and much more.

Genomic imprinting

Genomic imprinting is a special case of epigenetics. When traits are inherited on autosomal chromosomes, they're generally expressed equally in males and females. In some cases, the sex of the parent who contributes the particular allele may determine whether that allele is expressed; this is called *genomic imprinting.* If a gene is in an *imprinted* region on a chromosome, it may be either turned on or turned off depending on which parent that chromosome came from.

Sheep breeders in Oklahoma discovered an amusing example of genomic imprinting. A ram named Solid Gold had unusually large hindquarters for his breed. Eventually, Solid Gold sired other sheep, which also had very large . . . butts. The breed was named Callipyge, which is Greek for *beautiful butt.* It turns out that six genes affect rump size in sheep. As breeders mated Callipyge sheep, it quickly became clear that the trait didn't obey Mendel's rules. Eventually, researchers determined that the big rump phenotype resulted only when the father passed on the trait. Callipyge ewes can't pass their big rumps on to their offspring.

The reasons behind genomic imprinting are still unclear. However, it is known that changes in the DNA that affect imprinting can cause serious developmental disorders. For example, children with deletion of a certain region of chromosome 15 (a region we now know to be imprinted so that different genes are expressed based on which parent the gene came from) will develop either Prader-Willi syndrome or Angelman syndrome, depending on whether that deletion occurred in either the chromosome inherited from the father (Prader-Willi) or the chromosome inherited from the mother (Angelman). Prader-Willi syndrome is typically characterized by intellectual disabilities, compulsive eating leading to obesity, a shorter than average height, and problems with puberty and infertility. In contrast, Angelman syndrome is typically characterized by significant intellectual disabilities, seizures, severe speech problems, and difficulties with movement and balance.

Anticipation

Sometimes, traits seem to grow stronger and gain more expressivity from one generation to the next. The strengthening of a trait as it's inherited is called *anticipation.* One condition that shows anticipation is Huntington disease. The symptoms of Huntington disease typically appear in adulthood, with changes in mood and difficulties with movement. Adults with the condition have involuntary muscle movements, problems walking, increased forgetfulness, difficulties making decisions, and problems with eating and swallowing. The symptoms appear and get worse over time, and affected individuals always have a shortened lifespan. The age of onset of Huntington disease symptoms tends to be earlier with each generation, and the severity of the symptoms tend to increase from one generation to the next.

The reason behind anticipation in Huntington disease and similar disorders is that there is a region within the Huntington disease gene that contains a repeated sequence. In individuals with Huntington disease, this repeat is actually longer than that found in people without the disease. During replication (covered in Chapter 7), these sections of the DNA within the gene are easily duplicated by accident, leading to longer repeat sequences that can be passed down to children. Thus, in successive generations, the repeated sequence gets longer and the effect of the expansion on cells in the brain gets more severe. As a result, an individual that inherits an expansion that is longer than what their parent had will often show signs at an earlier age than their parent, and the symptoms may be more severe and progress more quickly.

Environmental effects

Most traits show little evidence of environmental effect. However, the environment that some organisms live in controls the phenotype that result from certain genes. For example, the gene that gives a Himalayan rabbit its characteristic phenotype of dark feet, ears, nose, and tail is a good example of a trait that varies in its expression based on the animal's environment. The pigment that produces dark fur in any animal results from the presence of an enzyme that the animal's body produces. But in this case, the enzyme's effect is deactivated at normal body temperatures. Thus, the allele that produces pigment in the rabbit's fur is expressed only in the cooler parts of the body. That's why Himalayan rabbits are all white when they're born (they've been kept warm inside their mother's body) but get dark feet, ears, noses, and tails later in life. (Himalayans also change color seasonally and get lighter during the warmer months.)

2 DNA: The Genetic Material

IN THIS PART . . .

See how DNA is put together and how it is copied during cell division.

Understand how chromosomes are packaged and where your genes are.

Discover how DNA is transcribed into messenger RNA and how the genetic code is used to translate messenger RNA into protein.

Find out how gene expression works and the factors that control it.

IN THIS CHAPTER

» Identifying the chemical components of DNA

» Understanding the structure of the double helix

» Looking at different DNA varieties

» Chronicling DNA's scientific history

Chapter 5
DNA: The Basis of Life

It's time to meet the star of the genetics show: *deoxyribonucleic acid,* otherwise known as DNA. If the title of this chapter hasn't impressed upon you the importance and magnitude of those three little letters, consider that DNA is also referred to as "*the* genetic material" or "*the* molecule of heredity." And you thought your title was impressive!

Every living thing on earth, from the smallest bacteria to the largest whale, uses DNA to store genetic information and transmits that info from one generation to the next; a copy of some (or all) of every creature's DNA is passed on to its offspring. The developing organism then uses DNA as a blueprint to make all its body parts. (Some nonliving things use DNA to transmit information, too; see the nearby sidebar "DNA and the undead: The world of viruses" for details.)

To get an idea of how much information DNA stores, think about how complex your body is. You have hundreds of kinds of tissues that all perform different functions. It takes a lot of DNA to catalog all that. See the section "Discovering DNA" later in this chapter to find out how scientists learned that DNA is the genetic material of all known life forms.

The structure of DNA provides a simple way for the molecule to copy itself and protects genetic messages from getting garbled. That structure is at the heart of many applications of DNA testing, too. But before you can start exploring genetic information and applications of DNA, you need to have a handle on its chemical makeup and physical structure. In this chapter, we explore the essential makeup of DNA and the various sorts of DNA present in living things.

DNA AND THE UNDEAD: THE WORLD OF VIRUSES

Viruses contain DNA (or RNA), but they aren't considered living things. To reproduce, a virus must attach itself to a living cell. As soon as the virus finds a host cell, the virus injects its DNA into the cell and forces that cell to reproduce the virus. A virus can't grow without stealing energy from a living cell, and it can't move from one organism to another on its own. Although viruses come in all sorts of fabulous shapes, they don't have all the components that cells do; in general, a virus is just DNA (or RNA) surrounded by a protein shell. So a virus isn't alive, but it's not quite dead either. Creepy, huh?

Chemical Ingredients of DNA

DNA is a remarkably durable molecule; it can be stored in ice or in a fossilized bone for thousands of years. DNA can even stay in one piece for as long as 100,000 years under the right conditions. This durability is why scientists can recover DNA from 14,000-year-old mammoths and learn that the mammoth is most closely related to today's Asian elephants. (Scientists have recovered ancient DNA from an amazing variety of organisms — check out the sidebar "Still around after all these years: Durable DNA" for more.) The root of DNA's extreme durability lies in its chemical and structural makeup.

REMEMBER

Chemically, DNA is really simple. It's made of three components: nitrogen-rich bases, deoxyribose sugars, and phosphates. The three components, which we explain in the following sections, combine to form a *nucleotide* (see the section "Assembling the double helix: The structure of DNA" later in this chapter). Thousands of nucleotides come together in pairs to form a single molecule of DNA.

Covering the bases

Each DNA molecule contains thousands of copies of four specific nitrogen-rich bases:

- Adenine (A)
- Guanine (G)
- Cytosine (C)
- Thymine (T)

STILL AROUND AFTER ALL THESE YEARS: DURABLE DNA

When an organism dies, it starts to decay and its DNA starts to break down (for DNA, this means breaking into smaller and smaller pieces). But if a dead organism dries out or freezes shortly after death, decay slows down or even stops. Because of this kind of interference with decay, scientists have been able to recover DNA from animals and humans that roamed the earth as many as 100,000 years ago. This recovered DNA tells scientists a lot about life and the conditions of the world long ago. But even this very durable molecule has its limits — about a million years or so.

In 1991, hikers in the Italian Alps discovered a human body frozen in a glacier. As the glacier melted, the retreating ice left behind a secret concealed for over 5,000 years: an ancient human. The Ice Man, renamed Otzi, has yielded amazing insight into what life was like in northern Italy thousands of years ago. Scientists have recovered DNA from this lonely shepherd, his clothing, and even the food in his stomach. Apparently, red deer and ibex meat were part of his last meal. His food was dusted with pollen from nearby trees, so even the forest he walked through can be identified!

By analyzing Otzi's mitochondrial (mt) DNA, which he inherited from his mother (see the "Mitochondrial DNA" section later in this chapter), scientists discovered that he wasn't related to any modern European population studied so far. A team of investigators from Australia, led by the late Thomas Loy, examined blood found on Otzi's clothing and possessions. Like modern forensic scientists, Loy's team determined that four different people's DNA fingerprints were present, in addition to Otzi's own (to find out how DNA fingerprints are used to solve modern crimes, check out Chapter 18). The team found blood from two different people on Otzi's arrow, a third person's blood on his knife, and a fourth person's blood on his clothing. These findings led people to speculate that he was involved in a fight shortly before he died.

As you can see in Figure 5-1, the bases are comprised of carbon (C), hydrogen (H), nitrogen (N), and oxygen (O) atoms.

The four bases come in two flavors:

- **Purines:** The two purine bases in DNA are adenine and guanine. If you were a chemist, you'd know that the word *purine* means a compound composed of two rings (check out adenine's and guanine's structures in Figure 5-1). If you're like us (not a chemist), you're likely still familiar with one common purine: caffeine.
- **Pyrimidines:** The two pyrimidine bases in DNA are cytosine and thymine. The term *pyrimidine* refers to chemicals that have a single six-sided ring structure (see cytosine's and thymine's structures in Figure 5-1)..

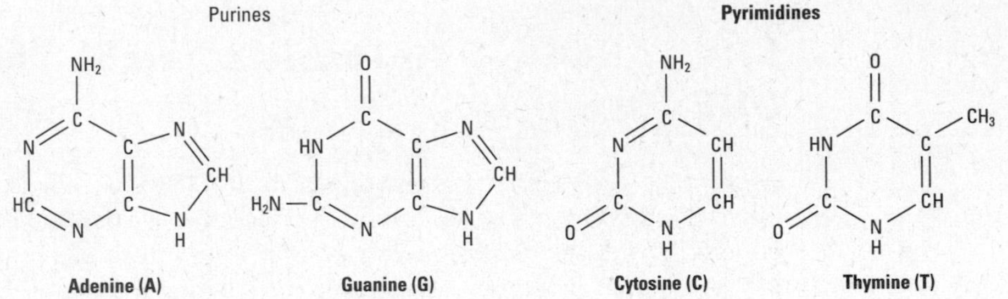

Purines

Pyrimidines

Adenine (A) Guanine (G) Cytosine (C) Thymine (T)

FIGURE 5-1: The four DNA bases.

REMEMBER

Because they're rings, all four bases are flat molecules. And as flat molecules, they're able to stack up in DNA much like a stack of coins. The stacking arrangement accomplishes two things: It makes the molecule both compact and very strong.

TIP

It's been our experience that students and other folks often get confused by spatial concepts where DNA is concerned. To see the chemical structures more easily, DNA is often drawn as if it were a flattened ladder. But in its true state, DNA isn't flat — it's three-dimensional. Because DNA is arranged in strands, it's also linear. One way to think about this structure is to look at a phone cord (that is, if you can find a phone that *has* a cord). A phone cord spirals in three dimensions, yet it's linear (rope-like) in form. That's sort of the shape DNA has, too.

Just like a sequence of letters can convey information in the form of words and sentences, the sequence of the bases carries the message of DNA, which provides the information necessary to produce the corresponding protein(s). However, the bases can't bond together by themselves. Two more ingredients are needed: a special kind of sugar and a phosphate.

Adding a spoonful of sugar and a little phosphate

To make a complete nucleotide (thousands of which combine to make one DNA molecule), the bases must attach to deoxyribose and a phosphate molecule. *Deoxyribose* is ribose sugar that has lost one of its oxygen atoms. When your body breaks down *adenosine triphosphate* (ATP), the molecule your body uses to power your cells, ribose is released with a phosphate molecule still attached to it. Ribose loses an oxygen atom to become deoxyribose (see Figure 5-2) and holds onto its phosphate molecule, which is needed to transform a lone base into a nucleotide.

Ribose is the precursor for deoxyribose and is the chemical basis for RNA (see Chapter 9 for more details about RNA). The only difference between ribose and deoxyribose sugars is the presence or absence of an oxygen atom at the 2′ site.

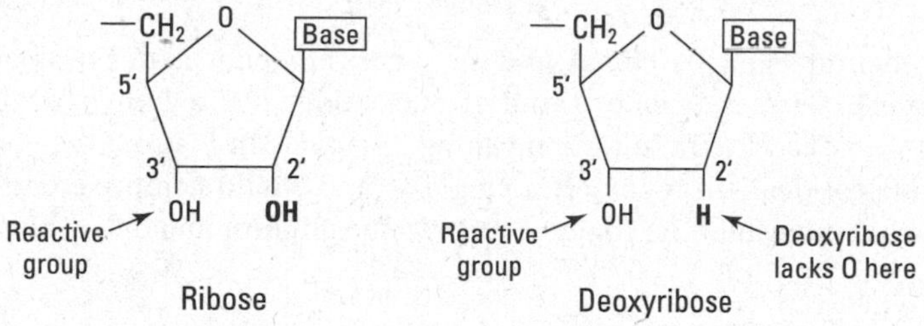

Ribose

Deoxyribose

Reactive group

Deoxyribose lacks O here

FIGURE 5-2: The chemical structure of ribose and deoxyribose.

TIP

Chemical structures are numbered so you can keep track of where atoms, branches, chains, and rings appear. On ribose sugars, numbers are followed by an apostrophe (') to indicate the designation "prime." The addition of "prime" prevents confusion with numbered sites on other molecules that bond with ribose.

Deoxy- means that an oxygen atom is missing from the sugar molecule and defines the *D* in DNA. As an added touch, some authors write "2-" before the "deoxy-" to indicate which site lacks the oxygen — the number 2 site, in this case (that is, the 2-prime site). The OH group at the 3′ site of both ribose and deoxyribose is a *reactive group.* That means the oxygen atom at that site is free to interact chemically with other molecules.

Assembling the Double Helix: The Structure of DNA

Nucleotides are the true building blocks of DNA. In Figure 5-3, you see the three components of a single nucleotide: one deoxyribose sugar, one phosphate, and one of the four bases. To make a complete DNA molecule, single nucleotides join to make chains that come together as matched pairs and form long double strands. This section walks you through the assembly process. To make the structure of DNA easier to understand, we start with how a single strand is put together.

REMEMBER

DNA normally exists as a double-stranded molecule. In living things, new DNA strands are *always* put together using a preexisting strand as a pattern (see Chapter 7).

Starting with one: Weaving a single strand

Hundreds of thousands of nucleotides link together to form a strand of DNA, but they don't hook up haphazardly. Nucleotides are a bit like coins in that they have two "sides" — a phosphate side and a sugar side. Nucleotides can only make a connection by joining phosphates to sugars. The bases wind up parallel to each other (stacked like coins), and the sugars and phosphates run perpendicular to the stack of bases. A long strand of nucleotides put together in this way is called a *polynucleotide* strand (*poly* meaning "many"). In Figure 5-4, you can see how the nucleotides join together; a single strand would comprise one-half of the two-sided molecule (the chain of sugars, phosphates, and one of the pair of bases).

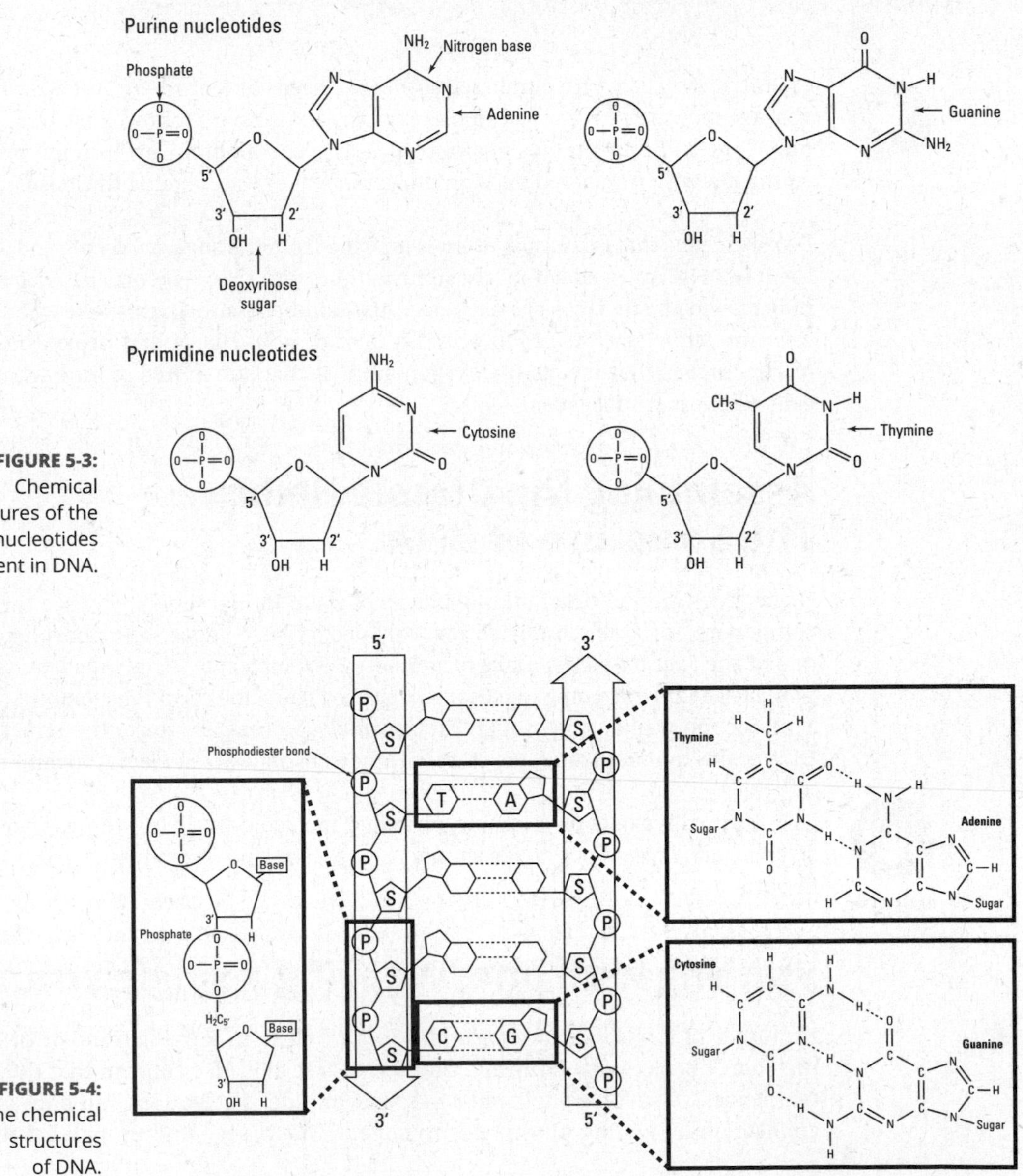

FIGURE 5-3: Chemical structures of the four nucleotides present in DNA.

FIGURE 5-4: The chemical structures of DNA.

Because of the way the chemical structures are numbered, DNA has numbered "ends." The phosphate end is referred to as the 5′ (5-prime) end, and the sugar end is referred to as the 3′ (3-prime) end. (For a discussion of how the chemical structure of deoxyribose is numbered, check out the earlier section "Adding a spoonful of sugar and a little phosphate.") The bonds between a phosphate and two sugar molecules in a nucleotide strand are collectively called a *phosphodiester bond.* This is a fancy way of saying that two sugars are linked together by a phosphate in between.

After they're formed, strands of DNA don't enjoy being single; they're always looking for a match. The arrangement in which strands of DNA match up is very, very important. A number of rules dictate how two lonely strands of DNA find their perfect matches and eventually form the star of the show, the molecule you've been waiting for — the double helix.

Doubling up: Adding the second strand

A complete DNA molecule has

- Two side-by-side polynucleotide strands twisted together
- Bases attached in pairs in the center of the molecule
- Sugars and phosphates on the outside, forming a backbone

If you were to untwist a DNA double helix and lay it flat, it would look a lot like a ladder (refer to Figure 5-4). The bases are attached to each other in the center to make the rungs, and the sugars are joined together by phosphates to form the sides of the ladder. It sounds pretty straightforward, but this ladder arrangement has some special characteristics.

REMEMBER

If you were to separate the ladder into two polynucleotide strands, you'd see that the strands are oriented in opposite directions (shown with arrows in Figure 5-5). The locations of the sugar and the phosphate give nucleotides heads and tails, two distinct ends. The heads-tails (or in this case, 5′ to 3′) orientation applies here. This head-to-tail arrangement is called *antiparallel,* which is a fancy way of saying that something is parallel and running in opposite directions. Part of the reason the strands must be oriented this way is to guarantee that the dimensions of the DNA molecule are even along its entire length. If the strands were put together in a parallel arrangement, the angles between the atoms would be all wrong, and the strands wouldn't fit together.

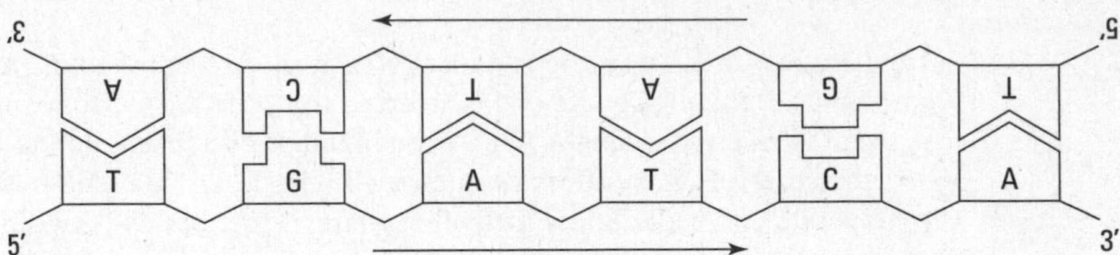

FIGURE 5-5: Antiparallel strands of DNA.

REMEMBER

The molecule is guaranteed to be the same size all over because the matching bases *complement* each other, making whole pieces that are all the same size. Adenine complements thymine, and guanine complements cytosine. The bases *always* match up in this complementary fashion. Therefore, in every DNA molecule, the amount of one base is equal to the amount of its complementary base. This condition is known as *Chargaff's rules*. You can find more on the discovery of these rules in the "Obeying Chargaff's rules" section later in the chapter.

TECHNICAL STUFF

Why can't the bases match up in other ways? First, purines have two rings and are larger than pyrimidines, which only have one ring (refer back to Figure 5-1). So matching like with like would introduce irregularities in the molecule's shape. Irregularities are bad because they can cause mistakes when the molecule is copied, as covered in Chapter 13.

An important result of the bases' complementary pairing is the way in which the strands bond to each other. Hydrogen bonds form between the base pairs. The number of bonds between the base pairs differs; G-C (guanine-cytosine) pairs have three bonds, and A-T (adenine-thymine) pairs have only two. Figure 5-4 illustrates the structure of the untwisted double helix — specifically, the bonds between base pairs. Every DNA molecule has hundreds of thousands of base pairs, and each base pair has multiple bonds, so the rungs of the ladder are very strongly bonded together.

When inside a cell, the two strands of DNA gently twist around each other like a spiral staircase (or a strand of licorice, or the stripes on a candy cane . . . anybody else have a sweet tooth?). The antiparallel arrangement of the two strands is what causes the twist. Because the strands run in opposite directions, they pull the sides of the molecule in opposite directions, causing the whole thing to twist around itself.

Most naturally occurring DNA spirals clockwise, as you can see in Figure 5-6. A full twist (or complete turn) occurs every ten base pairs or so, with the bases safely protected on the inside of the helix. The helical form is one way that the information that DNA carries is protected from damage that can result in mutation.

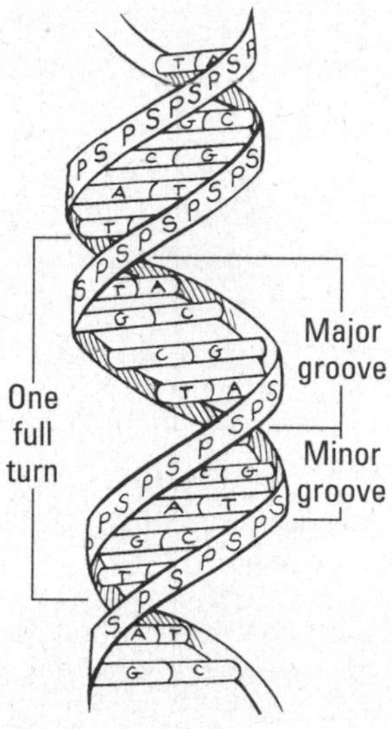

FIGURE 5-6: The DNA double helix.

TECHNICAL STUFF

The helical form creates two grooves on the outside of the molecule (refer to Figure 5-6). The major groove lets the bases peep out a little, which is important when it's time to read the information DNA contains, as covered in Chapter 9.

TECHNICAL STUFF

Because base pairs in DNA are stacked on top of each other, chemical interactions make the center of the molecule repel water. Molecules that repel water are called *hydrophobic* (Greek for "afraid of water"). The outside of the DNA molecule is just the opposite; it attracts water. The result is that the inside of the helix remains safe and dry while the outside is encased in a "shell" of water.

REMEMBER

Here are a few additional details about DNA that you need to know:

- **A DNA strand is measured by the number of base pairs it has.**
- **The sequence of bases in DNA isn't random.** The genetic information in DNA is carried in the order of the base pairs. That is, the order of the base pairs determines the structure of the protein that is generated from a given gene. Chapter 10 explains how the sequences are read and decoded.
- **DNA uses a preexisting DNA strand as a pattern or template in the assembly process.** DNA doesn't just form on its own. The process of making a new strand of DNA using a preexisting strand is called *replication.* We cover replication in detail in Chapter 7.

MOLECULAR MADNESS: EXTRACTING DNA AT HOME

Using this simple recipe, you can see DNA right in the comfort of your own home! You need a strawberry, salt, water, two clear jars or juice glasses, a sandwich bag, a measuring cup, a white coffee filter, clear liquid soap, and rubbing alcohol. (Other foods such as onions, bananas, kiwis, and tomatoes also work well if strawberries are unavailable.) After you've gathered these ingredients, follow these steps:

1. **Put slightly less than ⅜ cup of water into the measuring cup. Add ¼ teaspoon of salt and enough clear liquid soap to make ⅜ cup of liquid altogether. Stir gently until the salt dissolves into the solution.**

   The salt provides sodium ions needed for the chemical reaction that allows you to see the DNA in Step 6. The soap causes the cell walls to burst, freeing the DNA inside.

2. **Remove the stem from the strawberry, place the strawberry into the sandwich bag, and seal the bag. Mash the strawberry thoroughly until completely pulverized (I roll a juice glass repeatedly over my strawberry to pulverize it). Make sure you don't puncture the bag.**
3. **Add 2 teaspoons of the liquid soap-salt solution to the bag with the strawberry, and then reseal the bag. Mix gently by compressing the bag or rocking the bag back and forth for at least 45 seconds to one minute.**
4. **Pour the strawberry mixture through the coffee filter into a clean jar. Let the mixture drain into the jar for 10 minutes.**

   Straining gets rid of most of the cellular debris (a fancy word for gunk) and leaves behind the DNA in the clean solution.

5. **While the strawberry mixture is draining, pour ¼ cup of rubbing alcohol into a clean jar and put the jar in the freezer. After 10 minutes have elapsed, discard the coffee filter and pulverized strawberry remnants. Put the jar with the cold alcohol on a flat surface where it will be undisturbed, and pour the strained strawberry liquid into the alcohol.**
6. **Let the jar sit for at least 5 minutes, and then check out the result of your DNA experiment. The cloudy substance that forms in the alcohol layer is the DNA from the strawberry. The cold alcohol helps strip the water molecules from the outside of the DNA molecule, causing the molecule to collapse on itself and "fall out" of the solution.**

Examining Different Varieties of DNA

All DNA has the same four bases, obeys the same base pairing rules, and has the same double helix structure. No matter where it's found or what function it's carrying out, DNA is DNA. That said, different sets of DNA exist within a single organism. These sets carry out different genetic functions. In this section, we explain where the various DNAs are found and describe what they do.

Nuclear DNA

Nuclear DNA is DNA in cell nuclei, and it's responsible for the majority of functions that cells carry out. Most of our genes are located in the nuclear DNA. In addition, the physical traits of an organism (that is, physical *phenotypes*) are generally the result of the genes located in the nuclear DNA. Nuclear DNA is packaged into chromosomes and passed from parent to offspring, as covered in Chapter 2. When scientists talk about sequencing the human genome, they mean human nuclear DNA. A *genome* is a full set of genetic instructions. We get more into the human genome in Chapter 8. The nuclear genome of humans is comprised of the DNA from all 22 pairs of autosomal chromosomes plus two sex chromosomes (either two X chromosomes or one X and one Y).

Mitochondrial DNA

Animals, plants, and fungi all have *mitochondria* (for a review of cell parts, turn to Chapter 2). These powerhouses of the cell come with their own DNA, which is quite different in form (and inheritance) from nuclear DNA. Each *mitochondrion* (the singular word for *mitochondria*) has many molecules of mitochondrial DNA — *mtDNA*, for short.

Whereas human nuclear DNA is linear, mtDNA is circular (hoop-shaped). Human mtDNA is very short (slightly less than 17,000 base pairs) and has 37 genes, which account for almost the entire mtDNA molecule. These genes control cellular metabolism — the processing of energy inside the cell.

Half of your nuclear DNA came from your mom, and the other half came from your dad. But *all* your mtDNA came from your mom. All your mom's mtDNA came from her mom, and so on. All mtDNA is passed from mother to child in the cytoplasm of the egg cell.

TECHNICAL STUFF

Sperm cells have essentially no cytoplasm and thus, virtually no mitochondria. Special chemicals in the egg destroy the few mitochondria that sperm do possess.

MIGHTY MITOCHONDRIA

Mitochondrial DNA (mtDNA) bears a strong resemblance to a bacterial DNA. The striking similarities between mitochondria and a certain bacteria called *Rickettsia* have led scientists to believe that mitochondria originated from *Rickettsia. Rickettsia* causes typhus, a flu-like disease transmitted by flea bites (the flea first bites an infected rat or mouse and then bites a person). As for the similarities, neither *Rickettsia* nor mitochondria can live outside a cellular home, both have circular DNA, and both share similar DNA sequences (see Chapter 8 for how DNA sequences are compared between organisms). Instead of being parasitic like *Rickettsia,* however, mitochondria are considered *endosymbiotic,* meaning they must be inside a cell to work *(endo-)* and they provide something good to the cell *(-symbiotic).* In this case, the something good is energy.

Because mtDNA is passed only from mother to child, scientists have compared mtDNA from people all over the world to investigate the origins of modern humans. These comparisons have led some scientists to believe that all modern humans have one particular female ancestor in common, a woman who lived on the African continent about 200,000 years ago. This hypothetical woman has been called "Mitochondrial Eve," but she wasn't the only woman of her time. There were many women, but apparently, none of their descendants survived, making Eve what scientists refer to as our "most recent common ancestor," or MRCA. Some evidence suggests that all humans are descended from a rather small population of about 100,000 individuals, meaning that all people on earth have common ancestry.

Chloroplast DNA

Plants have three sets of DNA: nuclear in the form of chromosomes, mitochondrial, and *chloroplast DNA* (cpDNA). Chloroplasts are organelles found only in plants, and they're where *photosynthesis* (the conversion of light to chemical energy) occurs. To complicate matters, plants have mitochondria (and thus mtDNA) in their chloroplasts. Like mitochondria, chloroplasts probably originated from bacteria, as explained in the sidebar "Mighty mitochondria."

Chloroplast DNA molecules are circular and fairly large (120,000–160,000 base pairs) but only have about 120 genes. Most of those genes supply information used to carry out photosynthesis. Inheritance of cpDNA can be either maternal or paternal, and cpDNA, along with mtDNA, is transmitted to offspring in the cytoplasm of the seed.

Digging into the History of DNA

Back when Mendel was poking around his pea pods in the early 1860s (see Chapter 3), neither he nor anybody else knew about DNA. DNA was discovered in 1868, but its importance as *the* genetic material wasn't appreciated until nearly a century later. This section gives you a rundown on how DNA and its role in inheritance was revealed.

Discovering DNA

HISTORICAL STUFF

In 1868, a Swiss medical student named Johann Friedrich Miescher isolated DNA for the first time. Miescher was working with white blood cells that he obtained from the pus drained out of surgical wounds (yes, this man was dedicated to his work). Eventually, Miescher established that the substance he called *nuclein* was rich in phosphorus and was acidic. Thus, one of his students renamed the substance *nucleic acid,* a name DNA still carries today. Like Mendel's findings on the inheritance of various plant traits, Miescher's work wasn't recognized for its importance until long after his death, and it took 84 years for DNA to be recognized as *the* genetic material. Until the early 1950s, everyone was sure that protein had to be the genetic material because, with only four bases, DNA seemed too simple.

In 1928, Frederick Griffith recognized that bacteria could acquire something — he wasn't quite sure what — from each other to transform harmless bacteria into deadly bacteria (see Chapter 21 for the whole story). A team of scientists led by Oswald Avery followed up on Griffith's experiments and determined that the "transforming principle" was DNA. Even though Avery's results were solid, scientists of the time were skeptical about the significance of DNA's role in inheritance. It took another elegant set of experiments using a virus that infected bacteria to convince the scientific community that DNA was the real deal.

HISTORICAL STUFF

Alfred Chase and Martha Hershey worked with a virus called a *bacteriophage* (which means "eats bacteria," even though the virus technically ruptures the bacteria rather than eats it). Bacteriophages grab onto the bacteria's cell wall and inject something into the bacteria. At the time of Hershey and Chase's experiment, the injected substance was unidentified. The bacteriophage produces its offspring inside the cell and then bursts the cell wall open to free the viral "offspring." Offspring carry the same traits as the original attacking bacteriophage, so it was certain that whatever got injected must be the genetic material, given that most of the bacteriophage stays stuck on the outside of the cell. Hershey and Chase attached radioactive chemicals to track different parts of the bacteriophage; for example, they used sulfur to track protein, because proteins contain sulfur, and DNA was marked with phosphorus (because of the sugar-phosphate backbone). Hershey and Chase reasoned that offspring bacteriophages would get marked with one or the other, depending on which — DNA or protein — turned out to be the genetic material. The results showed that the viruses injected only DNA into the bacterial cell to infect it. All the protein stayed stuck on the outside of the bacterial cell. They published their findings in 1952, when Hershey was merely 24 years old!

Obeying Chargaff's rules

Long before Hershey and Chase published their pivotal findings, Erwin Chargaff read Oswald Avery's paper on DNA as the transforming principle (see Chapter 21 for more information on DNA transformation) and immediately changed the focus of his entire research program. Unlike many scientists of his day, Chargaff recognized that DNA was the genetic material.

HISTORICAL STUFF

Chargaff focused his research on learning as much as he could about the chemical components of DNA. Using DNA from a wide variety of organisms, he discovered that all DNA had something in common: When DNA was broken into its component bases, the amount of guanine fluctuated wildly from one organism to another, but the amount of guanine always equaled the amount of cytosine. Likewise, in every organism he studied, the amount of adenine equaled the amount of thymine. Published in 1949, these findings are so consistent that they're called *Chargaff's rules.* Unfortunately, Chargaff was unable to realize the meaning of his own work. He knew that the ratios said something important about the structure of DNA, but he couldn't figure out what that something was. It took a pair of young scientists named Watson and Crick — Chargaff called them "two pitchmen in search of a helix" — to make the breakthrough.

Hard feelings and the helix: Franklin, Wilkins, Watson, and Crick

HISTORICAL STUFF

If you don't know the name Rosalind Franklin, you should. Her data on the shape of the DNA molecule revealed its structure as a double helix. Watson and Crick get all the credit for identifying the double helix, but Franklin did much of the work. While researching the structure of DNA at King's College, London, in the early 1950s, Franklin bounced X-rays off the molecule to produce incredibly sharp, detailed photos of it. Franklin's photos show a DNA molecule from the end, not the side, so it's difficult to envision the side view of the double helix you normally see. Yet Franklin knew she was looking at a helix.

Meanwhile, James Watson, a 23-year-old postdoctoral fellow at Cambridge, England, was working with a 38-year-old graduate student named Francis Crick. Together, they were building an enormous model of metal sticks and wooden balls, trying to figure out the structure of the same molecule Franklin had photographed.

Franklin was supposed to be collaborating with Maurice Wilkins, another scientist in her research group, but she and Wilkins despised each other (because of a switch in research projects in which Franklin was instructed to take over Wilkins's project without his knowledge). As their antagonism grew, so did Wilkins's friendship with Watson. What happened next is the stuff of science infamy. Just a few weeks before Franklin was ready to publish her findings, Wilkins showed Franklin's photographs of the DNA molecule to Watson — without her knowledge or permission! By giving Watson access to Franklin's data, Wilkins gave Watson and Crick the scoop on the competition.

Watson and Crick cracked the mystery of DNA structure using Chargaff's rules (see the preceding section, "Obeying Chargaff's rules," for more on Chargaff's rules) and Franklin's measurements of the molecule. They deduced that the structure revealed by Franklin's photo, hastily drawn from memory by Watson, had to be a double helix, and Chargaff's rules pointed to bases in pairs. The rest of the structure came together like a big puzzle, and they rushed to publish their discovery in 1953. Franklin's paper, complete with the critical photos of the DNA molecule, was published in the same issue of the journal *Nature*.

In 1962, Watson, Crick, and Wilkins were honored with the Nobel Prize. Franklin wasn't properly credited for her part in their discovery but couldn't protest because she had died of ovarian cancer in 1957. It's quite possible that Franklin's cancer was the result of long-term exposure to X-rays during her scientific career. In a sense, Franklin sacrificed her life for science.

IN THIS CHAPTER

- How DNA is packaged into chromosomes
- Understanding chromosome structure
- How sex is determined in humans and other animals

Chapter 6

Chromosomes: The Big Picture

Nearly every cell of the human body has a complete set of chromosomes. *Chromosomes,* which means "colored bodies," were named as such because these structures were found to absorb certain dyes when cells undergoing mitosis were studied. Chromosomes were identified long before we knew that DNA was the genetic material and genes were the individual "units of heredity." Chromosomes were first observed in plant cells in the 1840s by the Swiss botanist Karl Wilhelm von Nägeli.

Now we know that all organisms have chromosomes. We also know that the location and number of chromosomes vary depending on the species. In prokaryotes like bacteria, there is a single ring-shaped chromosome inside the cell (there is no cell nucleus). In humans, there are 46 total chromosomes (23 pairs) in the nucleus of each cell. The number of chromosomes is generally consistent within a species, but it can vary significantly between types of organisms, ranging from 1 to more than 200!

Chromosome structure allows all an organism's DNA to be packaged nicely in the nuclei of its cells and allows for the efficient and accurate distribution of genes during mitosis and meiosis. In this chapter, we get into more detail about how DNA is packaged into chromosomes in *eukaryotes* (organisms with cell nuclei) and about chromosome structure, building on what we review in Chapter 2. We also discuss the chromosomal basis of sex determination in humans, as well as how sex determination varies in other eukaryotic organisms.

Deconstructing the Double Helix

If you're like most folks, when you think of DNA, you think of a double helix. But DNA isn't just a double helix; it's a *huge* molecule — so huge that it's called a *macromolecule.* It can even be seen with the naked eye! (Chapter 5 provides an experiment you can do to see actual DNA.) If you were to lay out, end to end, all the DNA from just one of your cells, the line would be a little over 6 feet long! You have roughly 100,000,000,000,000 cells in your body (that's 100 trillion, in case you don't feel like counting zeros). Put another way, laid out altogether, the DNA in your body would easily stretch to the sun and back nearly 100 times.

You're probably wondering how a huge DNA molecule can fit into a cell so small that you can't see it with the naked eye. Here's how: DNA is tightly packed into *chromatin.* Chromatin contains both DNA and proteins. First, the negatively charged DNA is wrapped around positively charged proteins called *histones* (see Figure 6-1). The DNA plus histones are referred to as *nucleosomes.* At this point, the DNA and proteins look like beads on a string. The DNA continues to twist and become more compact, in a process called *supercoiling* — much like a phone cord that's been twisted around and around on itself. The whole "necklace" twists around itself so tightly that over 6 feet of DNA is compressed into only a few thousandths of an inch.

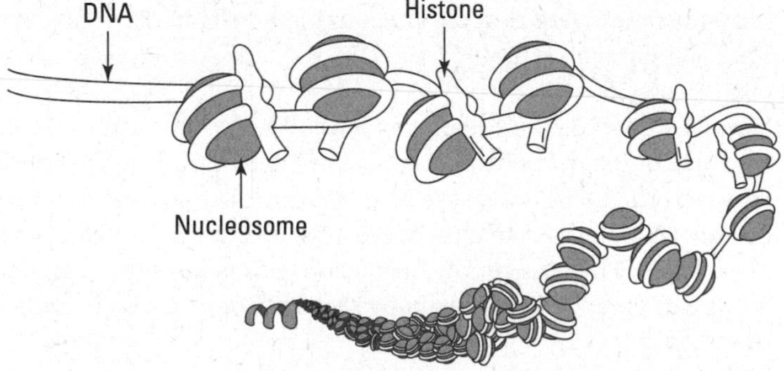

FIGURE 6-1: DNA is wrapped around nucleosomes and tightly coiled to fit into tiny cell nuclei.

Cell nuclei contain two types of chromatin: euchromatin and heterochromatin. *Euchromatin* is not as compact as *heterochromatin.* Genes that are to be expressed are generally located in the euchromatin. In contrast, heterochromatin, which is highly condensed, tends to contain more repetitive, non-coding DNA and genes that are poorly expressed.

Although the idea of a DNA path to the sun works great for visualizing the size of the DNA molecule, an organism's DNA usually doesn't exist as one long piece. Rather, strands of DNA are divided into *chromosomes*. In humans and all other *eukaryotes* (organisms whose cells have nuclei), a full set of chromosomes is stored in the nucleus of each cell. That means that practically every cell contains a complete set of instructions to build the entire organism! The instructions are packaged as *genes*, with each gene providing the instructions for a particular protein. Chapter 11 covers genes and how they work.

TECHNICAL STUFF

Cells with nuclei are found only in eukaryotes; however, not every eukaryotic cell has a nucleus. For example, humans are eukaryotes, but human red blood cells don't have nuclei. For more on cells and nuclei, see Chapter 2.

Anatomy of a Chromosome

During most of the cell cycle (discussed in Chapter 2 and illustrated in Figure 2-5), chromosomes are present as loose strings of DNA in the nuclei of cells. However, during metaphase, the chromosomes are present as the condensed, characteristic sausage shape. At this point, several distinctive features can be identified (see Figure 6-2).

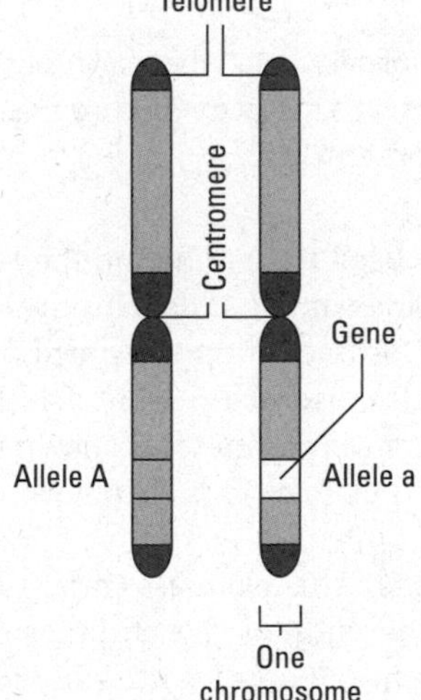

FIGURE 6-2: Chromosome features are evident in a metaphase chromosome.

First, each chromosome has a "pinched" region in the center, which is known as the *centromere*. As discussed in Chapter 2, the centromere plays a key role during both mitosis and meiosis. The spindles that attach to each chromosome and pull the chromosomes to each pole during cell division are attached to the chromosome at the centromere. Also, as shown in Figure 6-3, the position of the centromere along the chromosome is used to help classify chromosomes. Chromosomes that have a centromere right in the middle of the chromosome are referred to as *metacentric chromosomes*. Chromosomes that have the centromere toward one end of the chromosome are referred to as *acrocentric chromosomes*. Chromosomes that have the centromere somewhere between these two places are referred to as *submetacentric chromosomes*. The DNA found at the centromere is heterochromatic (highly condensed chromatin).

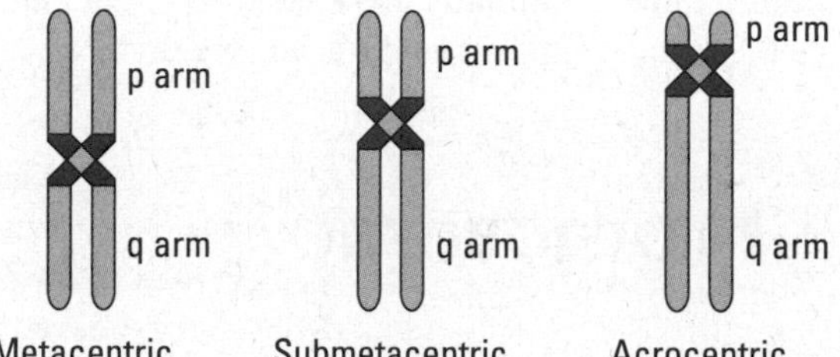

FIGURE 6-3: Classification of chromosomes based on centromere location.

On either side of the centromere are the chromosome *arms* (refer to Figure 6-3). The two chromosome arms are the:

- ***p* arm:** The shorter of the two arms (from the word *petit,* French for "small"), or the arm located above the centromere.
- ***q* arm:** The longer arm (because *q* follows *p* alphabetically), or the arm located below the centromere. The long arm in a metacentric chromosome may not actually appear longer than the short arm.

The designation of *p* or *q* arm is also used to help define the location of specific genes on each of the chromosomes. For example, if the chromosome in Figure 6-2 was chromosome number 1, the gene located on the long arm (the arm below the centromere) is said to be located on chromosome 1q. (Note that the gene would not be this large relative to the size of the chromosome; it is shown larger here for the purpose of illustrating the location of a gene on this chromosome.)

At the very ends of the chromosomes are the *telomeres* (refer to Figure 6-2). The role of the telomeres is to protect the ends of the chromosomes. The DNA at the telomeres does not code for any genes. Instead it is highly repetitive DNA that contains repeats of the sequence TTAGGG. In most cells, telomeres become shorter with each round of replication, as described in Chapter 7. After a certain number of cell divisions, the telomeres in these cells will have become so short that the cells become inactive or die. In cells that need to continually divide, such as blood cells, a special enzyme called *telomerase* comes to the rescue. Telomerase adds new telomere repeats to the ends of the chromosomes, which keeps the telomeres sufficiently long and the cell able to divide without losing precious coding DNA. Like the centromere, the DNA located at the telomeres is heterochromatic (highly condensed chromatin).

Two Chromosomes Are Better than One (or Three)

Cell nuclei contain two main types of chromosomes: *autosomes* and *sex chromosomes*. The autosomes are the same in males and females. Human cell nuclei contain 22 pairs of autosomes, which are numbered from 1 to 22 based on size, shape, and banding pattern that can be observed when the chromosomes are stained and viewed under a microscope (see Figure 6-4). The sex chromosomes include the X chromosome and the Y chromosome. Females should have two X chromosomes and males should have one X chromosome and one Y chromosome. Females get an X chromosome from both parents, while males get their X chromosome from their mother and their Y chromosome from their father.

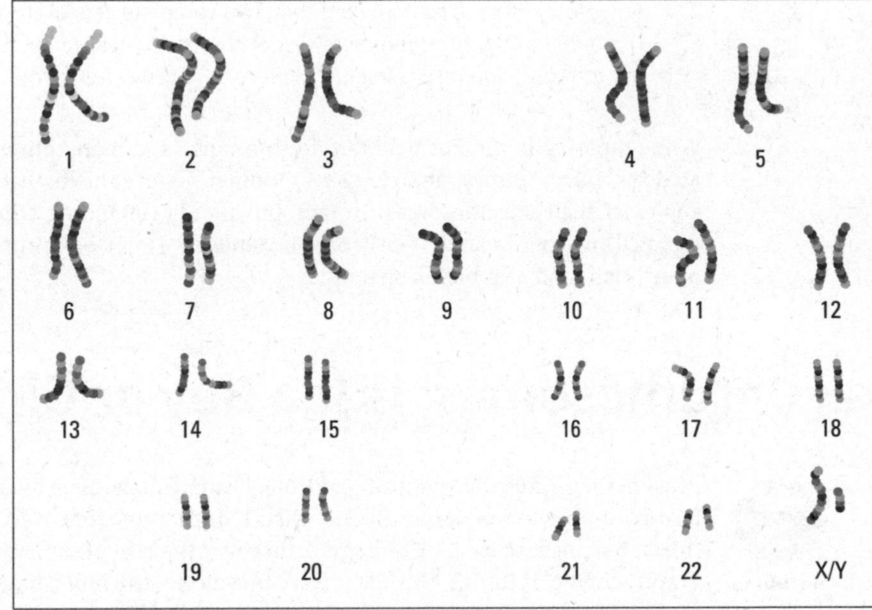

FIGURE 6-4: Twenty-two pairs of autosomes and two sex chromosomes from a single human cell.

The pairs of chromosomes are called *homologous chromosomes* (or *homologs*, introduced in Chapter 2). Each homolog of a pair should contain the same genes, but not necessarily the same versions (or *alleles*) of each gene (see Figure 6-2). For example, a child may inherit the allele for "dimples" from his or her mother, but the allele for "no dimples" from his or her father.

Ploidy (discussed in Chapter 2) refers to the number of chromosomes a particular organism possesses. Any deviation from this number can significantly affect the development of the individual. Several sorts of "ploidy" are commonly bandied about in genetics:

- **Euploid** refers to the number of *sets* of chromosomes an organism has. When an organism is euploid, its total number of chromosomes is an exact multiple of its haploid number *(n)*. All *diploid* organisms (like humans) have two complete sets of chromosomes (2*n*).
- **Aneuploid** refers to an imbalance in the number of chromosomes, with either missing or extra chromosomes. Situations involving aneuploidy include *monosomy* (when one chromosome is missing) and *trisomy* (when there is one extra copy of a chromosome). Chapter 13 covers the effects of monosomy and trisomy, either of which can have a significant impact on growth and development.
- **Polyploid** refers to having more than two complete sets of chromosomes. *Triploidy* (having three complete sets of chromosomes) can occur in humans, but the condition results in death before or shortly after birth.

While diploidy is the normal state for humans (and many other organisms, such as dogs, mice, horses, and chickens), numerous organisms have more than two copies of their chromosomes. In fact, polyploidy (including triploidy [3n], tetraploidy [4n], pentaploidy [5n], and hexaploidy [6n]) is common among many plant, fish, and amphibian species.

Sex Chromosomes: Is It a Boy or Girl?

Presumably, since the beginning of time, humans have been aware of the dissimilarities between the sexes. But it wasn't until 1905 that Nettie Stevens stared through a microscope long enough to discover the role of the Y chromosome in the grand scheme of things. Until Stevens came along, the much larger X chromosome was credited with creating all the celebrated differences between males and females.

From a genetics standpoint, the phenotypes of sex — male and female — depend on which type of gamete an individual produces. If an individual produces sperm (or has the potential to produce sperm when mature), it's considered male. If the individual can produce eggs, it's considered female. Some organisms are both male and female (that is, they're capable of producing viable eggs and sperm); this situation is referred to as *monoecy* (pronounced mo-*knee*-see, which means "one house"). Many plants, fish, and invertebrates (organisms lacking a bony spine like yours) are *monoecious* (mo-*knee*-shus).

Humans are *dioecious* (*di*-ee-shus; literally "two houses"), meaning that individuals have either functional male or female reproductive structures, but not both. (There are exceptions that result from genetic disorders of sex differentiation, but most humans have only one set of reproductive structures.) Most of the species you're familiar with are dioecious: Mammals, insects, birds, reptiles, and many plants all have separate sexes.

Organisms with separate sexes get their sex phenotypes in various ways.

- Chromosomal sex determination occurs when the presence or absence of certain chromosomes control sex phenotype.
- Genetic sex determination occurs when specific genes control sex phenotype.
- The environment an organism develops in may determine its sex.

This section examines how chromosomes, genetics, and the environment determine whether an organism is male or female.

Sex determination in humans

REMEMBER

In humans and most other mammals, males and females have the same number of chromosomes, which are found in pairs, making them *diploid*. Humans have 23 pairs of chromosomes, including 22 pairs of autosomes and one pair of sex chromosome. Sex phenotype is determined by the two sex chromosomes: X and Y.

Figure 6-5 shows the basic size and shape of the human X and Y chromosomes. Female humans have two X chromosomes, and male humans have one X and one Y. Check out the sidebar "The X (and Y) files" for how X and Y got their names.

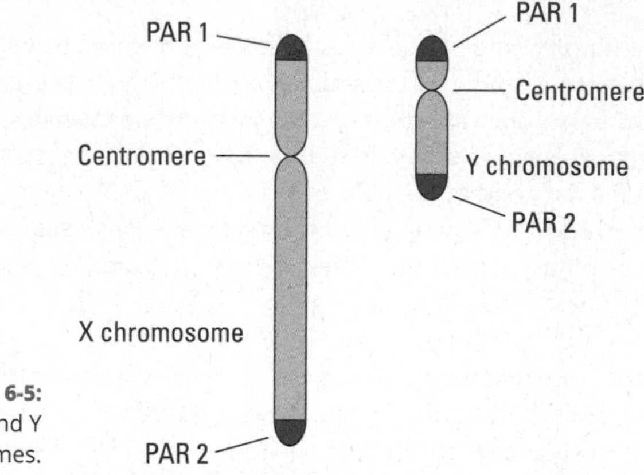

FIGURE 6-5: Human X and Y chromosomes.

The very important X

During metaphase, the X chromosome truly has an x-shape, with the centromere placed roughly in the middle (a *metacentric* chromosome). Genetically speaking, unlike the relatively puny Y chromosome, X is quite large. Of the 23 pairs of chromosomes ordered by size, X occupies the eighth place, weighing in at slightly over 150 million base pairs long. See Chapter 5 for more about how DNA is measured in base pairs.

THE X (AND Y) FILES

Hermann Henking discovered the X chromosome while studying insects in the early 1890s. He wasn't quite sure what the lonely, unpaired structure did, but it seemed different from the rest of the chromosomes he was looking at. So rather than assign it a number (chromosomes are generally numbered according to size, largest to smallest), he called it *X*. In the early 1900s, Clarence McClung decided, rightly, that Henking's X was a chromosome, but he wasn't quite sure of its role. McClung started calling X the *accessory chromosome.* At the time, what we know as the Y chromosome carried the cumbersome moniker of *small ideochromosome.* The prefix *ideo-* means "unknown" — in other words, McClung and other geneticists of the time had no idea what the little Y guy was for.

Edmund Wilson discovered XX-XY sex determination in insects in 1905 (independent of Nettie Stevens, who accomplished the same feat that year). Wilson seems to have had the honor of naming the Y chromosome. According to multiple genetics historians, Wilson first used the name Y in 1909. The Y designation was in no way romantic — it was just convenient shorthand. The new name caught on rapidly, and by 1914 or so, all geneticists were calling the two sex chromosomes X and Y.

The X chromosome is home to approximately 800 to 900 genes and is incredibly important for normal human development. When no X is present, the zygote (fertilized egg) can't begin development. Surprisingly, only one gene on X has a role in determining female phenotype; all the other genes that act to make females are on the autosomal (non-sex) chromosomes.

The not very significant Y

REMEMBER

In comparison to X, the Y chromosome is scrawny, antisocial, and surprisingly expendable. Y contains approximately 200 genes along its 50-million base pair length and is generally considered the smallest and least gene-rich human chromosome.

Most of Y doesn't seem to code for any genes at all. In addition, individuals with only one X and no Y can survive the condition (known as Turner syndrome, discussed in Chapter 13), demonstrating that Y supports no genes required for survival. Almost all the genes Y has are involved in male sex determination and sexual function.

Unlike the other chromosomes, most of Y doesn't recombine during meiosis because Y is so different from X — it has only small regions near the telomeres that allow X and Y to pair during meiosis.

As discussed in the section "Two Chromosomes Are Better than One (or Three)," pairs of human chromosomes are considered *homologous,* meaning the members of each pair are identical in structure and shape and contain similar (although not identical) genetic information.

X and Y aren't homologous — they're different in size and shape and carry different sets of genes. They do, however, contain a small subset of genes that overlap, located in regions called *pseudoautosomal regions* (or PARs). One PAR is located at the tips of the short arms of the sex chromosomes, and one PAR is located at the tips of the long arms of the sex chromosomes, as shown in Figure 6-5.

The genes in the PARs are the same on the X and Y chromosomes. This allows for proper pairing of the X and Y during male meiosis. Crossing over (or recombination) can occur between the PARs of the X and Y, but it does not occur anywhere else in the sex chromosomes.

One of the most important genes located on the Y chromosome is *SRY,* the Sex-determining Region Y gene, which was discovered in 1990. The *SRY* gene is necessary for males to become males. *SRY* codes for a mere 204 amino acids (to see how the genetic code works to make proteins from amino acids, see Chapter 10).

SRY's most important function is starting the development of testes. Embryos that have at least one Y chromosome differentiate into males when the *SRY* gene is turned on during week seven of development. *SRY* acts with at least one other gene (on chromosome 17) to stimulate the expression of the male phenotype in the form of testes. The testes secrete testosterone, the hormone responsible for the expression of most traits belonging to males.

Coping with X and Y inequality

In organisms such as humans that have chromosomal sex determination, male organisms normally have only one X, giving them one copy of each gene on the X and allowing some genes on the X chromosome to act like dominant genes when, in fact, they're recessive. Female organisms must cope with two copies, or doses, of the X chromosome and its attendant genes. If both copies of a female's X were active, she'd get twice as much X-linked gene product as a male. (*X-linked* means located on the X chromosome.) The extra protein produced by two copies of the gene acting at once derails normal development. The solution to this problem is a process called *dosage compensation,* when the amount of gene product is equalized in both sexes.

Dosage compensation is achieved in one of two ways:

- The organism increases gene expression on the X to get a double dose for males. This is what happens in fruit flies, for example.
- The female inactivates essentially all the genes on one X to get a "half" dose of gene expression. The genes located in the pseudoautosomal region of the X chromosome escape inactivation, since there are versions of these genes on the Y chromosome.

Both methods equalize the amount of gene product produced by each sex. In humans, dosage compensation is achieved by *X inactivation;* one X chromosome is permanently and irreversibly turned off in every cell of a female's body.

TECHNICAL STUFF

X inactivation in humans is controlled by a single gene, called *XIST* (*X Inactive-Specific Transcript*), that lies on the X chromosome. When a female zygote starts to develop, it goes through many rounds of cell division. When the zygote gets to be a little over 16 cells in size, X inactivation takes place. The *XIST* gene gets turned on and goes through the normal process of transcription, which we cover in Chapter 9. The RNA (a close cousin of DNA) produced when *XIST* is transcribed isn't translated into protein. Instead, the *XIST* transcript binds directly to one of the X chromosomes to inactivate its genes.

X inactivation causes the inactivated chromosome to change form; it becomes highly condensed and genetically inert. Highly condensed chromosomes are easy for geneticists to spot because they soak up a lot of dye, which geneticists use when studying chromosomes (see Chapter 13). Murray Barr was the first person to observe the highly condensed, inactivated X chromosomes in mammals. Therefore, these inactivated chromosomes are called *Barr bodies*.

REMEMBER

You should remember two very important things about X inactivation:

- In humans, X inactivation is random. Only one X remains turned on, but which X remains on is completely up to chance.
- If more than two Xs are present, only one remains completely active.

The ultimate result of X inactivation is that the tissues that arise from each embryonic cell have a "different" X. Because females get one X from their father and the other from their mother, their Xs are likely to carry different alleles of the same genes. Therefore, their tissues may express different phenotypes depending on which X (Mom's or Dad's) remains active. This random expression of X chromosomes is best illustrated in cats.

Calico and tortoiseshell cats both have patchy-colored fur (often orange and black, but other combinations are possible). The genes that control these fur colors are on the X chromosomes. Male cats are usually all one color because they always have only one active X chromosome (and are XY). Females (XX), on the other hand, also have one active X chromosome, but the identity of the active X (maternal or paternal) varies over the cat's body. Therefore, calico females get a patchy distribution of color depending on which X is active (that is, as long as her parents had different alleles on their Xs — each with an allele for a different color). If you have a calico male cat, he possesses an extra X and has the genotype XXY. XXY cats have normal phenotypes.

Sex determination in other organisms

In mammals, sex determination is directed by the presence of sex chromosomes that turn on the appropriate genes to make male or female phenotypes. In most other organisms, however, sex determination is highly variable. This section looks at how various arrangements of chromosomes, genes, and even temperature affect the determination of sex.

Insects

When geneticists first began studying chromosomes in the early 1900s, insects were the organisms of choice. Grasshopper, beetle, and especially fruit fly chromosomes were carefully stained and studied under microscopes. Much of what we now know about chromosomes in general and sex determination in particular comes from the work of these early geneticists.

In 1901, Clarence McClung determined that female grasshoppers had two X chromosomes, but males had only one (see the sidebar "The X (and Y) files" for more about McClung's role in discovering the sex chromosomes). This arrangement, now known as XX-XO, with the O representing a lack of a chromosome, occurs in many insects. For these organisms, the number of X chromosomes in relation to the autosomal chromosomes determines maleness or femaleness. Two doses of X produce a female. One X produces a male.

In the XX-XO system, females (XX) are *homogametic*, which means that every gamete (in this case, eggs) that the individual produces has the same set of chromosomes composed of one of each autosome and one X. Males (XO) are *heterogametic*; their sperm can come in two different types. Half of a male's gametes have one set of autosomes and an X; the other half have one set of autosomes and no sex chromosome at all. This imbalance in the number of chromosomes is what determines sex for XX-XO organisms.

A similar situation occurs in fruit flies. Male fruit flies are XY, but the Y doesn't have any sex-determining genes on it. Instead, sex is determined by the number of X chromosomes compared to the number of sets of autosomes. The number of X chromosomes an individual has is divided by the number of sets of autosomes (sometimes referred to as the haploid number, *n*; see Chapter 2). This equation is the X to autosome (A) ratio, or X:A ratio. If the X:A ratio is ½ or less, the fly is male. Otherwise, it would be female.

Bees and wasps have no sex chromosomes at all. Their sex is determined by whether the individual is diploid or haploid. Females develop from fertilized eggs and are diploid. Males develop from unfertilized eggs and are therefore haploid.

Organisms that are *diploid* have paired chromosomes. Those that are *haploid* have a single set of chromosomes.

PERPLEXING SEX DETERMINATION

Some organisms have *location-dependent* sex determination, meaning the organism becomes male or female depending on where it ends up. Take the slipper limpet, for example. Slipper limpets (otherwise known by their highly suggestive scientific name of *Crepidula fornicata*) have concave, unpaired shells and cling to rocks in shallow seawater environments. Basically, they look like half of an oyster.

All young slipper limpets start out as male, but a male can become female as a result of his (soon to be her) circumstances. If a young slipper limpet settles on bare rock, it becomes female. If a male settles on top of another male, the one on the bottom becomes a female to accommodate the new circumstances. If a male is removed from the top of a pile and placed on bare rock, he becomes a she and awaits the arrival of a male. After an individual becomes female, she's a female from then on.

Some fish also change sex depending on their locations or their social situations. Bluehead wrasse, a large reef fish familiar to many scuba divers, can change from female to male, depending on the other bluehead wrasse around. If no male is around, or if the local male disappears, large females change sex to become males. The fish's brain and nervous system control its ability to switch from one sex to another. An organ in the brain called the *hypothalamus* regulates sex hormones and controls growth of the needed reproductive tissues.

To add to the list of the truly bizarre, a parasitic critter that lives inside certain fish has an unusual way of changing gender: cannibalism. When a male *Ichthyoxenus fushanensis* (a sort of parasitic pill bug), eats a female, the diner changes sex — that is, he becomes a she. In the case of the isopod, the sex change is a form of hermaphroditism where the sexes are expressed sequentially and in response to some change in the environment or diet.

Birds

Like humans, birds have two sex chromosomes: Z and W. Female birds are ZW, and males are ZZ. Sex determination in birds isn't completely understood, but two genes, one on the Z and the other on the W, seem to play roles in whether an individual becomes male or female. The Z-linked gene suggests that, like the XX-XO system in insects (see the preceding section), the number of Z chromosomes may help determine sex (but with reversed results from XX-XO). On the other hand, the W-linked gene suggests the existence of a "female-determining" gene. The chicken genome sequence has provided critical information for geneticists to learn how sex is determined in birds.

Reptiles

Sex chromosomes determine the sex of most reptiles (like snakes and lizards). However, the sex of most turtles and all crocodiles and alligators is determined by the temperature the eggs experience during incubation. Female turtles and crocodilians dig nests and bury their eggs in the ground. Females usually choose nest sites in open areas likely to receive a lot of sunlight. Female turtles don't bother to guard their eggs; they lay 'em and forget 'em. Alligators and crocodiles, on the other hand, guard their nests (quite aggressively) but let the warmth of the sun do the work.

In turtles, lower temperatures (78–82 degrees Fahrenheit) produce all males. At temperatures over 86 degrees, all eggs become females. Intermediate temperatures produce both sexes. Male alligators, on the other hand, are produced only at intermediate temperatures (around 91 degrees). Cooler conditions (84–88 degrees) produce only females; really warm temperatures (95 degrees) produce all females also.

An enzyme called *aromatase* seems to be the key player in organisms with temperature-dependent sex determination. Aromatase converts testosterone into estrogen. When estrogen levels are high, the embryo becomes a female. When estrogen levels are low, the embryo becomes a male. Aromatase activity varies with temperature. In some turtles, for example, aromatase is essentially inactive at 77 degrees, and all eggs in that environment hatch as males. When temperatures around the eggs get to 86 degrees, aromatase activity increases dramatically, and all the eggs become females.

IN THIS CHAPTER

» Uncovering the pattern for copying DNA

» Putting together a new DNA molecule

» Revealing how circular DNA versus molecules replicate

Chapter 7

Replication: A Copy Machine for DNA

Everything in genetics relies on *replication* — the process of copying DNA accurately, quickly, and efficiently. Replication is part of reproduction (producing eggs and sperm), development (making all the cells needed by a growing embryo), and maintaining normal life (replacing skin, blood, and muscle cells).

Before meiosis can occur, the entire genome must be replicated so that a potential parent can make the eggs or sperm necessary for creating offspring (see Chapter 2 for a review of meiosis). After fertilization occurs, the growing embryo must have the right genetic instructions in every cell to make all the tissues needed for life. As life outside the womb goes on, almost every cell in your body needs a copy of the entire genome to ensure that the genes that carry out the business of living are present and ready for action. For example, because you're constantly replacing your skin cells and white blood cells, your DNA is being replicated right now so that your cells have the genes they need to work properly.

This chapter explains all the details of the fantastic molecular photocopier that allows DNA — the stuff of life — to do its job. First, you tackle the basics of how DNA's structure provides a pattern for copying itself. Then you find out about all the enzymes — those helpful protein workhorses — that do the labor of opening the double-stranded DNA and assembling the building blocks of DNA into a new strand. Finally, you see how the copying process works, from beginning (origins) to ends (telomeres).

Unzipped: Creating the Pattern for More DNA

DNA is the ideal material for carrying genetic information because it:

- Stores vast amounts of complex information *(genotype)* that can be "translated" into physical characteristics *(phenotype)*.
- Can be copied quickly and accurately.
- Is passed down from one generation to the next (in other words, it's *heritable*).

HISTORICAL STUFF

When James D. Watson and Francis Crick proposed the double helix as the structure of DNA, they ended their 1953 paper with a pithy sentence about replication. That one little sentence paved the way for their next major publication, which hypothesized how replication may work. It's no accident that Watson and Crick won the Nobel Prize; their genius was uncanny and amazingly accurate. Without their discovery of the double helix, they never could have figured out replication; the trick that DNA pulls off during replication depends entirely on how DNA is put together in the first place.

REMEMBER

If you skipped Chapter 5, which focuses on how DNA is put together, you may want to skim over that material now. The main points about DNA you need to know to understand replication are:

- DNA is double-stranded.
- The nucleotide building blocks of DNA always match up in a complementary fashion — A (adenosine) with T (thymine) and C (cytosine) with G (guanine).
- DNA strands run antiparallel (that is, in opposite directions) to each other.

REMEMBER

If you were to unzip a DNA molecule by breaking all the hydrogen bonds between the bases, you'd have two strands, and each would provide the pattern to create the other. During replication, special helper proteins called *enzymes* bring matching (complementary) nucleotide building blocks to pair with the bases on each strand. The result is two exact copies built on the *templates* that the unzipped original strands provide.

Figure 7-1 shows how the original double-stranded DNA supplies a template to make copies of itself. This mode of replication is called *semiconservative*. Semiconservative means that only half the molecule is "conserved," or left in its original state. (*Conservative*, in the genetic sense, means keeping something protected in its original state.)

FIGURE 7-1: DNA provides its own pattern for copying itself using semiconservative replication.

HISTORICAL STUFF

At Columbia University in 1957, J. Herbert Taylor, Philip Woods, and Walter Hughes used the cell cycle to determine how DNA is copied. They came up with two possibilities: conservative or semiconservative replication.

Figure 7-2 shows how conservative replication may work. For both conservative and semiconservative replication, the original, double-stranded molecule comes apart and provides the template for building new strands. The result of semiconservative replication is two complete, double-stranded molecules, each composed of half "new" and half "old" DNA (which is what you see in Figure 7-1). Following conservative replication, the complete, double-stranded copies are composed of all "new" DNA, and the templates come back together to make one molecule composed of "old" DNA, as shown in Figure 7-2.

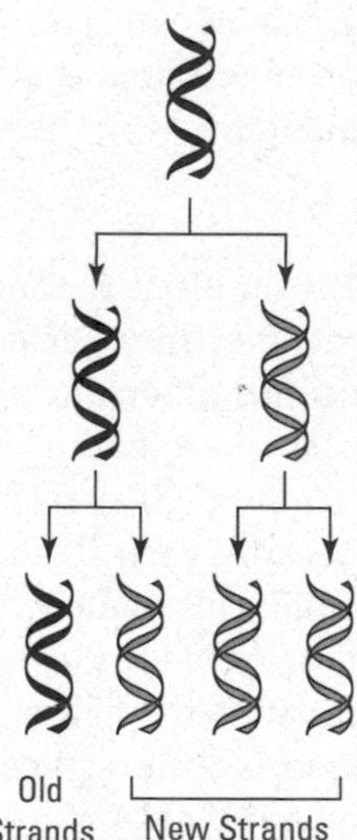

FIGURE 7-2: Conservative replication.

To sort out replication, Taylor and his colleagues exposed the tips of a plant's roots to water that contained a radioactive chemical. This chemical was a form of the nucleotide building block *thymine,* which is found in DNA. Before cells in the root tips divided, their chromosomes incorporated the radioactive thymine as part of newly replicated DNA. In the first step of the experiment, Taylor and his team let the root tips grow for eight hours. That was just long enough for the DNA of the cells in the growing tips to replicate. The researchers collected some cells after this first step to see whether one or both sister chromatids of each chromosome were radioactive. Then, for the second step, they put the root tips in water with no radioactive chemical in it. After the cells started dividing, Taylor and his team examined the replicated chromosomes while they were in *metaphase* (when the replicated chromosomes, called *sister chromatids,* are all lined up together in the center of the cell, before they're pulled apart to opposite ends of the soon-to-divide cell).

The radioactivity allowed Taylor and his team to trace the fate of the template strands after replication was completed and determine whether the strands stayed together with their copies (semiconservative) or not (conservative). They examined the results of both steps of the experiment to ensure that their conclusions were accurate.

If replication was semiconservative, Taylor, Woods, and Hughes expected to find that one sister chromatid of the replicated chromosome would be radioactive and the other would be radiation-free — and that's what they got. Figure 7-3 shows how their results ended up as they did. The shaded chromosomes represent the ones containing the radioactive thymine. After one round of replication in the presence of the radioactive thymine (Step 1 in Figure 7-3), the entire chromosome appears radioactive.

If Taylor and his team could have seen the DNA molecules themselves (as you do figuratively here), they would have known that one strand of each double-stranded molecule contained radioactive thymine and the other did not (the radioactive strands are depicted with a thicker line). After one round of replication without access to the radioactive thymine (Step 2 in Figure 7-3), one sister chromatid was radioactive, and the other was not. That's because each strand from Step 1 provided a template for semiconservative replication: The radioactive strand provided one template, and the nonradioactive strand provided the other. After replication was completed, the templates remained paired with the new strands. This experiment showed conclusively that DNA replication is truly semiconservative — each replicated molecule of DNA is half "new" and half "old."

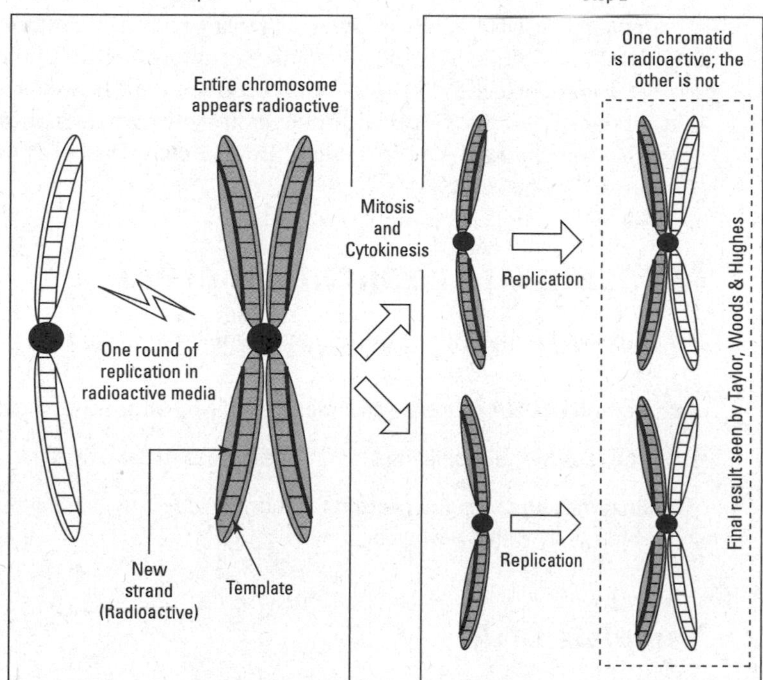

FIGURE 7-3: The results of Taylor, Woods, and Hughes' experiment show that DNA replication is semiconservative.

How DNA Copies Itself

Replication occurs during interphase of each cell cycle, just before prophase in both mitosis and meiosis. If you skipped over Chapter 2, you may want to take a quick glance at it to get an idea of when replication occurs with respect to the life of a cell.

The process of replication follows a very specific order:

1. The helix is opened to expose single strands of DNA.
2. Nucleotides are strung together to make new partner strands for the two original strands.

REMEMBER

DNA replication was first studied in bacteria, which are *prokaryotic* (lacking cell nuclei). *Eukaryotes* are organisms composed of cells with nuclei, like fungi, plants, and animals. Prokaryotic and eukaryotic DNA replication differ in a few ways. Basically, bacteria use slightly different versions of the same enzymes that eukaryotic cells use, and most of those enzymes have similar names. If you understand prokaryotic replication, which we explain in this section, you have enough background to understand the details of eukaryotic replication, too.

Most eukaryotic DNA is linear, whereas most bacterial DNA (and your mitochondrial DNA) is circular. The shape of the chromosome (an endless loop versus a string) doesn't affect the process of replication at all. However, the shape means that circular DNAs have special problems to solve when replicating their hoop-shaped chromosomes. See the section "How Circular DNAs Replicate" later in this chapter to find out more.

Meeting the replication crew

For successful replication, several players must be present:

- **Template DNA,** a double-stranded molecule that provides a pattern to copy
- **Nucleotides,** the building blocks necessary to make new DNA
- **Enzymes and various proteins** that do the unzipping and assembly work of replication, called *DNA synthesis*

Template DNA

In addition to the material earlier in this chapter detailing how the template DNA is replicated in a semiconservative manner (see the section "Unzipped: Creating the Pattern for More DNA"), it's vitally important for you to understand all the meanings of the term *template.*

- Every organism's DNA exists in the form of chromosomes. Therefore, the chromosomes undergoing replication and the template that DNA uses during replication are one and the same.
- Both strands of each double-stranded original molecule are copied, and therefore, each of the two strands serves as a template (that is, a pattern) for replication.

REMEMBER

The bases of the template DNA provide critical information needed for replication. Each new base of the newly replicated strand must be *complementary* to the base opposite it on the template strand (see Chapter 5 for more about the complementary nature of DNA). For example, if an A is present in the template strand, then a T must be added opposite in the newly replicated strand. Together, template and replicated DNA (like you see in Figure 7-1) make two identical copies of the original, double-stranded molecule.

Nucleotides

DNA is made up of nucleotides linked together in paired strands. The nucleotide building blocks of DNA that come together during replication start in the form of *deoxyribonucleoside triphosphates*, or *dNTPs*, which are made up of:

- A sugar (deoxyribose)
- One of four bases (adenine, guanine, thymine, or cytosine)
- Three phosphates

Figure 7-4 shows a dNTP being incorporated into a double-stranded DNA molecule. The dNTPs used in replication are very similar in chemical structure to the ones in double-stranded DNA (you can refer to Figure 5-3 in Chapter 5 to compare a nucleotide to the dNTP in Figure 7-4). The key difference is the number of phosphate groups — each dNTP has three phosphates, and each nucleotide has one.

Take a look at the blowup of the dNTP in Figure 7-4. The three phosphate groups (the "tri-" part of the name) are at the top end (usually referred to as the 5-prime end, or 5′ end) of the molecule. At the bottom left of the molecule, also known as the 3-prime (3′) spot, is a little tail made of an oxygen atom attached to a hydrogen atom (collectively called an *OH group* or a *reactive group*). The oxygen atom in the OH tail is present to allow a nucleotide in an existing DNA strand to hook up with a dNTP; multiple connections like this one eventually produce the long chain of DNA.

When DNA is being replicated, the OH tail on the 3′ end of the last nucleotide in the chain reacts with the phosphates of a newly arrived dNTP (as shown in the right-hand part of Figure 7-4). Two of the dNTP's three phosphates get chopped off, and the remaining phosphate forms a phosphodiester bond with the previously incorporated nucleotide. Hydrogen bonds form between the base of the template strand and the complementary base of the dNTP. This reaction — losing two phosphates to form a phosphodiester bond and hydrogen bonding — converts the dNTP into a nucleotide. The only real difference between dNTP and the nucleotide it becomes is the number of phosphates each carries. Remember, the template DNA must be single-stranded for these reactions to occur (see the section "Splitting the helix," later in this chapter).

Each dNTP incorporated during replication must be complementary to the base it's hooked up with on the template strand.

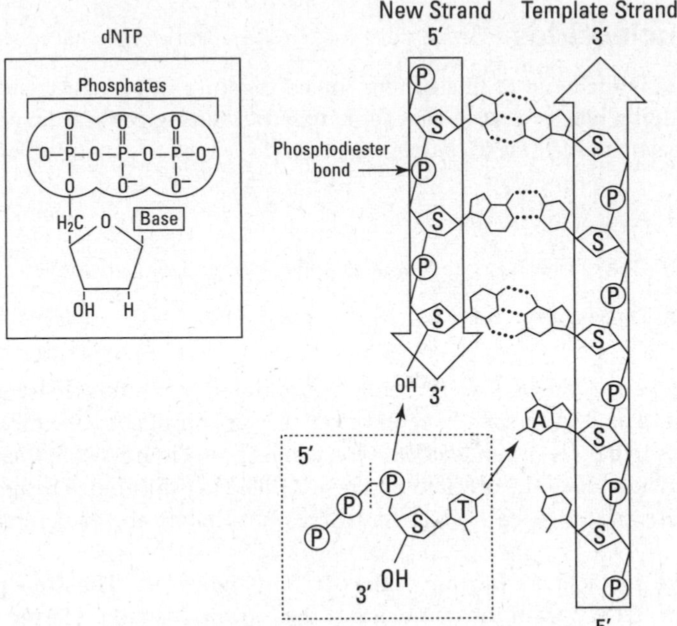

FIGURE 7-4: Connecting the chemical building blocks (nucleotides as dNTPs) during DNA synthesis.

REMEMBER

A nucleotide is a deoxyribose sugar, a base, and a phosphate joined together as a unit. A nucleotide is a nucleotide regardless of whether it's part of a whole DNA molecule or not. A dNTP is also a nucleotide, just a special sort: a nucleotide triphosphate.

Enzymes

Replication can't occur without the help of a huge suite of enzymes. *Enzymes* are proteins that cause reactions. Generally, enzymes come in two flavors: those that put things together and those that take things apart. Both types are used during replication.

TIP

Although you can't always tell the function of an enzyme (building or destroying) by its name, you can always identify enzymes because they end in *-ase*. The *-ase* suffix usually follows a reference to what the enzyme acts on. For example, the enzyme helicase acts on the helix of DNA to make it single-stranded (helix + ase = helicase).

So many enzymes are used in replication that it's hard to keep up with them all. However, the main players and their roles are:

- **Helicase:** Opens the double helix so replication can occur.
- **Gyrase:** Prevents the helix from forming knots.

» **Primase:** Lays down a short piece of RNA (a primer) to get replication started (see Chapter 9 for more on RNA).

» **DNA polymerase:** Adds dNTPs to build the new strand of DNA.

» **Ligase:** Seals the gaps between newly replicated pieces of DNA.

» **Telomerase:** Replicates the ends of chromosomes (the telomeres) — a very special job.

TECHNICAL STUFF

Prokaryotes have 5 forms of DNA polymerase, and eukaryotes have at least 13 forms. In prokaryotes, DNA polymerase III is the enzyme that performs replication. DNA polymerase I removes RNA primers and replaces them with DNA. DNA polymerases II, IV, and V all work to repair damaged DNA and carry out proofreading activities. Eukaryotes use a whole different set of DNA polymerases. (For more details on eukaryotic DNA replication, see the section "Replication in Eukaryotes," later in the chapter.)

Splitting the helix

DNA replication starts at very specific spots, called *origins*, along the double-stranded template molecule. Bacterial chromosomes are so short (only about 4 million base pairs) that only one origin for replication is needed. Copying larger genomes would take far too long if each chromosome had only one origin, so to make the process of copying very rapid, human chromosomes each have thousands of origins. (See the section "Replication in Eukaryotes" later in this chapter for more details on how human DNA is replicated.)

Special proteins called *initiators* move along the double-stranded template DNA until they encounter a group of bases that are in a specific order. These bases represent the origin for replication; think of them as a road sign with the message: "Start replication here." The initiator proteins latch onto the template at the origin by looping the helix around themselves like looping a string around your finger. The initiator proteins then make a very small opening in the double helix.

Helicase (the enzyme that separates the DNA strands so replication can occur) finds this opening and starts breaking the hydrogen bonds between the complementary template strands to expose a few hundred bases and splits the helix open even wider. DNA has such a strong tendency to form double-strands that if another protein didn't come along to hold the single strands exposed by helicase apart, they would snap right back together again. These proteins, called *single-stranded-binding* (SSB) proteins, prop the two strands apart so replication can occur. Figure 7-5 shows the whole process of replication. For now, focus on the part that shows how helicase breaks the strands apart as it moves along the double helix and how the strands are kept separated and untwisted.

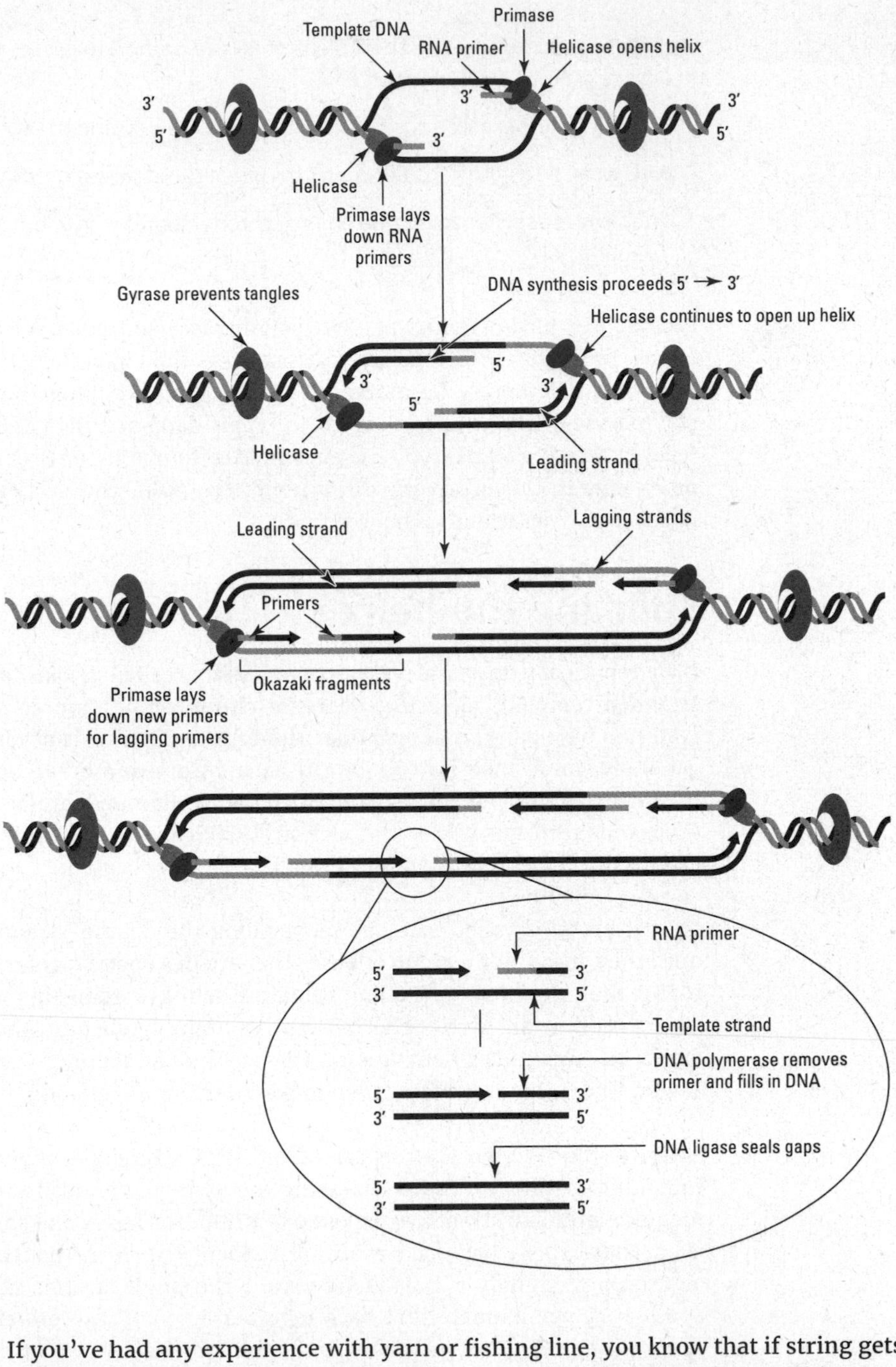

FIGURE 7-5: The process of replication.

If you've had any experience with yarn or fishing line, you know that if string gets twisted together and you try to pull the strands apart, a knot forms. This same problem occurs when opening the double helix of DNA. When helicase starts pulling the two strands apart, the opening of the helix sends extra turns along the intact helix. To prevent DNA from ending up a knotty mess, an enzyme called *gyrase* comes along to relieve the tension. Exactly how gyrase does this is unclear, but some researchers think that gyrase actually snips the DNA apart temporarily to let the twisted parts relax and then seals the molecule back together again.

Priming the pump

When helicase opens the DNA molecule, a Y forms at the opening. This Y is called a *replication fork*. You can see a replication fork in Figure 7-5, where the helicase has split the DNA helix apart. For every opening in the double-stranded molecule, two forks form on opposite sides of the opening. DNA replication is very particular in that it can only proceed in one direction: 5-prime to 3-prime ($5' \rightarrow 3'$). In Figure 7-5, the top strand runs $3' \rightarrow 5'$ from left to right, and the bottom strand runs $5' \rightarrow 3'$ (that is, the template strands are *antiparallel;* see Chapter 5 to review the importance of the antiparallel arrangement of DNA strands).

REMEMBER

Replication must proceed antiparallel to the template, running 5′ to 3′. Therefore, replication on the top strand runs right to left; on the bottom strand, replication runs left to right. Basically, replication occurs in opposite directions on the two DNA strands.

After helicase splits the molecule open (as we explain in the preceding section), two naked strands of template DNA are left. Replication can't start on the naked template strands because no new strand has formed yet. (That sounds a bit like Yogi Berra saying "It ain't over 'til it's over," doesn't it?) All funny business aside, nucleotides can only form chains if a nucleotide is already present with a free reactive tail on which to attach the incoming dNTP. DNA solves the problem of starting replication by inserting *primers*, little complementary starter strands made of RNA (refer to Figure 7-5; to find out more about RNA, see Chapter 9).

Primase, the enzyme that manufactures the RNA primers for replication, lays down primers at each replication fork so that DNA synthesis can proceed from $5' \rightarrow 3'$ on both strands. The RNA primers made by primase are only about 10 or 12 nucleotides long. They're complementary to the single strands of DNA and end with the same sort of OH tail found on a nucleotide of DNA. DNA uses the primers' free OH tails to add nucleotides in the form of dNTPs; the primers are later snipped out and replaced with DNA.

Leading and lagging

As soon as the primers are in place, actual replication can get underway. *DNA polymerase* is the enzyme that does all the work of replication. At the OH tail of each primer, DNA polymerase tacks on dNTPs by snipping off two phosphates and forming phosphodiester bonds. Meanwhile, helicase opens the helix ahead of the growing chain to expose more template strand. From Figure 7-5, it's easy to see that replication can just zoom along this way — but only on one strand (in this case, the top strand in Figure 7-5). The replicated strands keep growing continuously 5′ → 3′ as helicase makes the template available.

At the same time, on the opposite strand, new primers need to be added to take advantage of the newly available template. The new primers are necessary because a naked strand (the bottom one in Figure 7-5) lacking the necessary free nucleotide for chain-building is created by the ongoing splitting of the helix.

Thus, the interaction of opening the helix and synthesizing DNA 5′ → 3′ on one strand while laying down new primers on the other leads to the formation of *leading* and *lagging strands* (see Figure 7-6).

- **Leading strands:** The strands formed in one bout of uninterrupted DNA synthesis. Leading strands follow the lead, so to speak, of helicase.
- **Lagging strands:** The strands that are begun over and over as new primers are laid down. Synthesis of the lagging strands stops when they reach the 5' end of a primer elsewhere on the strand. Lagging strands "lag behind" leading strands in the sense of frequent starting and stopping, versus continuous replication. Replication happens so rapidly that there's no difference in the amount of time it takes to replicate leading and lagging strands. The short pieces of DNA formed by lagging DNA synthesis have a special name: *Okazaki fragments,* named for the scientist, Reiji Okazaki, who discovered them.

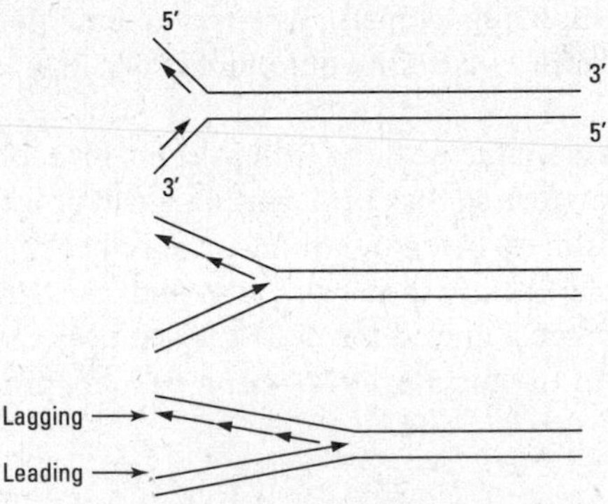

FIGURE 7-6: Leading and lagging strands.

Joining all the pieces

After the template strands are replicated, the newly synthesized strands need to be modified to be complete and whole:

- The RNA primers must be removed and replaced with DNA.
- The Okazaki fragments formed by lagging DNA synthesis must be joined together.

A special kind of DNA polymerase moves along the newly synthesized strands seeking out the RNA primers. When DNA polymerase encounters the short bits of RNA, it snips them out and replaces them with DNA. (Refer to Figure 7-5 for an illustration of this process.) The snipping out and replacing of RNA primers proceeds in the usual $5' \rightarrow 3'$ direction of replication and follows the same procedures as normal DNA synthesis (adding dNTPs and forming phosphodiester bonds).

After the primers are removed and replaced, one phosphodiester bond is missing between the Okazaki fragments. *Ligase* is the enzyme that seals these little gaps (*ligate* means to join things together). Ligase has the special ability to form phosphodiester bonds without adding a new nucleotide.

Proofreading replication

Despite its complexity, replication is unbelievably fast. In humans, replication speeds along at about 2,000 bases a minute. Bacterial replication is even faster at about 1,000 bases per *second!* Working at that speed, it's really no surprise that DNA polymerase makes mistakes — about one in every 100,000 bases is incorrect. Fortunately, DNA polymerase can use the backspace key!

DNA polymerase constantly checks its work through a process called *proofreading* — the same way we proofread our work as we wrote this book. DNA polymerase looks over its shoulder, so to speak, and keeps track of how well the newly added bases fit with the template strand. If an incorrect base is added, DNA polymerase backs up and cuts the incorrect base out. The snipping process is called *exonuclease activity,* and the correction process requires DNA polymerase to move $3' \rightarrow 5'$ instead of the usual $5' \rightarrow 3'$ direction. DNA proofreading eliminates most of the mistakes made by DNA polymerase, and the result is nearly error-free DNA synthesis. Generally, replication (after proofreading) has an astonishingly low error rate of one in 10 million base pairs.

If DNA polymerase misses an incorrect base, special enzymes come along after replication is complete to carry out another process, called *mismatch repair* (much like our editors checked our proofreading). The mismatch repair enzymes detect the bulges that occur along the helix when noncomplementary bases are paired up, and the enzymes snip the incorrect base out of the newly synthesized strand. These enzymes replace the incorrect base with the correct one and, like ligase, seal up the gaps to finish the repair job.

Replication is a complicated process that uses a dizzying array of enzymes. The key points to remember are:

- Replication always starts at an origin.
- Replication can only occur when template DNA is single-stranded.
- RNA primers must be put down before replication can proceed.
- Replication always moves 5' → 3'.
- Newly synthesized strands are exact complementary matches to template ("old") strands.

Replication in Eukaryotes

Although replication in prokaryotes and eukaryotes is very similar, you need to know about four differences:

- For each of their chromosomes, eukaryotes have many, many origins for replication. Prokaryotes generally have one origin per circular chromosome.
- The enzymes that prokaryotes and eukaryotes use for replication are similar but not identical. Compared to prokaryotes, eukaryotes have many more DNA polymerases, and these DNA polymerases carry out other functions besides replication.
- Linear chromosomes, found in eukaryotes, require special enzymes to replicate the *telomeres* — the ends of chromosomes.
- Eukaryotic chromosomes are tightly wound around special proteins in order to package large amounts of DNA into very small cell nuclei (explained in Chapter 6).

Pulling up short: Telomeres

When linear chromosomes replicate, the ends of the chromosomes, called *telomeres*, present special challenges. These challenges are handled in different ways depending on what kind of cell division is taking place (that is, mitosis versus meiosis).

At the completion of replication for cells in mitosis, a short part of the telomere tip is left single-stranded and unreplicated. A special enzyme comes along and snips off this unreplicated part of the telomere. Losing this bit of DNA at the end of the chromosome isn't as big a deal as it may seem, because telomeres are long strings of *non-coding DNA*, which contains repeats of the sequence TTAGGG. Non-coding DNA doesn't contain genes but may have other important functions, like the regulation of gene expression (which will be reviewed in Chapter 11).

For telomeres, being non-coding DNA is good because when telomeres get snipped off, the chromosomes aren't damaged too much and the genes still work just fine — up to a point. After many rounds of replication, all the non-coding, repetitive DNA at the ends of the chromosomes is snipped off (essentially, the chromosomes run out of telomeric DNA), and actual genes themselves are affected. Therefore, when the chromosomes of a mitotic cell (like a skin cell, for example) get too short, the cell dies through a process called *apoptosis*. Paradoxically, cell death through apoptosis is a good thing because it protects you from the ravages of mutations, which can cause cancer.

If the cell is being divided as part of meiosis, telomere snipping is not okay. The telomeres must be replicated completely so that perfectly complete, full-size chromosomes are passed on to offspring. An enzyme called *telomerase* takes care of replicating the ends of the chromosomes. Figure 7-7 gives you an idea of how telomerase replicates telomeres. Primase lays down a primer at the very tip of the chromosome as part of the normal replication process. DNA synthesis proceeds from $5' \rightarrow 3'$ as usual, and then, a DNA polymerase comes along and snips out the RNA primer from $5' \rightarrow 3'$. Without telomerase, the process stops, leaving a tail of unreplicated, single-stranded DNA flapping around (this is what happens during mitosis).

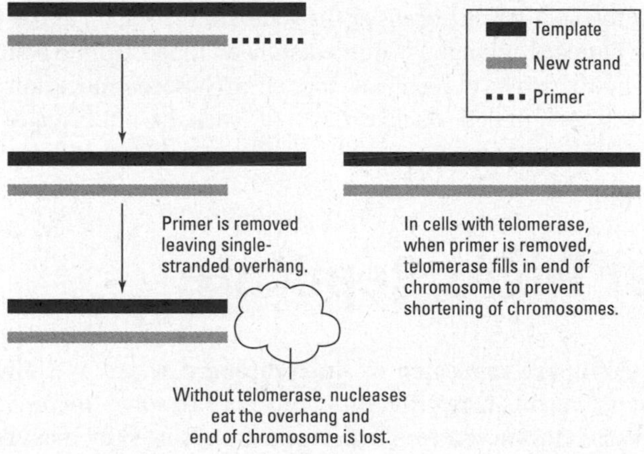

FIGURE 7-7: Telomeres require special help to replicate during meiosis.

Telomerase easily detects the unreplicated telomere because telomeres have long sections of guanines, or Gs. Telomerase contains a section of cytosine-rich RNA, allowing the enzyme to bind to the unreplicated, guanine-rich telomere. Telomerase then uses its own RNA to extend the unreplicated DNA template by about 15 nucleotides. Scientists suspect that the single-stranded template then folds back on itself to provide a free OH tail to replicate the rest of the telomere in the absence of a primer (see "Priming the pump," earlier in this chapter, for more details).

Finishing the job

Your DNA (and that of all eukaryotes) is tightly wound around special proteins to form structures called *nucleosomes* (see Chapter 6 for a review of how DNA is packaged in the cell nuclei). This allows the enormous molecule to fit neatly into the cell nucleus. Like replication, packaging DNA is a very rapid process.

In the packaging stage, DNA is normally twisted tightly around hundreds of thousands of nucleosomes, much like string wrapped around beads. The whole "necklace" gets wound very tightly around itself in a process called *supercoiling.* In each nucleosome, about 150 base pairs of DNA are wrapped around a set of special proteins called *histones.*

DNA is packaged in this manner both before and after replication. Because only 30 or 40 base pairs of DNA are exposed between nucleosomes, the DNA must be removed from the nucleosomes in order to replicate. If it isn't removed from the nucleosomes, the enzymes used in replication aren't able to access the entire molecule.

TECHNICAL STUFF

As helicase opens the DNA molecule during replication, an unidentified enzyme strips off the nucleosome beads at the same time. As soon as the DNA is replicated, the DNA (both old and new) is immediately wrapped around waiting nucleosomes. Studies show that the old nucleosomes (from before replication) are reused along with newly assembled nucleosomes to package the freshly replicated DNA molecule.

How Circular DNAs Replicate

Circular DNAs are replicated in three different ways, as shown in Figure 7-8. Different organisms take different approaches to solve the problem of replicating hoop-shaped chromosomes. Theta replication is used by most bacteria, including *E. coli.* Viruses use rolling circle replication to rapidly manufacture vast numbers of copies of their genomes. Finally, human mitochondrial DNA and the chloroplast DNA of plants both use D-loop replication.

Theta

Theta replication refers to the shape the chromosome takes on during the replication process. After the helix splits apart, a bubble forms, giving the chromosome a shape reminiscent of the Greek letter theta (Θ; see Figure 7-8). Bacterial chromosomes have only one origin of replication (see the earlier section, "Splitting the helix"), so after helicase opens the double helix, replication proceeds in both directions simultaneously, rapidly copying the entire molecule. As we describe previously in the section "Leading and lagging," leading and lagging strands form, and ligase seals the gaps in the newly synthesized DNA to complete the strands. Ultimately, theta replication produces two intact, double-stranded molecules.

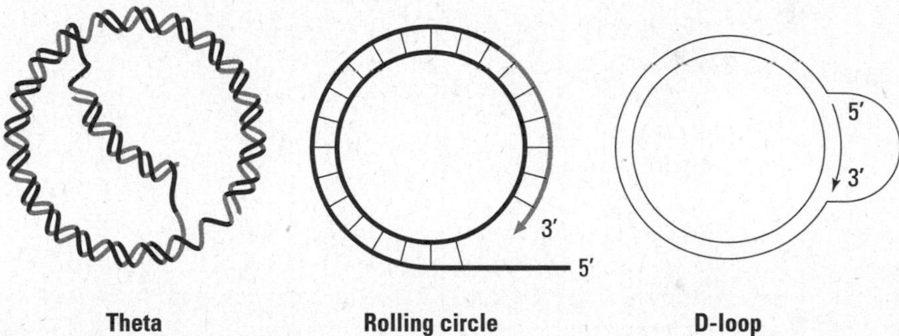

FIGURE 7-8: Circular DNA can be replicated in one of three ways.

Rolling circle

Rolling circle replication creates an odd situation. No primer is needed because the double-stranded template is broken at the origin to provide a free OH tail to start replication. As replication proceeds, the inner strand is copied continuously as a leading strand (refer to Figure 7-8). Meanwhile, the broken strand is stripped off. As soon as enough of the broken strand is freed, a primer is laid down so replication can occur as the broken strand is stripped away from its complement. Thus, rolling circle replication is continuous on one strand and lagging on the other. As soon as replication is completed for one copy of the genome, the new copies are used as templates for additional rounds of replication. Viral genomes are often very small (only a few thousand base pairs), so rolling circle replication is an extremely rapid process that produces hundreds of thousands of copies of viral DNA in only a few minutes.

D-loop

Like rolling circle replication, *D-loop replication* creates a displaced, single strand (refer to Figure 7-8). Helicase opens the double-stranded molecule, and an RNA primer is laid down, displacing one strand. Replication then proceeds around the circle, pushing the displaced strand off as it goes. The intact, single strand is released and used as a template to synthesize a complementary strand.

IN THIS CHAPTER

» Sequencing DNA to determine the order of the bases

» Appreciating the contributions of the Human Genome Project

» Discovering the genomes of other species

Chapter 8

DNA Sequencing: Decoding the Genome

Imagine owning a library of 22,000 books. We don't mean just any books; this collection contains unimaginable knowledge, such as solutions to diseases that have plagued humankind for centuries, basic building instructions for just about every creature on earth, and even the explanation of how thoughts are formed inside your brain. This fabulous library has only one problem — it's written in a mysterious language, a code made up of only four letters that are repeated in arcane patterns. The very secrets of life on earth have been contained within this library since the dawn of time, but no one could read the books — until now.

The 22,000 books are the genes that carry the information that make you. The library storing these books is the human genome. Sequencing *genomes* (that is, all the DNA in one set of chromosomes of an organism), both human genomes and those of other organisms, means discovering the order of the four bases (C, G, A, and T) that make up DNA. The order of the bases in DNA is incredibly important because it's the key to DNA's language, and understanding the language is the first step in reading the books of the library. Most of your genes are identical to those in other species, so sequencing the DNA of other organisms, such as fruit flies, roundworms, chickens, and even yeast, supplies scientists with a lot of information about the human genome and how human genes function.

Sequencing: Reading the Language of DNA

DNA sequencing involves determining the order of nucleotides in DNA. A wealth of information is contained in the sequence of those nucleotides. The sequence can be used to identify genes and locate segments of DNA that involve the regulation of gene expression. In addition, DNA sequencing can be used to determine the cause of genetic syndromes and identify specific changes in genes that lead to disease.

The chemical nature of DNA, which we covered in Chapter 5, and the replication process, which we covered in Chapter 7, are essential to DNA sequencing. If you skipped either of these chapters, you may want to flip back for a brief review of these topics.

Identifying the players in DNA sequencing

New technologies are rapidly changing the way DNA sequencing is done. However, the old tried-and-true approach we describe here, called the Sanger method (named after inventor Frederick Sanger), still provides the basis for many of the newer methods.

The key ingredients for DNA sequencing using the Sanger method are:

- **DNA:** From a single individual of the organism to be sequenced.
- **Primers:** Several thousand copies of short sequences of DNA that are complementary to the part of the DNA to be sequenced.
- **dNTPs:** Many As, Gs, Cs, and Ts, put together with sugars and phosphates as *nucleotides,* the normal building blocks of DNA.
- **ddNTPs:** Many As, Gs, Cs, and Ts as nucleotides that each lack an oxygen atom at the 3' spot. These are fluorescently labeled with different colored dyes so that they can be "read" by a computer.
- **Taq polymerase:** The enzyme that puts the DNA molecule together (see Chapter 18 for more details on Taq).

The DNA is typically extracted from an individual's white blood cells or skin cells found in their saliva. However, cells from any tissue can be used (such as cells from a biopsy). The remaining ingredients can either be manufactured in the laboratory or purchased from a supplier.

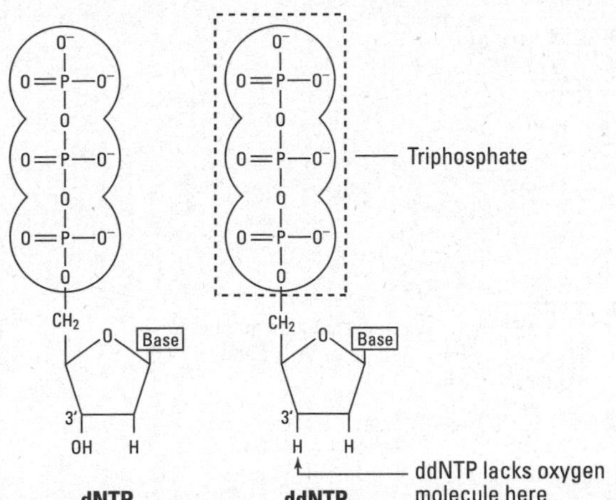

FIGURE 8-1: Comparison of the chemical structure of a generic dNTP (left) and a ddNTP (right).

The use of ddNTPs is the key to how this sequencing process works. Take a careful look at Figure 8-1. On the left is a generic dNTP, the basic building block of DNA used during replication (if you don't remember all the details, see Chapter 5 for more on dNTPs). The molecule on the right is ddNTP *(di-deoxyribonucleoside triphosphate)*. The ddNTP is identical to the dNTP in every way except that it has no oxygen atom at the 3′ spot. No oxygen means no reaction, because the phosphate group of the next nucleotide can't form a phosphodiester bond without that extra oxygen atom to aid the reaction. The next nucleotide can't hook up to ddNTP at the end of the chain, and the replication process stops. So how does *stopping* the reaction help the sequencing process? The idea is to create thousands of pieces of DNA of varying lengths that give the identity of each and every base along the sequence (see Figure 8-2).

In the Sanger method, four parallel reactions are performed at the same time — one for each nucleotide. Each of the reactions has one ddNTP (ddATP, ddTTP, ddGTP, or ddCTP) and all four dNTPs (needed to generate the DNA fragments that result from the reactions). They also contain the template DNA (the DNA fragment to be sequenced), which is heated to separate the two strands of the DNA. Last, primers that can bind to the template provide a place for the sequencing reaction to start. With the addition of Taq polymerase (pronounced "*tack*"), nucleotides are then added to the 3′ end of the primer (using the free OH group). When dNTP is added, the reaction can continue and more nucleotides can be added to the growing fragment. However, when a ddNTP is added, no more nucleotides can be added to that particular fragment (since there is no free 3′ OH to add on to).

When the reactions are complete, the result is DNA fragments of varying lengths. By knowing which ddNTP stopped the reaction and compiling the data from all four reactions, one can determine the order of the nucleotides in the sequence generated (refer to Figure 8-2).

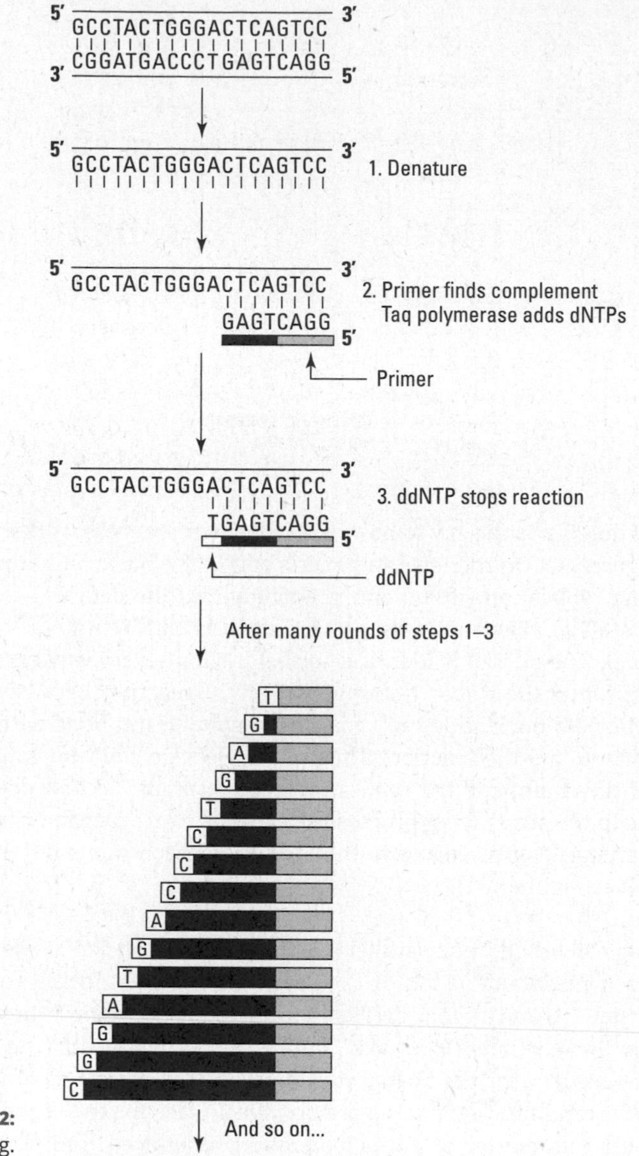

FIGURE 8-2: DNA sequencing.

The result of a typical Sanger sequencing reaction is a few hundred fragments representing a few hundred bases of the template strand. The shortest fragment is made up of a primer and one ddNTP representing the complement of the first base of the template. The next shortest fragment is made up of the primer, one nucleotide (from a dNTP), and a ddNTP — and so on, with the largest fragment being around a thousand bases long.

TIP

Sanger sequencing is a complicated process. If you are more of a visual learner, you can watch a brief animation of the process at `https://www.biointeractive.org/classroom-resources/sanger-sequencing`. This may help you visualize the steps and better understand the process from beginning to end.

Finding the message in sequencing results

To see the results of the sequencing reaction, scientists put the DNA fragments through a process called *gel electrophoresis*. Electrophoresis is the movement of charged particles (in this case, DNA) through a gel with small pores under the influence of electricity. The purpose of electrophoresis is to sort the fragments of DNA by size, with the smallest going the farthest and the longest going the shortest distance. The smallest fragment gives the first base in the sequence, the second-smallest fragment gives the second base, and so on, until the largest fragment gives the last base in the sequence. This arrangement of fragments allows researchers to read the sequence in its proper order.

TECHNICAL STUFF

In the automated version of this process, a computer-driven machine called a *sequencer* uses a laser to see the colored dyes of the ddNTPs at the end of each fragment. The laser shines into the gel and reads the color of each fragment as it passes by. Fragments pass the laser in order of size, from smallest to largest. Each dye color signals a different letter. For example, adenines (A) show up green, thymines (T) are red, cytosines (C) are blue, and guanines (G) are yellow. The computer automatically translates the colors into letters and stores all the information for later analysis.

The resulting picture is a series of peaks, as shown in Figure 8-3. Each peak represents a different base and is shown in a different color, although the color is not shown in Figure 8-3. The sequence indicated by the peaks is the *complement* of the template strand (see Chapter 5 for more on the complementary nature of DNA). When you know the complement of the template, you know the template sequence itself. You can then mine this information for the location of genes (see Chapter 10) and compare it to the sequences of other organisms.

Newer, cheaper, faster

Since Sanger sequencing was first introduced, sequencing methods have come a long way. New sequencing technologies have not only made sequencing much faster, but also much cheaper. Initially, Sanger sequencing required the use of radioactive nucleotides (ddNTPs) to visualize the fragments generated, and it required someone in the laboratory to "read" the results manually. That means that someone had to look at a picture of the results from the four sequencing reactions, determine the order of the bases from that picture, and type them in to a computer file manually. To determine the order of several hundred bases in a single fragment took several days.

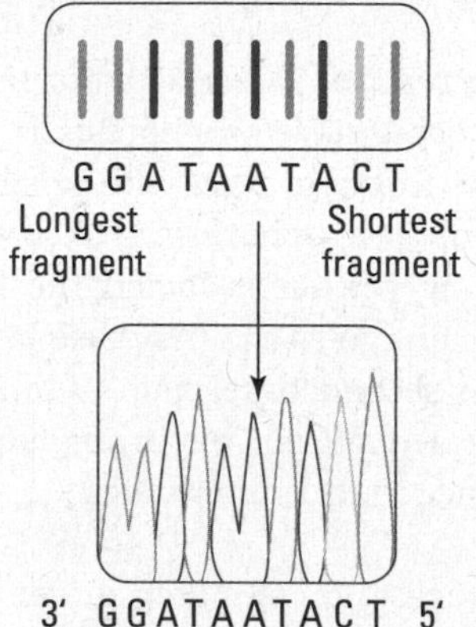

FIGURE 8-3: Results of a typical sequencing reaction.

TECHNICAL STUFF

More modern sequencing technologies, often referred to as *next-generation sequencing methods*, are almost completely automated. One of these newer methods involves identifying each base in a segment of DNA as it is added to a growing DNA strand. Each base emits a unique fluorescent signal which can then be identified by a computer. Another newer technique that uses a fluorescent signal to detect the bases as they are added to a growing DNA strand is called *pyrosequencing*. However, in this method, what is detected is pyrophosphate release as the new nucleotides are incorporated by DNA polymerase. A third method is *ion torrent sequencing*, but this method does not utilize fluorescence to identify bases. Instead, ion torrent sequencing measures the direct release of hydrogen ions as the nucleotides are added by DNA polymerase.

What is common about these newer techniques is that they can be used to sequence larger amounts of DNA much faster than the older methods used in the laboratory. The first reference genome sequenced as a part of the Human Genome Project (described in the next section) took approximately 13 years and around 2.7 billion dollars. By the end of 2015, the cost of sequencing a genome dropped to less than $1,500, and it could be done in just a matter of days!

Sequencing Your Way to the Human Genome

The DNA of all organisms holds a vast amount of information. Amazingly, most cell functions work the same, regardless of which animal the cell comes from. Yeast, elephants, and humans all replicate DNA in the same way, using almost identical genes. Because nature uses the same genetic machinery over and over, finding out about the DNA sequences in other organisms tells us a lot about the human genome (and it's far easier to experiment with yeast and roundworms than with humans). Table 8-1 is a timeline of the major milestones of DNA sequencing projects so far. In this section, you find out about several of these projects, including the granddaddy of them all, the Human Genome Project.

TABLE 8-1 **Major Milestones in DNA Sequencing**

Year	Event
1985	Human Genome Project is proposed.
1990	Human Genome Project officially begins.
1992	First map of known genes is published (the location and order of these genes on their respective chromosomes has been established).
1995	First sequence of an entire living organism — *Haemophilus influenzae,* a flu bacterium — is completed.
1997	Genome of *Escherichia coli,* the most common intestinal bacteria, is completed.
1999	First human chromosome, chromosome 22, is completely sequenced. Human Genome Project passes the 1 billion base pairs milestone.
2000	Fruit fly genome is completed. First entire plant genome — *Arabidopsis thaliana,* the common mustard plant — is sequenced.
2001	First working "draft" sequence of the human genome is published.
2002	Mouse genome is completed.
2004	Chicken genome is completed, as is the *euchromatin* (gene-containing) sequence of the human genome.
2006	Cancer Genome Atlas project launched.
2008	First high-resolution map of genetic variation among humans is published.

The Human Genome Project

In 2001, the triumphant publication of the human genome sequence was heralded as one of the great feats of modern science. The sequence was considered a draft, and indeed, it was a really *rough* draft. The 2001 sequence was woefully incomplete (it represented only about 60 percent of the total human genome) and was full of errors that limited its utility. In 2004, the *euchromatic* (or gene-containing) sequence had only a few gaps, and most of the errors had been corrected. By 2008, new technologies allowed comparisons between individual humans, laying the foundation for a better understanding of how genes vary to create the endless phenotypes you see around you.

HISTORICAL STUFF

The Human Genome Project (HGP) is akin to some of the greatest adventures of all time — it's not unlike putting a person on the moon. However, unlike many proposed projects of this size, the HGP was completed on time and *under* budget! It was estimated that the project would cost $3 billion. It was completed for $2.7 billion, and was completed almost two years earlier than expected. When first proposed in 1985, the HGP was considered completely impossible. At that time, sequencing technology was slow, requiring several days to generate only a few hundred base pairs of data. James Watson, one of the discoverers of DNA structure way back in the 1950s, was one of the first to push the project (in 1988) from idea to reality during his tenure as director of the National Institutes of Health. When the project got off the ground in 1990, a global team of scientists from 20 institutions participated. (The 2001 human genome sequence paper had a staggering 273 authors.)

REMEMBER

The enormous benefits of the HGP remain underappreciated. Most genetic applications wouldn't exist without the HGP. Here are just a few:

- Knowledge of which genes control what functions and how those genes are turned on and off.
- Understanding of the causes of cancer.
- Diagnosis of genetic disorders.
- Development of better drugs and gene therapy.
- Identification of bacteria and viruses to allow for targeted treatment of disease. Some antibiotics, for example, target some strains of bacteria better than others. Genetic identification of bacteria is quick and inexpensive, allowing physicians to rapidly identify and prescribe the right antibiotic.
- Development of bioinformatics, an entirely new field focused on advancing technological capability to generate genetic data, catalog results, and compare

genomes. For more on bioinformatics, see the sidebar "Bioinformatics: Mining the Data."

» Forensics applications, such as identification of criminals and determination of identity after mass disasters.

Listing and explaining all the HGP's discoveries would fill this book and then some. As you can see in Table 8-1, all other genome projects — mouse, fruit fly, yeast, roundworm, mustard weed, and so on — were started as a result of the HGP.

As the HGP progressed, the gene count in the human genome steadily declined. Originally, researchers thought that humans had as many as 100,000 genes. But as new and more accurate information has become available over the years, they've determined that the human genome has only about 22,000 genes. Genes are often relatively small, from a base-pair standpoint (roughly 3,000 base pairs), meaning that less than 2 percent of your DNA actually codes for some protein. The number of genes on different chromosomes varies enormously, from nearly 3,000 genes on chromosome 1 (the largest) to less than 300 genes on the Y chromosome (the smallest).

The Human Genome Project has revealed the surprisingly dynamic nature of the human genome. One of the surprising discoveries of the HGP is that the human genome is still changing. Genes, which we discuss in detail in Chapter 11, can get duplicated and then gain new functions, adding new genes to the genome. Other genes can be altered and lose function, removing genes from the genome. Some genes that were once functional now exist as *pseudogenes*, which have the sequence structure of normal genes but no longer code for proteins.

REMEMBER

Of the human genes that researchers have identified, they only understand what a portion of them do. Comparisons with genomes of other organisms help identify what genes do, because most of the proteins that human genes produce have counterparts in other organisms. Thus, humans share many genes in common with even the simplest organisms, such as bacteria and worms. Over 99 percent of your DNA is identical to that of any other human on earth, and as much as 98 percent of your DNA is identical to the sequences found in the mouse genome. Perhaps the greatest take-home message of the HGP is how alike all life on earth really is.

One of the ways scientists figure out what functions various kinds of sequences carry out is by comparing genomes of different organisms. To make these comparisons, the projects we describe in this section use the methods explained in the section "Sequencing: Reading the Language of DNA," earlier in this chapter. The results of these comparisons tell us a lot about ourselves and the world around us.

BIOINFORMATICS: MINING THE DATA

With the introduction of newer sequencing methods that are both faster and cheaper, a vast amount of data has been generated. Along with this has come a huge growth in the field of bioinformatics, which combines the computer analysis of biological data with the development of programs to store, retrieve, and distribute this information. The field of bioinformatics involves a combination of biological sciences, mathematics, statistics, and computer science.

Scientists working in bioinformatics use computers to develop algorithms that allow them to analyze large amounts of sequencing data and measure the similarity between sequences from different people and different organisms. Computer programs may also be used to predict protein structure based on amino acid sequences and to predict which proteins may interact with each other or with certain DNA sequences. In addition, bioinformatics is being used in the development of new drugs by helping to identify potential drug targets and to predict possible drug interactions or drug resistance.

Many software programs have been developed to analyze the data generated by the Human Genome Project and other large projects that have followed. In addition, large databases have been created and are used to store the data generated. Many of these are the result of large international collaborations, and the result of a large movement toward open access (meaning the data is open to all who wish to use it). Consequently, there are now public databases that allow anyone to see the reference sequence for a variety of genomes (human, mouse, dog, and many other plants and animals). There are also public databases that catalog variations found in human DNA sequences and that classify them as harmless (benign), potentially disease-causing (pathogenic), or of unknown consequence. In addition, there are multiple projects that have sequenced the genomes of thousands of individuals of different ethnic backgrounds and that report their findings regarding the frequency of DNA sequence variations in publicly accessible databases. These include the 1000 Genomes Project (now supported by the International Genome Sample Resource; `www.internationalgenome.org`), the Exome Aggregation Consortium (ExAC; `http://exac.broadinstitute.org/about`), and the Genome Aggregation Database (gnomAD; `https://gnomad.broadinstitute.org/`).

Trying on a Few Genomes

Humans are incredibly complex organisms, but when it comes to genetics, they're not at the top of the heap. Other organisms have vastly larger genomes than humans do. Genomes are usually measured in the number of base pairs they contain (you can flip back to Chapter 5 for more about how DNA is put together in base pairs). Table 8-2 lists the genome sizes and estimated number of genes for various organisms (for some genomes, the numbers of genes are still unknown). The size of the human genome is smaller than several organisms, including a fish, a salamander, and a plant! It's humbling, but true — a Japanese herb has a genome that is 50 times the size of the human genome. If genome size and complexity were related (and they obviously aren't), you'd expect less complex organisms to have smaller genomes than those of more complex organisms. On the flip side, it doesn't take a lot of DNA to have a big impact on the world. For example, the human immunodeficiency virus (HIV), which causes acquired immunodeficiency syndrome (AIDS), is a mere 9,700 bases long and is responsible for the deaths of over 25 million people worldwide. With only nine genes, HIV isn't very complex, but it's still very dangerous.

Even organisms that are similar have vastly different genome sizes. *Arabadopsis thaliana*, a small plant in the mustard family, has roughly 120 million base pairs of DNA. Compare that to the genome of the herb plant *Paris japonica*, which weighs in at a whopping 150 *billion* base pairs. But these two plants aren't *that* different. So if it isn't organism complexity, what causes the differences in genome size among organisms?

TABLE 8-2 Genome Sizes of Various Organisms

Species	Number of Base Pairs	Number of Genes
HIV virus	9,700	9
Influenza A virus	14,000	11
E. coli	4,600,000	~3,200
Yeast	12,000,000	~6,600
Roundworm	103,000,000	~20,000
Arabidopsis thaliana (mustard weed)	120,000,000	~27,000
Fruit fly	180,000,000	~14,000
Chicken	1,000,000,000	~16,000
Mouse	2,700,000,000	~23,000
Human	3,000,000,000	~22,000
Axolotl (salamander)	32,000,000,000	Unknown
Marbled lungfish	130,000,000,000	Unknown
Paris japonica (herb plant)	150,000,000,000	Unknown

Neither the number of chromosomes nor the number of genes tell the whole story. What may be the difference between the genomes of the various organisms is how much of the various *types* of DNA each organism has.

DNA sequences fall into two major categories:

- Unique sequences found in the genes
- Repetitive sequences found in noncoding DNA

The presence of repetitive sequences of DNA in some organisms seems to best explain genome size — that is, large genomes have many repeated sequences that smaller genomes lack. Repetitive sequences vary from 150 to 300 base pairs in length and are repeated thousands and thousands of times. Initially, since all this repetitive DNA didn't seem to do anything, it was dubbed *junk DNA*.

Junk DNA has suffered a bum rap. For years, it was touted as a genetic loser, just along for the ride, doing nothing except getting passed on from one generation to the next. But no more. At long last, so-called junk DNA is getting proper respect. Recent evidence has shown that non-coding DNA is filled with sequences that are involved in the regulation of gene activity — playing a role in deciding when and where different genes get turned off or on. We delve into the regulation of gene expression in Chapter 11.

The yeast genome

Brewer's yeast (scientific name *Saccharomyces cerevisiae*) was the first eukaryotic genome to be fully sequenced. (*Eukaryotes* have cells with nuclei; see Chapter 2.) Yeast has an established track record as one of the most useful organisms known to humankind. It's responsible for making bread rise and for the fermentation that results in beer and wine. It's also a favorite organism for genetic study. Much of what we know about the eukaryotic cell cycle came from yeast research. Yeast has also provided information about how genes are inherited together (called linkage; see Chapter 4) and how genes are turned on and off (see Chapter 11). Because many human genes have yeast counterparts, yeast is extremely valuable for finding out how our own genes work.

The elegant roundworm genome

The genome of the lowly roundworm, more properly referred to by its full name *Caenorhabditis elegans*, was the first genome of a multicellular organism to be fully sequenced. Weighing in at roughly 97 million base pairs, the roundworm boasts nearly 20,000 genes — only a few thousand fewer than the human genome — on just six chromosomes. Like humans, roundworms have lots of junk DNA; only 25 percent of the roundworm genome is made up of genes.

Roundworms are a fabulous species to study because they reproduce sexually and have organ systems, such as digestive and nervous systems, similar to those in much more complex organisms. Additionally, roundworms have a sense of taste, can detect odors, and react to light and temperature, so they're ideal for studying all sorts of processes, including behavior. Full-grown roundworms have exactly 959 cells and are transparent, so figuring out how their cells work was relatively easy. Scientists determined the exact function of each of the 959 roundworm cells! Although roundworms live in soil, these microscopic organisms have contributed to our understanding of many human diseases.

One of the ways to discover what a gene does is to stop it from functioning and observe the effect. In 2003, a group of researchers fed roundworms a particular kind of RNA that temporarily puts gene function on hold (see Chapter 10 for how this effect on gene function works). By briefly turning genes off, the scientists were able to determine the functions of roughly 16,000 of the roundworms' genes. Another study using the same technique identified how fat storage and obesity are controlled in roundworms. Given that an amazing 70 percent of proteins that humans produce have roundworm counterparts, these gene function studies have obvious implications for human medicine.

The chicken genome

Chickens don't get enough respect. The study of chicken biology has revealed much about how organisms develop from embryos to adults. For example, a study of how a chicken's wings and legs are formed in the egg greatly enhanced a study of human limb formation. Chickens have contributed to our understanding of diseases such as muscular dystrophy and epilepsy, and chicken eggs are the principal ingredient used to produce vaccinations to fight human disease epidemics. So, when the chicken genome was sequenced in 2004, there should have been a lot of crowing about the underappreciated chicken.

The chicken genome is very different from mouse and human genomes. It's much smaller (about a third the size of the human genome), with fewer chromosomes (39 compared to our 46), and a similar number of genes (23,000 or so). Roughly 60 percent of chicken genes have human counterparts. Unlike mammals, some chicken chromosomes are tiny (only about 5 million base pairs). These microchromosomes are unique because they have a very high content of guanine and cytosine (see Chapter 5 for more about the bases that make up DNA) and very few repetitive sequences.

IN THIS CHAPTER

- » Picking out the chemical components of RNA
- » Meeting the various RNA molecules
- » Transcribing DNA's message into RNA

Chapter 9
RNA: DNA's Close Cousin

DNA is the stuff of life. Practically every organism on earth relies on DNA to store genetic information and transmit it from one generation to the next. The road from *genotype* (building plans) to *phenotype* (physical traits) begins with *transcription* — making a special kind of copy of a segment of DNA (a gene). DNA is so precious and vital to *eukaryotes* (organisms made up of cells with nuclei) that it's kept packaged in the cell nucleus, like a rare document that's copied but never removed from storage. Because it can't leave the safety of the nucleus, DNA directs all the cell's activity by delegating responsibility to another chemical, RNA. RNA carries messages out of the cell nucleus into the cytoplasm (visit Chapter 2 for more about navigating the cell) to direct the production of proteins during *translation*, a process you find out more about in Chapter 10.

You Already Know a Lot about RNA

If you read Chapter 5, in which we cover DNA at length, you already know a lot about *ribonucleic acid*, or RNA. From a chemical standpoint, RNA is very simple. It's composed of:

- » Ribose sugar (instead of deoxyribose, which is found in DNA).
- » Four nucleotide bases (three you know from DNA — adenine, guanine, and cytosine — plus an unfamiliar one called *uracil*).
- » Phosphate (the same phosphate found in DNA).

REMEMBER

RNA has three major characteristics that make it different from DNA:

- RNA is very unstable and decomposes rapidly.
- RNA contains uracil in place of thymine.
- RNA is almost always single-stranded.

Using a slightly different sugar

Both RNA and DNA use a *ribose* sugar as a main element of their chemical structures. The ribose sugar used in DNA is deoxyribose (see Chapter 5 if you need a review). RNA, on the other hand, uses unmodified ribose. Take a careful look at Figure 9-1. You can see that three spots on ribose are marked with numbers. On ribose sugars, numbers are followed by a single prime (′). Ribose and deoxyribose both have an oxygen (O) atom and a hydrogen (H) atom (an OH group) at their 3′ sites.

OH groups are also called *reactive groups* because oxygen atoms are very aggressive from a chemical standpoint (so aggressive that some chemists say they "attack" incoming atoms). The 3′ OH tail is required for phosphodiester bonds to form between nucleotides in both ribose and deoxyribose atoms, thanks to their aggressive oxygen atoms. (To review how phosphodiester bonds form during replication, see Chapter 7.)

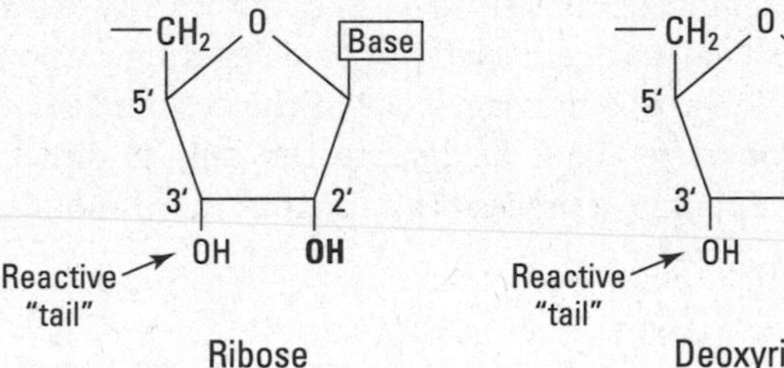

FIGURE 9-1: The ribose sugar is part of RNA.

REMEMBER

The difference between the two molecules is an oxygen atom at the 2′ spot: absent (with deoxyribose) or present (with ribose). This one oxygen atom has a huge hand in the differing purposes and roles of DNA and RNA:

- **DNA:** DNA must be protected from decomposition. The absence of one oxygen atom is part of the key to extending DNA's longevity. When the 2' oxygen is missing, as in deoxyribose, the sugar molecule is less likely to get involved in chemical reactions (because oxygen is chemically aggressive); by being aloof, DNA avoids being broken down.

» **RNA:** RNA easily decomposes because its reactive 2' OH tail introduces RNA into chemical interactions that break up the molecule. Unlike DNA, RNA is a short-term tool the cell uses to send messages and manufacture proteins as part of gene expression, which we cover in Chapter 11. Messenger RNAs (mRNAs) carry out the actions of genes. Put simply, to turn a gene "on," mRNAs have to be made, and to turn a gene "off," the mRNAs that turned it "on" have to be removed. So the 2' OH tail is a built-in mechanism that allows RNA to be decomposed, or *removed,* rapidly and easily when the message is no longer needed and the gene needs to be turned "off." Chapter 11 covers how genes are turned off and on.

Meeting a new base: Uracil

RNA is composed of four nucleotide bases. Three of the four bases may be quite familiar to you because they're also part of DNA: adenine (A), guanine (G), and cytosine (C). The fourth base, uracil (U), is found only in RNA. In DNA, the fourth base is thymine. RNA's bases are pictured in Figure 9-2.

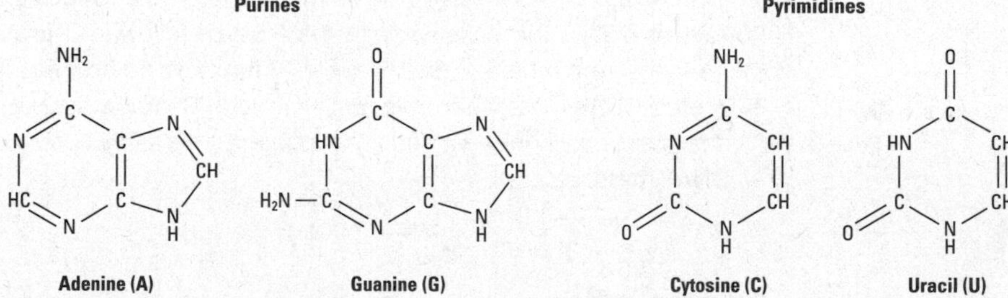

FIGURE 9-2: The four bases found in RNA.

Uracil may be new to you, but it's actually the precursor of DNA's thymine. When your body produces nucleotides, uracil is hooked up with a ribose and three phosphates to form a ribonucleoside triphosphate (rNTP). (Check out Figure 9-5 later in the chapter to see an rNTP.) If DNA is being replicated, as discussed in Chapter 7, deoxyribonucleotide triphosphates (dNTPs) of thymine — not uracil — are needed, meaning that a few things have to happen:

» The 2' oxygen must be removed from ribose to make deoxyribose.

» A chemical group must be added to uracil's ring structure (all the bases are rings; see Chapter 5 to review how these rings stack up). Folic acid, otherwise known as vitamin B9, helps add a carbon and three hydrogen atoms (CH_3, referred to as a *methyl group*) to uracil to convert it to thymine.

Uracil carries genetic information in the same way thymine does, as part of sequences of bases. In fact, the genetic code that's translated into protein is written using uracil. We discuss more about the genetic code in Chapter 10.

REMEMBER

The complementary base pairing rules that apply to DNA also apply to RNA: purines with pyrimidines (that is, G with C and A with U). So why are there two versions of essentially the same base (uracil and thymine)?

- Thymine protects the DNA molecule better than uracil can because that little methyl group (CH_3) helps make DNA less obvious to chemicals called *nucleases* that chew up both DNA and RNA. Nucleases are *enzymes* (chemicals that cause reactions to occur) that act on nucleic acids. Your body uses nucleases to attack unwanted RNA and DNA molecules (such as viruses and bacteria), but if methyl groups are present, nucleases can't bond as easily with the nucleic acid to break its chains. The methyl group also makes DNA hydrophobic, or afraid of water, in order to keep the inside of the DNA helix safe.
- Uracil is a very friendly base; it easily bonds with the other three bases to form pairs. Uracil's amorous nature is great for RNA, which needs to form all sorts of interesting turns, twists, and knots to do its job (see the next section, "Stranded!"). DNA's message is too important to trust to such an easygoing base as uracil; strict base pairing rules must be followed to protect DNA's message from mutation (see Chapter 12 for more on how base pair rules protect DNA's message from getting garbled). Thymine, as uracil's less friendly near-twin, only bonds with adenine, making it perfectly suited to protect DNA's message.

Stranded!

RNA is almost always single-stranded, and DNA is always double-stranded. The double-stranded nature of DNA helps protect its message and provides a simple way for the molecule to be copied during replication. Like DNA, RNA loves to hook up with complementary bases. But RNA is a bit narcissistic; it likes to form bonds with itself (see Figure 9-3), creating what's called a *secondary structure.* The primary structure of RNA is the single-stranded molecule; when the molecule bonds with itself and gets all twisted and folded up, the result is the secondary structure.

Three major types of RNA carry out the business of expressing DNA's message. Although all three RNAs function as a team during translation, which we cover in Chapter 10, the individual types carry out very specific functions.

- **mRNA (messenger RNA):** Carries the message of the gene out of the cell nucleus so it can be translated into a protein.

- **tRNA (transfer RNA)** Carries amino acids around during translation. Amino acids are the building blocks of proteins.
- **rRNA (ribosomal RNA):** Puts amino acids together in chains during translation.

Primary Structure

5′ AUGCGGCUACGUAACGAGCUUAGCGCGUAUACCGAAAGGGUAGAAC 3′

Complementary regions bond to form secondary structure

FIGURE 9-3: Single-stranded RNAs form interesting shapes in order to carry out various functions.

Transcription: Copying DNA's Message into RNA's Language

A *transcript* is a record of something, not an exact copy. In genetics, *transcription* is the process of recording part of the DNA message in a related, but different, language — the language of RNA. Transcription is necessary because DNA is too valuable to be moved or tampered with. The DNA molecule is *the* plan, and any error that's introduced into the plan (as a mutation, which we address in Chapter 12) can cause lots of problems. Transcription keeps DNA safe by letting a temporary RNA copy take the risk of leaving the cell nucleus and going out into the cytoplasm.

With *transcription*, the DNA inside the nucleus goes through a process similar to replication, which we discuss in Chapter 7, in order to get the message out as RNA. When DNA is replicated, the result is another DNA molecule that's exactly like the original in every way. But in transcription, many mRNAs are created because, instead of transcribing the entire DNA molecule, only genes are transcribed into mRNA.

REMEMBER

Transcription is the process by which a messenger RNA is created from a gene. *Messenger RNAs* (mRNAs) are the specific type of RNA responsible for carrying DNA's message from the cell nucleus into the cytoplasm (see Figure 9-4). *Translation* is the process of taking the message found in the mRNA and turning it into the protein product of the gene. This process occurs in the cytoplasm of the cell, after the mRNA has been transported out of the nucleus (also shown in Figure 9-4).

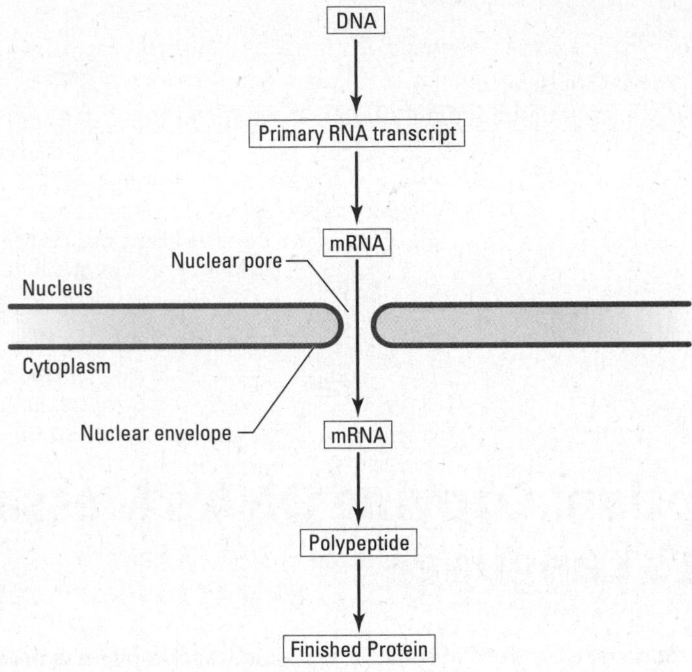

FIGURE 9-4: Transcription results in the mRNA transcript, while translation results in the final protein product.

Transcription has several steps:

1. Enzymes identify the right part of the DNA molecule to transcribe.
2. The DNA molecule is opened to make the message accessible.
3. Enzymes build the mRNA strand.
4. The DNA molecule snaps shut to release the newly synthesized mRNA.

Getting ready to transcribe

In preparing to transcribe DNA into mRNA, three things need to be completed:

- Locate the proper gene sequence within the billions of bases that make up DNA.
- Determine which of the two strands of DNA to transcribe.

» Gather up the nucleotides of RNA and the enzymes needed to carry out transcription.

Locating the gene

Your chromosomes are made up of roughly 3 billion base pairs of DNA and contain roughly 22,000 genes. But less than 2 percent of your DNA gets transcribed into mRNA. Genes, the sequences that do get transcribed, vary in size. The average gene is only about 3,000 base pairs long, but the human genome also has some gigantic genes — for example, the gene that's implicated in a particular form of muscular dystrophy (Duchenne muscular dystrophy) is a whopping 2.5 million base pairs.

REMEMBER

Before a gene of any size can be transcribed, it must be located. The cue that says "start transcription here" is written right into the DNA in regions called *promoters*. (The promoter also controls how often the process takes place; see the "Initiation" section later in the chapter.) The sequence that indicates where to stop transcribing is called a *terminator*. The gene, the promoter, and the terminator together are called the *transcription unit* (see Figure 9-5).

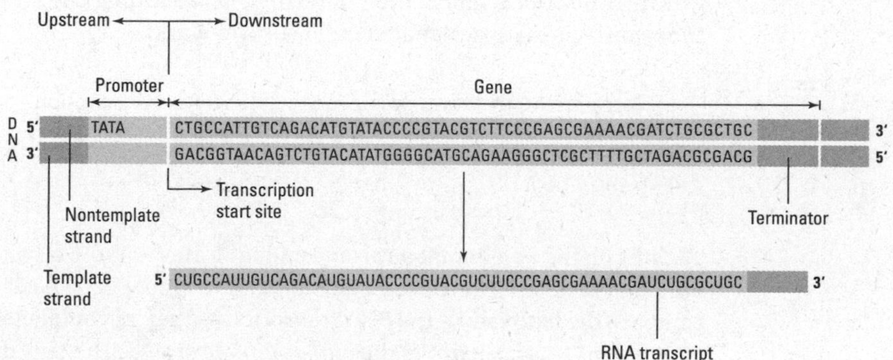

FIGURE 9-5: The transcription unit is made up of a promoter, the gene, and a terminator sequence.

The promoter sequences tell the enzymes of transcription where to start work and are located just upstream of the genes they control (see Figure 9-5). Each gene has its own promoter. In eukaryotes, most genes contain a sequence called the *TATA box* (so called because the sequence of the bases is most commonly TATAAA). The presence of TATA tells the transcription-starting enzyme that the gene to transcribe is about 30 base pairs away. Sequences like the TATA box that are the same in many (if not all) organisms are called *consensus sequences*, indicating that the sequences agree or mean the same thing everywhere they appear.

Locating the right strand

By now you've (hopefully) picked up on the fact that DNA is double-stranded. Those double strands aren't identical, though; they're complementary, meaning that the sequence of bases matches up, but it doesn't spell the same words of the genetic code (see Chapter 10 for more details about the genetic code). The genetic code of DNA works like this: Bases of genes are read in three base sets, like words. For example, three adenines in a row (AAA) are transcribed into mRNA as three uracils (UUU). During translation, UUU tells the ribosome to use an amino acid called phenylalanine as part of the protein it's making. If the complementary DNA, TTT, were transcribed, you'd wind up with an mRNA saying AAA, which specifies lysine. A protein containing lysine will function differently than one containing phenylalanine.

REMEMBER

Because complements don't spell the same genetic words, you can get two different messages depending on which strand of DNA is transcribed into mRNA. Therefore, genes can only be read from *one* of the two strands of the double-stranded DNA molecule — but which one? The TATA box not only indicates where a gene is but also tells which strand holds the gene's information. TATA boxes indicate that a gene is about 30 bases away going in the 3′ direction (sometimes referred to as *downstream*). Genes along the DNA molecule run in both directions, but any given gene is transcribed only in the 3′ direction. Because only one strand is transcribed, the two strands are designated in one of two ways:

- **Template:** This strand provides the pattern for transcription.
- **Non-template:** This strand is the original message that's actually being transcribed.

TATA is on the non-template strand and indicates that the other (complementary) strand is to be used as the template for transcription. Look back at Figure 9-5 and compare the template to the RNA transcript — they're complementary. Now compare the mRNA transcript to the non-template strand. The only difference between the two is that uracil appears in place of thymine. The RNA is the transcript of the non-template strand.

Gathering building blocks and enzymes

In addition to template DNA, the following ingredients are needed for successful transcription:

- **Ribonucleotides,** the building blocks of RNA.
- **Enzymes and other proteins,** to assemble the growing RNA strand in the process of *RNA synthesis.*

The building blocks of RNA are nearly identical to those used in DNA replication, which we discussed in Chapter 7. The differences, of course, are that for RNA, ribose is used in place of deoxyribose, and uracil replaces thymine. Otherwise, the rNTPs (ribonucleoside triphosphates; see Figure 9-6) look very much like the dNTPs you're hopefully already familiar with.

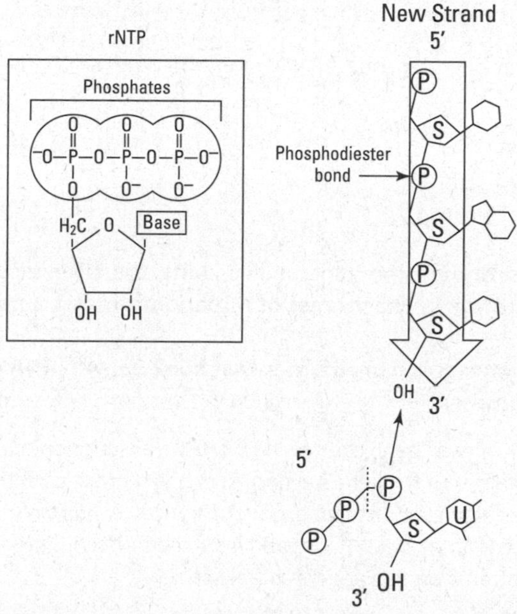

FIGURE 9-6: The basic building blocks of RNA and the chemical structure of an RNA strand.

In a process similar to replication, transcription requires the services of various enzymes to:

- Find the promoter.
- Open the DNA molecule.
- Assemble the growing strand of RNA.

Unlike replication, though, transcription has fewer enzymes to keep track of. The main player is RNA polymerase. Like DNA polymerase, which we cover in Chapter 7, *RNA polymerase* recognizes each base on the template and adds the appropriate complementary base to the growing RNA strand, substituting uracil where DNA polymerase would supply thymine. RNA polymerase hooks up with a large group of enzymes — called a *holoenzyme* — to carry out this process. The individual enzymes making up the holoenzyme vary between prokaryotes and eukaryotes, but their functions remain the same: to recognize and latch onto the promoter and to call RNA polymerase over to join the party.

TECHNICAL STUFF

Eukaryotes have three kinds of RNA polymerase, which vary only in which genes they transcribe.

- RNA polymerase I takes care of long rRNA molecules.
- RNA polymerase II carries out the synthesis of most mRNA and some tiny, specialized types of RNA molecules that are used in RNA editing after transcription is over.
- RNA polymerase III transcribes tRNA genes and other small RNAs used in RNA editing.

Initiation

Initiation includes finding the gene and opening the DNA molecule so that the enzymes can get to work. The process of initiation is pretty simple:

1. **The holoenzyme (group of enzymes that hook up with RNA polymerase) finds the promoter.**

   The promoter of each gene controls how often transcription makes an mRNA transcript to carry out the gene's action. RNA polymerase can't bind to a gene that isn't scheduled for transcription. In eukaryotes, *enhancers,* which are sequences sometimes distantly located from the transcription unit, also control how often a particular gene is transcribed.

2. **RNA polymerase opens up the double-stranded DNA molecule to expose a very short section of the template strand.**

   When the promoter "boots up" to initiate transcription, the holoenzyme complex binds to the promoter site and signals RNA polymerase. RNA polymerase binds to the template at the start site for transcription. RNA polymerase can't "see" past the sugar-phosphate backbone of DNA, so transcription can't occur if the molecule isn't first opened up to expose single strands. RNA polymerase breaks the hydrogen bonds between the double-stranded DNA molecule and opens up a short stretch of the helix to expose the template. The opening created by RNA polymerase when it wedges its way between the two strands of the helix is called the *transcription bubble* (see Figure 9-7).

3. **RNA polymerase strings together rNTPs to form mRNA (or one of the other types of RNA, such as tRNA or rRNA).**

   RNA polymerase doesn't need a primer to begin synthesis of a new mRNA molecule (unlike DNA replication). RNA polymerase simply reads the first base of the transcription unit and lays down the appropriate complementary rNTP. This first rNTP doesn't lose its three phosphate molecules because no phosphodiester bond is formed at the 5' side. Those two extra phosphates remain until the mRNA is edited later in the transcription process (see "Post-transcription Processing" later in this chapter).

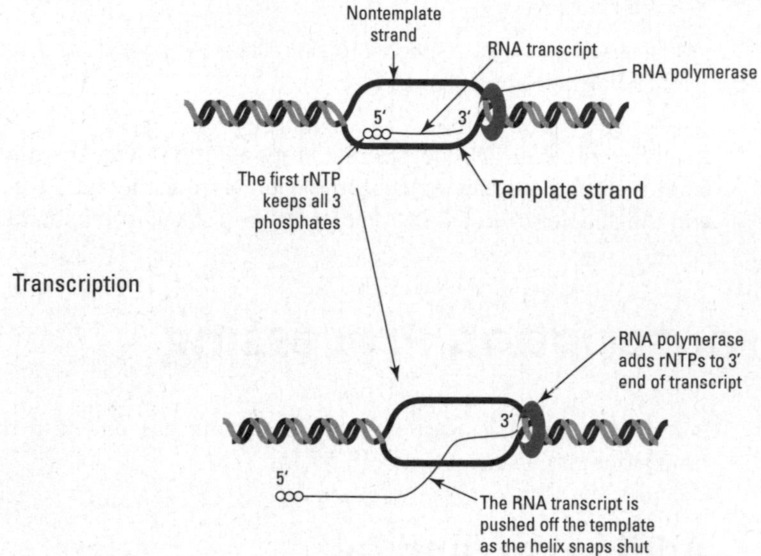

FIGURE 9-7: Transcribing DNA's message into RNA.

Elongation

After RNA polymerase puts down the first rNTP, it continues opening the DNA helix and synthesizing mRNA by adding rNTPs until the entire transcriptional unit is transcribed. The transcription bubble (the opening between DNA strands) itself is very small; only about 20 bases of DNA are exposed at a time. So, as RNA polymerase moves down the transcription unit, only the part of the template that's actively being transcribed is exposed. The helix snaps shut as RNA polymerase steams ahead to push the newly synthesized mRNA molecule off the template (refer to Figure 9-7).

Termination

When RNA polymerase encounters the terminator (as a sequence in the DNA, not the scary, gun-toting movie character), it transcribes the terminator sequence and then stops transcription. What happens next varies depending on the organism.

- In prokaryotic cells, some terminator sequences have a series of bases that are complementary and cause the mRNA to fold back on itself. The folding stops RNA polymerase from moving forward and pulls the mRNA off the template.
- In eukaryotic cells, a special protein called a *termination factor* aids RNA in finding the right stopping place.

In any event, after RNA polymerase stops adding rNTPs, the mRNA gets detached from the template. The holoenzyme and RNA polymerase let go of the template, and the double-stranded DNA molecule snaps back into its natural helix shape.

Post-transcription Processing

Before mRNA can venture out of the cell nucleus and into the cytoplasm for translation, it needs a few modifications.

Adding cap and tail

The "naked" mRNA that's produced by transcription needs to get dressed before translation:

- A 5' cap is added to the transcript.
- A long tail of adenine bases is tacked on the 3' end of the transcript.

RNA polymerase starts the process of transcription by using an unmodified rNTP (see the section "Initiation" earlier in this chapter). But a 5′ cap needs to be added to the mRNA to allow the ribosome to recognize it during translation (which will be covered in Chapter 10). The first part of adding the cap is the removal of one of the three phosphates from the leading end of the mRNA strand. A guanine, in the form of a ribonucleotide, is then attached to the lead base of the mRNA. Figure 9-8 illustrates the process of cap and tail attachment to the mRNA. Several groups composed of a carbon atom with three hydrogen atoms (CH_3, called a *methyl group*) attach at various sites — on the guanine and on the first and second nucleotides of the mRNA. Like the methyl groups that protect the thymine-bearing DNA molecule, the methyl groups at the 5′ end of the mRNA protect it from decomposition and allow the ribosome to recognize the mRNA as ready for translation.

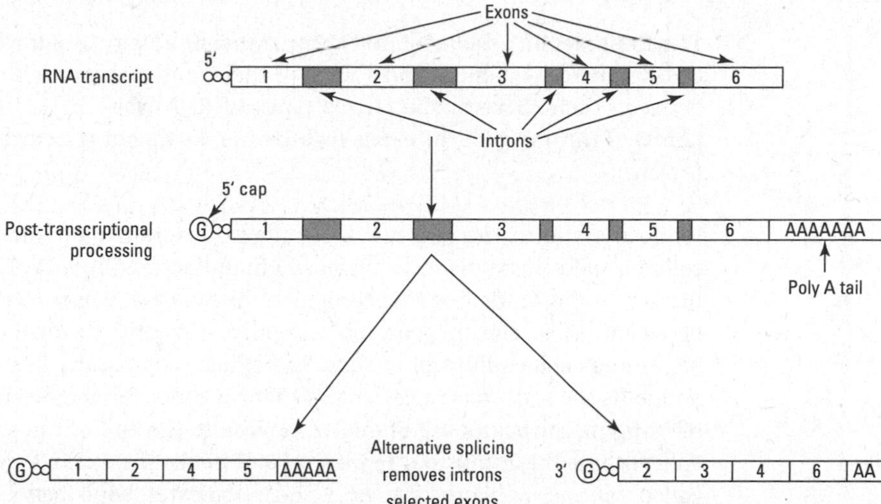

FIGURE 9-8: A 5' cap, 3' poly-A tail, and alternative splicing.

In eukaryotes, a long string of adenines is added to the 3′ end of the mRNA to further protect the mRNA from natural nuclease activity long enough to get translated (refer to Figure 9-8). This string is called the *poly-A tail.* RNA molecules are easily degraded and destroyed because of their chemical structure. Like memos, RNA molecules are linked to a specific task, and when the task is over, the memo is discarded. But the message needs to last long enough to be read, sometimes more than once, before it hits the shredder (in this case, nucleases do the shredding instead of guilty business executives). The length of the poly-A tail determines how long the message lasts and how many times it can be translated by the ribosomes before nucleases eat the tail and destroy the message.

Editing the message

The transcriptional units of genes contain sequences that aren't translated into protein. However, these sequences are important because they may help control how genes are expressed. As you may expect, geneticists have come up with terms for the parts that are translated and those that aren't:

- **Introns:** Non-coding sequences that get their name from their *in*tervening presence. Genes often have many introns that fall between the parts of the gene that code for the corresponding protein.
- **Exons:** Coding sequences that get their name from their *ex*pressed nature. Exons are translated into the protein product of the gene.

The entire gene — introns and exons — is transcribed (refer to Figure 9-8). After transcription has terminated, however, part of the editing process is the removal of introns.

The final step in preparing mRNA for translation is twofold: removing the non-coding intron sequences and stringing the exons together without interruptions between them. Several specialized types of RNA work to find the start and end points of introns, pull the exons together, and snip out the extra RNA (that is, the intron).

While it's still in the nucleus, a complex of proteins and small RNA molecules called a *spliceosome* inspects the newly manufactured mRNA. The spliceosome is like a roaming workshop that recognizes introns and works to remove them from between exons. The spliceosome recognizes consensus sequences that mark the beginnings and endings of introns. The spliceosome grabs each end of the intron and pulls the ends toward each other to form a loop. This movement has the effect of bringing the beginning of one exon close to the end of the preceding one. The spliceosome then snips out the intron and hooks the exons together in a process called *splicing.* Splicing creates a phosphodiester bond between the two exon sequences, which seals them together as one strand of mRNA.

Introns can be spliced out leaving all the exons in their original order, or introns *and* exons can be spliced out to create a new sequence of exons (refer to Figure 9-8 for a couple of examples). The splicing of introns and exons is called *alternative splicing* and results in the possibility for one gene to be expressed in different ways. Thanks to alternative splicing, the 22,000 or so genes in humans are able to produce around 90,000 different proteins. New evidence suggests that practically all multi-exon genes (which make up roughly 86 percent of the human genome) can be sliced and diced in multiple ways, thanks to alternative splicing. The enormous versatility of RNA editing has led some scientists to think of RNA as "the" genetic material instead of DNA.

After the introns are spliced and all the exons are strung together, the mRNA molecule is complete and ready for action. It migrates out of the cell nucleus, encounters an army of ribosomes, and goes through the process of translation — the final step in converting the genetic message from DNA to protein.

TECHNICAL STUFF

Prokaryotes don't have introns because prokaryotic genes are all coding, or exon. Only eukaryotes have genes interrupted by intron sequences, and almost all eukaryotic genes have at least one intron. Scientists continue to explore the function of introns, which in part control how different mRNAs are edited.

IN THIS CHAPTER

» Exploring the features of the genetic code

» Understanding the process of translation

» Molding polypeptides into functional proteins

Chapter 10
Translating the Genetic Code

From building instructions to implementation, the message that DNA carries follows a predictable path. First, DNA provides the template for transcription of the message into RNA. Then, RNA (in the form of messenger RNA) moves out of the cell nucleus and into the cytoplasm to provide the building plans for *proteins.* Every living thing is made of proteins, which are long chains of amino acids called *polypeptides* that are folded into complex shapes and hooked together in intricate ways.

All the physical characteristics of your body (that is, the *phenotypes*) are the result of thousands of different proteins. Of course, your body is also composed of other things, too, like water, minerals, and fats. But proteins supply the framework to organize all those other building blocks, and proteins carry out all your necessary bodily functions, like digestion, respiration, and elimination. In this chapter, we explain how RNA provides the blueprint for manufacturing proteins, the final step in the transformation from *genotype* (genetic information) to phenotype (physical trait).

Translation is the process of converting information from one language into another. In this case, the genetic language of nucleic acid is translated into the language of protein. However, before you dive into the translation process, you need to know a few things about the genetic code — the information that mRNA carries — and how the code is read. If you skipped over Chapter 9, you may want to go back and review its material on RNA before moving on.

Discovering the Good in a Degenerate

When Watson and Crick (along with Rosalind Franklin; see Chapter 5 for the full scoop) discovered that DNA is made up of two strands composed of four bases, the big question they faced was: How can only four bases contain enough information to encode complex phenotypes?

REMEMBER

Complex phenotypes (such as your bone structure, eye color, and ability to digest spicy food) are the result of combinations of proteins. The genetic code (that is, DNA transcribed as RNA) provides the instructions to make these proteins (via translation). Proteins are made up of amino acids strung together in various combinations to create chains called polypeptides (which is a fancy way of saying "protein"). Polypeptide chains generally vary from 50 to 1,000 amino acids in length. Because there are 20 different amino acids, and because chains are often more than 100 amino acids in length, the variety of combinations is enormous. For example, a polypeptide that's only 5 amino acids long has 3,200,000 combinations!

After experiments showed that DNA was truly the genetic material (see Chapter 5 for details), skeptics continued to point to the simplicity of the four bases in RNA and argued that a code of four bases wouldn't work to encode complex peptides. Reading the genetic code one base at a time — U, C, A, and G — would mean that there simply weren't enough bases to make 20 amino acids. So, it was obvious to scientists that the code must be made up of multiple bases read together. A two-base code didn't work because it only produced 16 combinations — too few to account for 20 amino acids. A three-base code (referred to as a *triplet code*) looked like overkill, because a *codon*, which is a combination of three nucleotides in a row, that chooses from four bases at each position produces 64 possible combinations. Skeptics argued that a triplet code contains too much redundancy — after all, there are only 20 amino acids.

As it turns out, the genetic code is *degenerate*, which is a fancy way of saying "too much information." Normally, degenerate is a negative term, like when referring to a person (which we are *not* doing). In the genetic sense, however, the degeneracy of the triplet code means that the code is highly flexible and tolerates some mistakes — which is a good thing.

REMEMBER

Several features of the genetic code are important to keep in mind. The code is:

- **Triplet,** meaning that bases are read three at a time in codons.
- **Degenerate,** meaning that 18 of the 20 amino acids are specified by two or more codons.
- **Orderly,** meaning that each codon is read in only one way and in only one direction, just as English is read left to right.
- **Nearly universal,** meaning that just about every organism on earth interprets the language of the code in exactly the same way.

Considering the combinations

Only 61 of the 64 codons are used to specify the 20 amino acids found in proteins. The three codons that don't code for any amino acid simply mean "stop," telling the ribosome to cease the translation process. In contrast, the one codon that tells the ribosome that an mRNA is ripe for translating — the "start" codon — codes for the amino acid *methionine.* In Figure 10-1, you can see the entire code with all the alternative spellings for the 20 amino acids. To read this table, you need to find the first nucleotide of the codon along the left side of the chart, and then find the second nucleotide along the top of the chart. Last, by finding the nucleotide along the right side of the chart in the row where the first two intersect, you can determine what amino acid the codon codes for. For example, the codon UUU codes for the amino acid phenylalanine.

For many of the amino acids, the alternative spellings differ only by one base — the third base of the codon. For example, four of the six codons for leucine start with the bases CU. This flexibility at the third position of the codon is called a *wobble.* The third base of the mRNA can vary, or wobble, without changing the meaning of the codon (and thus the amino acid it codes for). The wobble is possible because of the way *tRNAs (transfer RNAs)* and mRNAs pair up during the process of translation. The first two bases of the code on the mRNA and the partner tRNA (which is carrying the amino acid specified by the codon) must be exact matches. However, the third base of the tRNA can break the base-pairing rules, allowing bonds with mRNA bases other than the usual complements. This rule violation, or wobble, allows different spellings to code for the same amino acid. However, some codons, like one of the three stop codons (spelled UGA), have only one meaning; wobbles in this stop codon change the meaning from stop to either cysteine (spelled UGU or UGC) or tryptophan (UGG).

First Letter ↓	Second Letter U	Second Letter C	Second Letter A	Second Letter G	Third Letter ↓
U	phenylalanine	serine	tyrosine	cysteine	U
	phenylalanine	serine	tyrosine	cysteine	C
	leucine	serine	STOP	STOP	A
	leucine	serine	STOP	tryptophan	G
C	leucine	proline	histidine	arginine	U
	leucine	proline	histidine	arginine	C
	leucine	proline	glutamine	arginine	A
	leucine	proline	glutamine	arginine	G
A	isoleucine	threonine	asparagine	serine	U
	isoleucine	threonine	asparagine	serine	C
	isoleucine	threonine	lysine	arginine	A
	methionine & START	threonine	lysine	arginine	G
G	valine	alanine	aspartate	glycine	U
	valine	alanine	aspartate	glycine	C
	valine	alanine	glutamate	glycine	A
	valine	alanine	glutamate	glycine	G

FIGURE 10-1: The 64 codons of the genetic code, as written by mRNA.

Framed! Reading the code

Besides its combination possibilities, another important feature of the genetic code is the way in which the codons are read. Each codon is separate, with no overlap. And the code doesn't have any punctuation — it's read straight through without pauses.

The codons of the genetic code run sequentially, as you can see in Figure 10-2. Each codon is read only once using a *reading frame*, a series of sequential, non-overlapping codons. The start codon defines the position of the reading frame. In the mRNA in Figure 10-2, the sequence AUG, which is the codon methionine, is used as the start codon (but it can also be used elsewhere in a protein). After the start codon, the bases are read three at a time without a break until the stop codon is reached.

FIGURE 10-2: The genetic code is nonoverlapping and uses a reading frame.

Nucleotide sequence	A U G C G A G U C U U G C A G . . .
Nonoverlapping code	[A U G] [C G A] [G U C] [U U G] [C A G] [. . .]
	1 2 3 4 5

Not quite universal

The meaning of the genetic code is nearly universal. That means nearly every organism on earth uses the same spellings in the triplet code. Mitochondrial DNA spells a few words differently from nuclear DNA, which may explain (or at least relate back to) mitochondria's unusual origins. Plants, bacteria, and a few micro-organisms also use unusual spellings for one or more amino acids. Otherwise, the way the code is read — influenced by its degenerate nature, with wobbles, without punctuation, and using a specific reading frame — is the same. However, as scientists tackle DNA sequencing for various creatures, more unusual spellings are likely to pop up.

Meeting the Translating Team

Translation takes place in the cytoplasm of cells. After messenger RNAs (mRNAs) are created through transcription and move into the cytoplasm, the protein production process begins. The players involved in protein production include:

- **Ribosomes:** This large protein-making factory reads mRNA's message and carries out the message's instructions. Ribosomes are made up of *ribosomal RNA* (rRNA) and are capable of constructing any sort of protein.
- **Messenger RNA (mRNA):** The message of each gene is carried by mRNA in the sequence of its nucleotides. The mRNA is "read" by the ribosomes in order to translate the sequence into a protein.
- **Amino acids:** These complex chemical compounds, which contain nitrogen and carbon, are strung together in thousands of unique combinations to make proteins. There are 20 different amino acids.
- **Transfer RNA (tRNA):** This molecule runs a courier service to provide amino-acid building blocks to the working ribosome; each tRNA summoned by the ribosome grabs the amino acid specified by the codon in the mRNA.

Taking the Translation Trip

Translation proceeds in a series of predictable steps (which are expanded upon in the next several sections):

1. A ribosome recognizes an mRNA and latches onto its 5' cap (see Chapter 9 for an explanation of how and why mRNAs get caps). The ribosome slurps up the mRNA and carefully scrutinizes it, looking for a start codon (AUG).
2. tRNAs supply the amino acids dictated by each codon when the ribosome reads the instructions. The ribosome assembles the polypeptide chain with the help of various enzymes and other proteins.
3. The ribosome continues to assemble the polypeptide chain until it reaches a stop codon. The completed polypeptide chain is released.

After it's released from the ribosome, the polypeptide chain is modified and folded to become a mature protein.

Initiation

Preparation for translation consists of two major events:

- The tRNA molecules must be hooked up with the right amino acids in a process called *charging.*
- The ribosome, which comes in two pieces, must assemble itself at the 5' end of the mRNA. For a review of 5' and 3' ends, flip back to Chapter 5.

Charge! tRNA hooks up with a nice amino acid

Transfer RNA (tRNA) molecules are small, specialized RNAs produced by transcription. However, unlike mRNAs, tRNAs are never translated into protein; tRNA's whole function is ferrying amino acids to the ribosomes for assembly into polypeptides. tRNAs are uniquely shaped to carry out their job. In Figure 10-3, you see two depictions of tRNA. The illustration on the left shows you tRNA's true form. The illustration on the right is a simplified version that makes tRNA's parts easier to identify. The cloverleaf shape is one of the keys to the way tRNA works. tRNA gets its unusual configuration because many of the bases in its sequence are complements; the strand folds, and the complementary bases form bonds, resulting in the loops and arms of a typical tRNA.

The two key elements of tRNA are:

- **Anticodon:** A three-base sequence on one loop of each tRNA, which is complementary to one of the codons spelled out by the mRNA sequence.
- **Acceptor arm:** The single-stranded tail of the tRNA, where the amino acid corresponding to the codon is attached to the tRNA.

REMEMBER

The codon of mRNA specifies the amino acid used during translation. The anticodon of the tRNA is complementary to the codon of mRNA and specifies which amino acid each tRNA is built to carry.

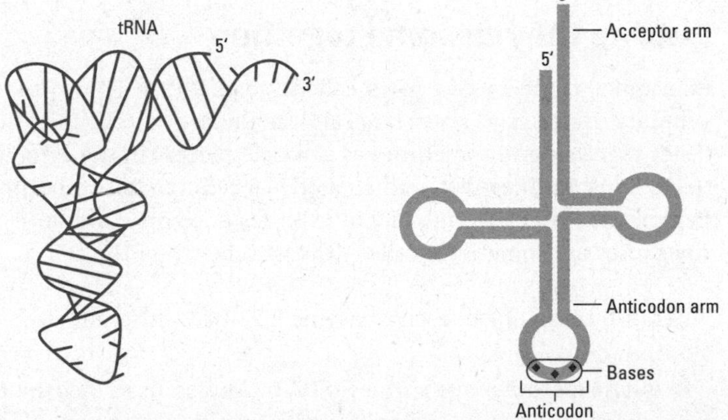

FIGURE 10-3: tRNA has a unique shape that helps it ferry amino acids to the ribosomes.

Like a battery, tRNAs must be charged in order to work. tRNAs get charged with the help of a special group of enzymes called *aminoacyl-tRNA synthetases.* Twenty synthetases exist, one for each amino acid specified by the codons of mRNA. Take a look at the illustration on the right in Figure 10-3, the schematic of tRNA. The aminoacyl-tRNA synthetases recognize sequences of bases in the anticodon of the tRNA that announce which amino acid that particular tRNA is built to carry. When the aminoacyl-tRNA synthetase encounters the tRNA molecule that matches its amino acid, the synthetase binds the amino acid to the tRNA at the acceptor arm — this is the charging part. Figure 10-4 shows the connection of amino acid and tRNA. The synthetases proofread to make sure that each amino acid is on the appropriate tRNA. This proofreading ensures that errors in tRNA charging are very rare and prevents errors later in translation. With the amino acid attached to it, the tRNA is charged and ready to make the trip to the ribosome.

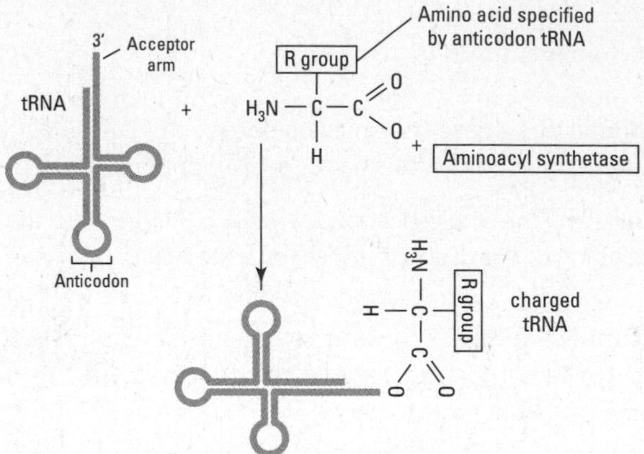

FIGURE 10-4:
tRNA charging.

Putting the ribosome together

Ribosomes come in two parts called *subunits* (see Figure 10-5), and ribosomal subunits come in two sizes: large and small. The two subunits float around (sometimes together and sometimes as separate pieces) in the cytoplasm until translation begins. Unlike tRNAs, which match specific codons, ribosomes are completely flexible and can work with any mRNA they encounter. Because of their versatility, ribosomes are sometimes called "the workbench of the cell."

When fully assembled, each ribosome has three different sites:

- **A-site (acceptor site):** Where tRNA molecules insert their anticodon arms to match up with the codon of the mRNA molecule.
- **P-site (peptidyl site):** Where amino acids get hooked together using peptide bonds.
- **Exit site:** Where tRNAs are released from the ribosome after their amino acids become part of the growing polypeptide chain.

Before translation can begin, the smaller of the two ribosome subunits attaches to the 5′ cap of the mRNA with the help of proteins called *initiation factors.* The small subunit then scoots along the mRNA until it hits the start codon (AUG). The P-site on the small ribosome subunit lines up with the start codon, and the small subunit is joined by the tRNA carrying methionine (with the anticodon UAC), the amino acid that matches the start codon. The tRNA uses its anticodon, which is complementary to the codon of the mRNA, to hook up to the mRNA. The large ribosome subunit joins with the small subunit to begin the process of hooking together all the amino acids specified by the mRNA (refer to Figure 10-5).

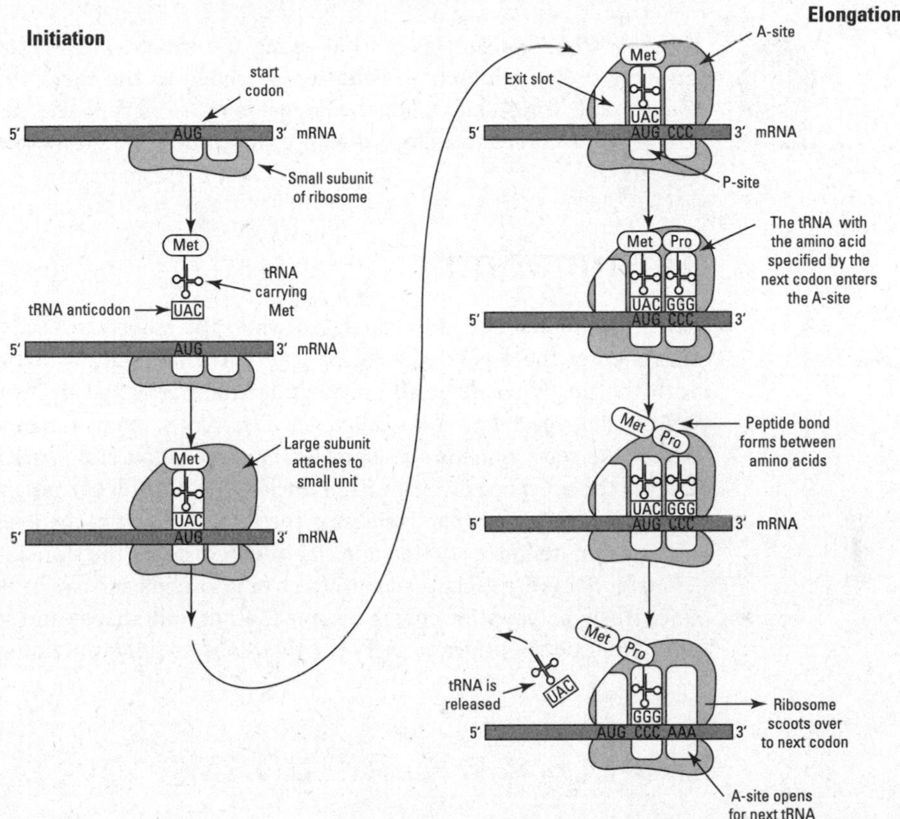

FIGURE 10-5: Initiation and elongation during translation.

Elongation

When the initiation process is complete, translation proceeds in several steps called *elongation*, which you can follow in Figure 10-5.

1. The ribosome calls for the tRNA carrying the amino acid specified by the codon residing in the A-site. The appropriate charged tRNA inserts its anticodon arm into the A-site.
2. Enzymes bond the two amino acids attached to the acceptor arms of the tRNAs in the P- and A-sites.
3. As soon as the two amino acids are linked together, the ribosome scoots over to the next codon of the mRNA. The tRNA that was formerly in the P-site now enters the exit site, and because it's no longer charged with an amino acid, the empty tRNA is released from the ribosome. The A-site is left empty, and the P-site is occupied by a tRNA holding its own amino acid and the amino acid of the preceding tRNA. The process of moving from one codon to the next is called *translocation* (not to be confused with the chromosomal translocations we describe in Chapter 13, where pieces of whole chromosomes are inappropriately swapped).

REMEMBER

The ribosome continues to scoot along the mRNA in a 5′ to 3′ direction. The growing polypeptide chain is always attached to the tRNA that's sitting in the P-site, and the A-site is opened repeatedly to accept the next charged tRNA. The process comes to a stop when the ribosome encounters one of the three stop codons.

Termination

No tRNAs match the stop codon, so when the ribosome reads "stop," no more tRNAs enter the A-site (see Figure 10-6). At this point, a tRNA sits in the P-site with the newly constructed polypeptide chain attached to it by the tRNA's own amino acid. Special proteins called *release factors* move in and bind to the ribosome; one of the release factors recognizes the stop codon and sparks the reaction that cleaves the polypeptide chain from the last tRNA. After the polypeptide is released, the ribosome comes apart, releasing the final tRNA from the P-site. The ribosomal subunits are then free to find another mRNA to begin the translation process anew. Transfer RNAs are recharged with fresh amino acids and can be used over and over. Once freed, polypeptide chains assume their unique shapes and sometimes hook up with other polypeptides to carry out their jobs as fully functioning proteins.

CHALLENGING THE DOGMA

The *Central Dogma of Genetics* (coined by our old friend Francis Crick, of DNA-discovery fame; see Chapter 5) posited that the trip from genotype to phenotype is a one-way information highway. Genes located in the DNA are transcribed into mRNAs, which are then translated into proteins. The dogma presumes that the starting point is always DNA and the endpoint is protein. While a dogma is a more or less universally accepted opinion about how the world works, there are always exceptions, just like with the laws of inheritance that we discussed earlier in Chapter 3.

Reverse transcription is one exception. Reverse transcription, which is basically transmitting RNA's message back into DNA, occurs in a type of virus known as a retrovirus. Retroviruses, such as HIV (the virus that causes AIDS) have an RNA genome that includes three genes needed for retroviral infection. An enzyme from these viruses, which is known as *reverse transcriptase,* is able to convert RNA into DNA.

Another exception to the central dogma is the case of RNAs that are not translated into protein. There have been many new discoveries about the powerful roles that RNA has outside of translation. It turns out that many non-coding RNAs exist — that is, RNAs that don't code for proteins but play important roles in how genes are expressed. Both ribosomal RNA (rRNA) and transfer RNA (tRNA) function in the process of translation but are not translated into protein themselves.

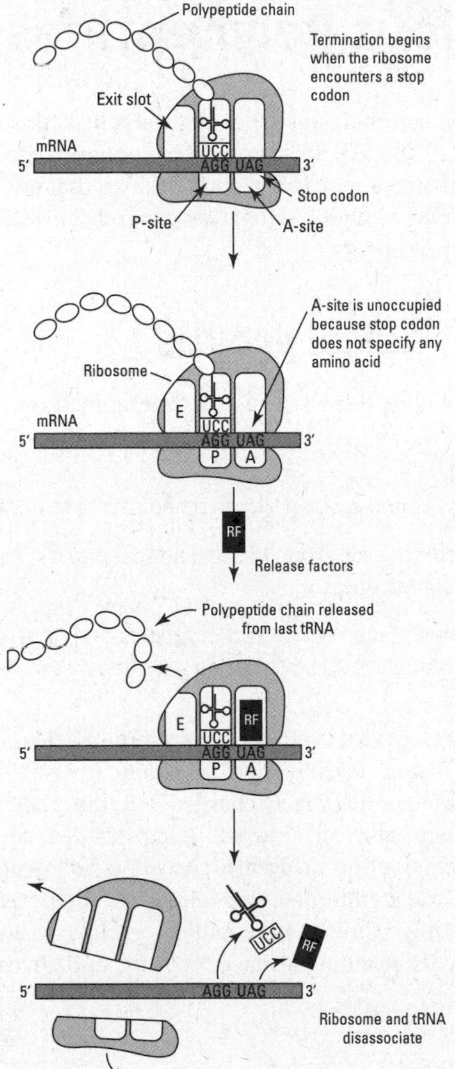

FIGURE 10-6: Termination of translation.

Messenger RNAs may be translated more than once and, in fact, may be translated by more than one ribosome at a time. As soon as the start codon emerges from the ribosome after the initiation of translation, another ribosome may recognize the mRNA's 5′ cap, latch on, and start translating. Thus, many polypeptide chains can be manufactured very rapidly.

Proteins Are Precious Polypeptides

REMEMBER

Besides water, the most common substance in your cells is protein. Proteins carry out the business of life. The key to a protein's function is its shape; completed proteins can be made of one or more polypeptide chains that are folded and hooked together. The way proteins fit and fold together depends on which amino acids are present in the polypeptide chains.

Recognizing radical groups

Every amino acid in a polypeptide chain shares several features, which you can see in Figure 10-7:

- A positively charged amino group (NH_2) attached to a central carbon atom.
- A negatively charged carboxyl group (COOH) attached to the central carbon atom opposite the amino group.
- A unique combination of atoms that form branches and rings, called *radical groups,* that differentiate the 20 amino acids specified by the genetic code.

There are several characteristics used to classify amino acids, including whether they are *hydrophobic* ("water-fearing") or *hydrophilic* ("water-loving"), whether they are positively charged or negatively charged, whether they are acidic or basic, and whether or not they have an aromatic (ring-shaped) side chain (refer to Figure 10-7). When their amino acids are part of a polypeptide chain, radical groups of adjacent amino acids alternate sides along the chain (refer to Figure 10-8). Because of their differing affinities, the radical groups either repel or attract neighboring groups. This reaction leads to folding and gives each protein its shape.

Giving the protein its shape

Proteins are folded into complex and often beautiful shapes, as you can see in Figure 10-8. These arrangements are partly the result of spontaneous attractions between radical groups (see the preceding section for details) and partly the result of certain regions of polypeptide chains that naturally form spirals (also called *helices,* not to be confused with DNA's double helix in Chapter 5). The spirals may weave back and forth to form sheets. These spirals and sheets are referred to as a *secondary structure* (the simple, unfolded polypeptide chain is the *primary structure*).

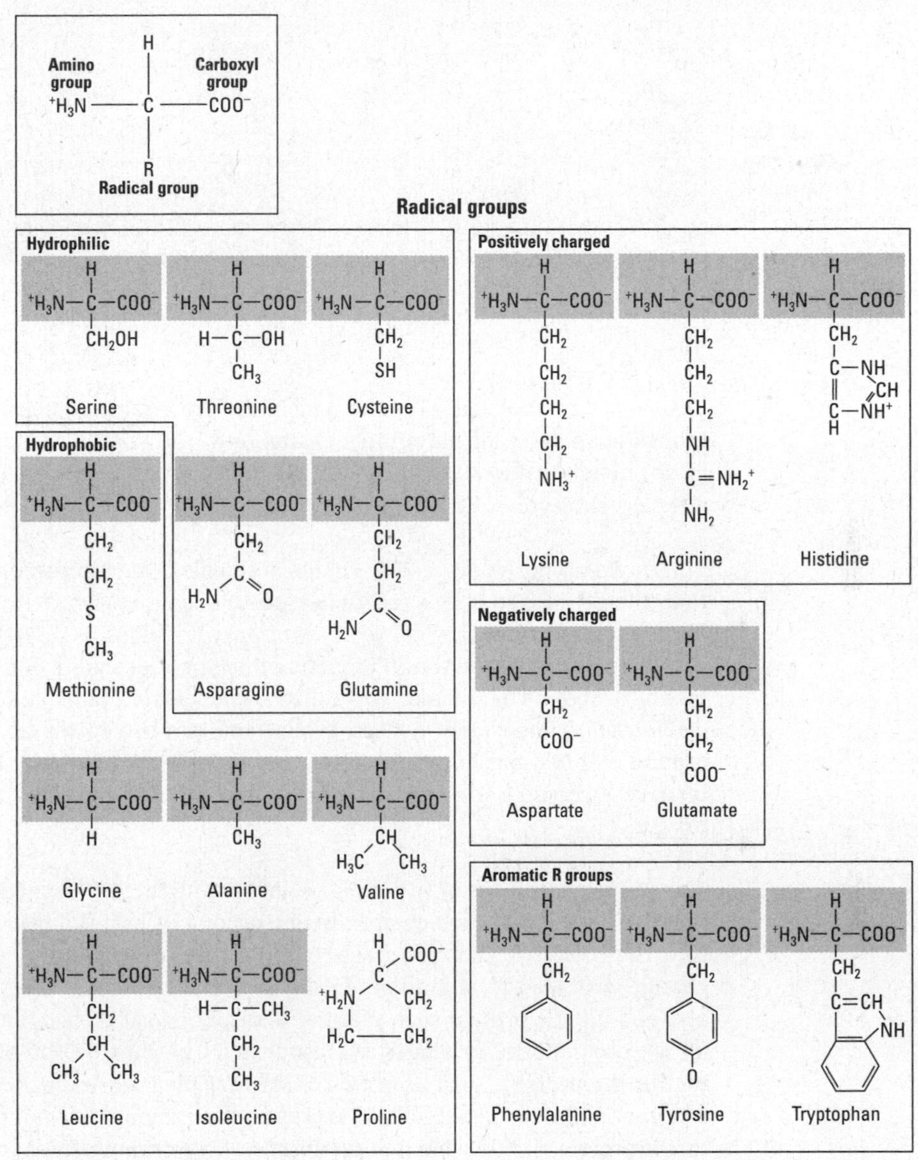

FIGURE 10-7: The 20 amino acids used to construct proteins.

Proteins are often modified after translation and may get hooked up with various chemical groups (such as phosphate groups) and metals (such as iron). In a process similar to the post-transcription modification of mRNA, proteins may also be sliced and spliced. Some protein modifications result in natural folds, twists, and turns, but sometimes the protein needs help forming its correct conformation. That's what chaperones are for.

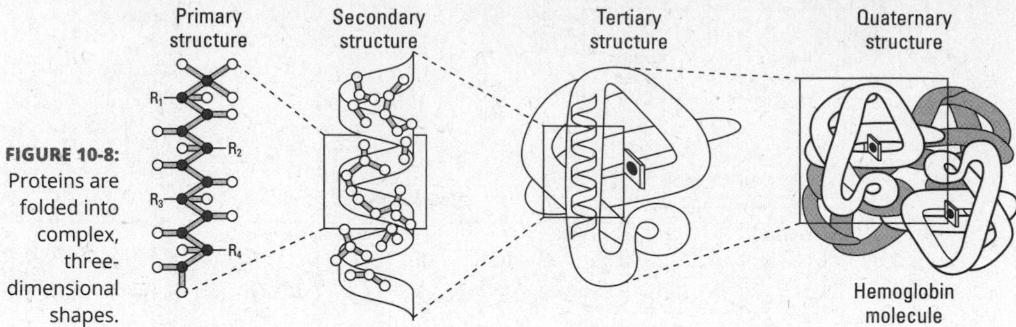

FIGURE 10-8: Proteins are folded into complex, three-dimensional shapes.

Chaperones are molecules that help mold the protein into the proper shape. Chaperones push and pull the protein chains until the appropriate radical groups are close enough to one another to form chemical bonds. This sort of folding is called a *tertiary structure.*

When two or more polypeptide chains are hooked together to make a single protein, they're said to have a fourth degree, or *quaternary structure.* For example, the hemoglobin protein that carries oxygen in your blood is a well-studied protein with a quaternary structure. Two pairs of polypeptide chains form a single hemoglobin protein. The chains, two called *alpha-globin chains* and two called *beta-globin chains,* each form helices, which you can see in Figure 10-8, that wind around and fold back on themselves into tertiary structures. Associated with the tertiary structures are iron-rich *heme* groups that have a strong affinity for oxygen.

The proper structure of a protein is important for it to function as necessary. Changes in the DNA, which are then incorporated into the mRNA and translated into the protein, can have drastic effects on the function of the protein. DNA changes can lead to alterations in the overall structure of the protein. They can also change the protein such that the necessary post-translational modifications do not occur. Certain changes can also lead to the substitution of key amino acids within the protein, such as amino acids that play a role in the active site of an enzyme. When something like this occurs, the enzyme may not be able to bind its target protein and do what it is supposed to do. For more on how good proteins go bad, see Chapter 12.

IN THIS CHAPTER

» Defining gene expression

» Confining gene activities to the right places

» Controlling genes before and after transcription

Chapter 11

Gene Expression: What a Cute Pair of Genes

Every cell in your body (with very few exceptions) carries the entire set of genetic instructions that make everything about you. Your eye cells contain the genes for growing hair. Your nerve cells contain the genes that turn on cell division — yet your nerve cells don't divide (under normal conditions). So why, then, aren't your eyeballs hairy?

It boils down to gene expression. *Gene expression* is basically the transcription and translation of a gene – the production of the protein that is encoded by the gene. Genes that are supposed to be active in certain cells are turned on only when needed and then turned off again, like turning off the light in a room when you leave. This process is highly regulated; your cells need to make sure the right genes make their products at the right time and in the right place. This chapter examines how your genes work and what controls them. Most of the chapter will focus on gene expression in eukaryotes (organisms with cell nuclei), with only a brief discussion of gene expression in prokaryotes (organism without cell nuclei).

Getting Your Genes Under Control

Gene expression occurs throughout an organism's life, starting at the very beginning. When an organism develops — first as a *zygote* (the fertilized egg) and later as an embryo and fetus — genes turn on to regulate the process. At first, all the cells are exactly alike, but that characteristic quickly changes. Cells that have the ability to turn into any kind of tissue (including the placenta) are *totipotent*. Cells that have the potential to turn into a variety of specialized cell types within the embryo are referred to as *pluripotent*. Eventually, cells become specialized based on gene expression in the cell, signals from the cells around it, and cues from the environment. Once the cell has turned into a specific cell type (such as a skin cell or a muscle cell), the cell is referred to as *differentiated*. After the tissue type is decided, certain genes in each cell become active, and others remain permanently turned off.

The default state of your genes is off, not on. Starting in the off position makes sense when you remember that almost every cell in your body contains a complete set of all your genes. You just can't have every gene in every cell flipped on and running amok all the time; you want specific genes acting only in the tissues where their actions are needed. Therefore, keeping genes turned off is every bit as important as turning them on.

To Be Expressed or Not To Be Expressed?

The control of gene expression is tightly regulated. Where genes are expressed is highly tissue-specific, which prevents the wrong genes from being turned on in the wrong cells. In part, the tissue-specific nature of gene expression is because of location — genes in cells respond to cues from the cells around them. These cues may be in the form of small molecules or proteins produced by neighboring cells. They may also be changes in the overall environment of the organism. See the sidebar "Heat and light" for a great example of the environment playing a role in gene expression.

Gene expression may also be regulated in a time-dependent manner, where some genes are set up to turn on (or off) at a certain stage of development. Take the genes that code for hemoglobin, for example.

Your *genome* (your complete set of genetic information) contains a large group of genes that all code for various components that make up the big protein, called *hemoglobin*, that carries oxygen in your blood. Hemoglobin is a complex structure comprised of two different types of proteins that are folded and joined together in pairs. During your development, nine different hemoglobin genes interacted at different times to make three kinds of hemoglobin. Changing conditions make it necessary for you to have three different sorts of hemoglobin at different stages of your life.

TECHNICAL STUFF

When you were still an embryo, your hemoglobin was composed mostly of epsilon-hemoglobin (Greek letters are used to identify the various types of hemoglobin). After about three months of development, the epsilon-hemoglobin gene was turned off in favor of two fetal hemoglobin genes, alpha and gamma. (*Fetal hemoglobin* is comprised of two proteins — two alphas and two gammas — folded and joined together as one functional piece.) When you were born, the gene producing the gamma-hemoglobin was shut off, and the beta-hemoglobin gene, which works for the rest of your life, kicked in.

The genes controlling the production of all these hemoglobins are on two chromosomes, 11 and 16 (see Figure 11-1). The genes on both chromosomes are turned on in order, starting at the 5′ end of the group for embryonic hemoglobin (see Chapter 5 for how DNA is set up with numbered ends). Adult hemoglobin is produced by the last set of genes on the 3′ end.

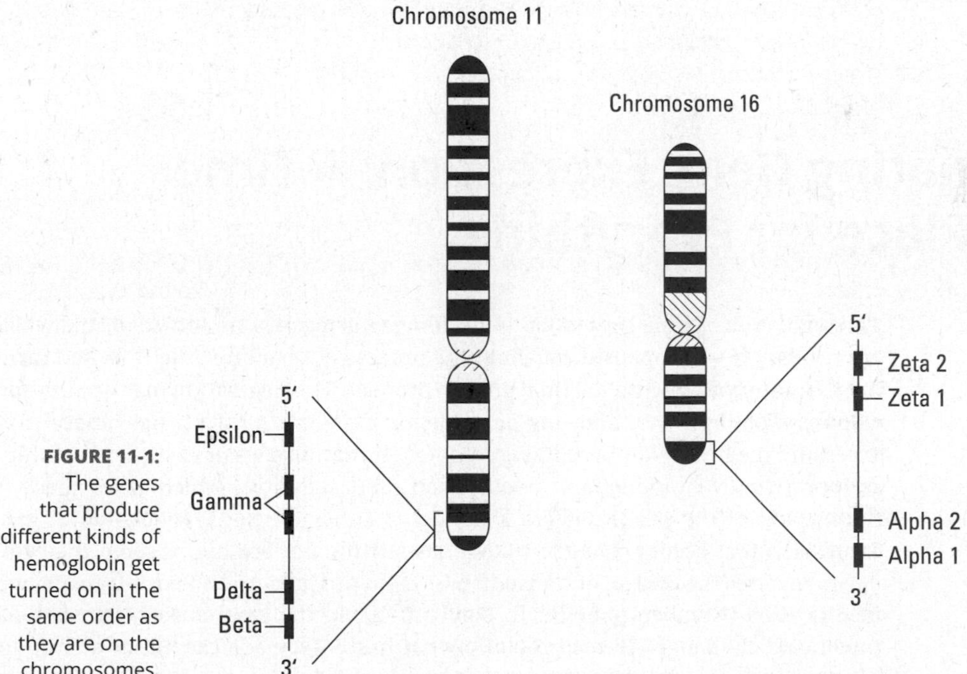

FIGURE 11-1: The genes that produce different kinds of hemoglobin get turned on in the same order as they are on the chromosomes.

HEAT AND LIGHT

Organisms need to respond quickly to changing conditions in order to survive. When external conditions turn on genes, it's called *induction.* Responses to heat and light are two types of induction that scientists understand particularly well.

When an organism is exposed to high temperatures, a suite of genes immediately kicks into action to produce *heat-shock proteins.* Heat has the nasty effect of mangling proteins so that they're unable to function properly, referred to as *denaturing.* Heat-shock proteins are produced by roughly 20 different genes and act to prevent other proteins from becoming denatured. Heat-shock proteins can also repair protein damage and refold proteins to bring them back to life. Heat-shock responses are best studied in fruit flies, but humans have a large number of heat-shock genes, too. These genes also protect you from the effects of stress and pollutants.

The expression of certain genes is also affected by the presence or absence of light. Your daily rhythms of sleeping and waking are controlled, in part, by light. When you're exposed to light during nighttime, your normal production of melatonin (a hormone that regulates sleep, among other things) is disrupted. In turn, a gene called *period* (so named because it controls circadian rhythms) is inactivated. Altered activity by the period gene is linked to breast cancer as well as depressed immune function. The increased incidence of breast cancer in women working the night shift was so dramatic that the researchers deemed night-shift work as a probable carcinogen.

Regulating Gene Expression: A Time and Place for Everything

The regulation of gene expression — deciding which genes are turned on and which are turned off — happens throughout the process of "reading" the DNA and turning that information into the final protein product. This regulation may occur *before* transcription, by either allowing or preventing the transcription machinery from accessing the DNA (Number 1 in Figure 11-2). It may also occur at the level of transcription (mRNA production, modification, and stability), which takes place in the nucleus of the cell (Numbers 2 through 4 in Figure 11-2). Regulation of gene expression that occurs once the mRNA has left the nucleus and entered the cytoplasm involves the control of translation (protein production) and post-translational modification (Numbers 5 and 6 in Figure 11-2). In the case of post-translational modification, changes to the protein once it has been made can either activate or inactivate the protein (depending on what is needed for the protein to function).

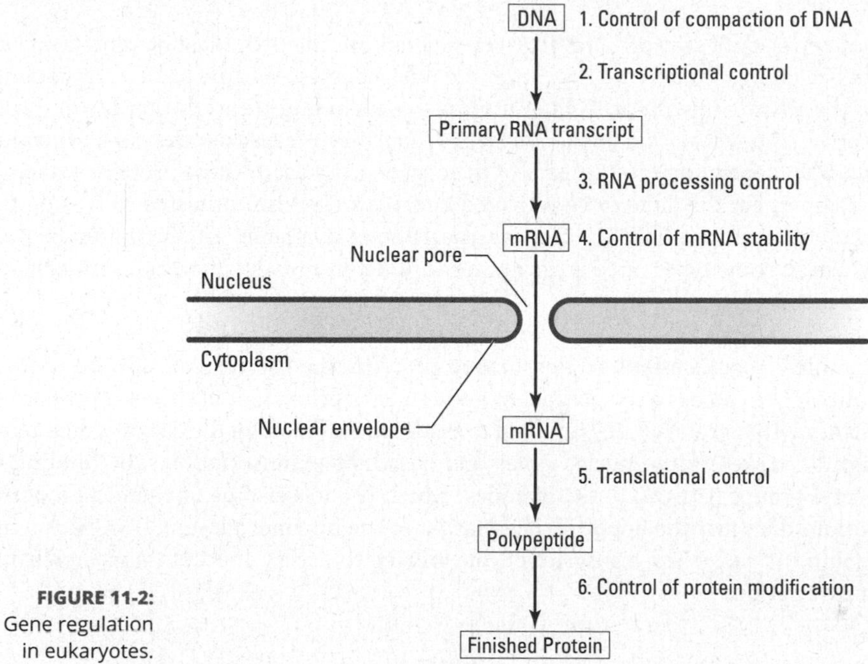

FIGURE 11-2: Gene regulation in eukaryotes.

Controlling Transcription Before It Starts

Whether a gene can be turned on and transcribed depends on access to the DNA that contains that gene. One way to control DNA access is through *epigenetics*. Epigenetics involves changes to the chemical structure of the DNA molecule, as opposed to changes in the DNA sequence. Small chemical tags called *methyl groups* (CH_3) are attached to the DNA. These chemical modifications can occur in or near gene promoters, preventing the transcription machinery from binding the DNA and effectively turning the gene "off."

DNA packaging is another way that gene transcription can be controlled. Tight packing of the DNA is a highly effective mechanism to make sure that most genes are off most of the time. Tightly wound DNA also prevents transcription from occurring by preventing the proteins involved in transcription from getting access to the genes. DNA is an enormous molecule, and the only way it can be scrunched down small enough to fit into the cell nucleus is by being tightly wound and supercoiled (see Chapter 6 to review DNA packaging). When DNA is wrapped up this way, it can't be transcribed, because the DNA cannot be accessed.

TECHNICAL STUFF

But genes can't stay off forever. To turn genes on, the DNA must be removed from its packaging. To unwrap DNA from the nucleosomes, specific proteins must bind to the DNA to unwind it. Lots of proteins — including transcription factors, collectively known as *chromatin-remodeling complexes* — carry out the job of unwinding DNA depending on the needs of the organism. Most of these proteins attach to a region near the gene to be activated and push the histones aside to free up the DNA for transcription. As soon as the DNA is available, transcription factors, which in some types of cells are always lurking around, latch on and immediately get to work.

Chemical modifications to the histone proteins themselves can also be used to control DNA access and subsequent gene transcription. Just like those that bind to DNA, methyl groups (CH_3) can bind to the histones, making the DNA become more tightly packed in the nucleosomes and decreasing transcription. The binding of acetyl groups (CH_3CO) to the histones, which is another type of chemical modification, does just the opposite. Acetylation of the histones loosens DNA packaging making it easier for transcription machinery to access the DNA and increasing transcription.

Regulation of Gene Transcription: Flipping the Switch

The status of a gene (that is, whether it is turned on or turned off) can also be controlled by the actions of certain regulatory sequences within the DNA or by the binding of certain proteins to specific sequences in the DNA.

Sequences controlling genes

The main sequences involved in transcription control are located in the promoter, which is just upstream of the gene, located at the 5′ end of the gene. (See Chapter 5 for a review of the 5 prime and 3 prime ends of DNA.) One of the key sequences in the promoter is the TATA box, which indicates that the gene to be transcribed in about 30 bases downstream (that is, going in the 3′ direction). When the appropriate regulatory proteins bind to the promoter sequences, RNA polymerase is recruited and transcription can be turned on, as discussed in Chapter 9.

Three additional types of DNA sequences act as regulatory agents to turn transcription up *(enhancers)*, turn it down *(silencers)*, or drown out the effects of enhancing or silencing elements *(insulators)*.

- **Enhancers:** This type of gene sequence makes transcription happen faster and more often. Enhancers can be upstream, downstream, or even smack in the middle of the transcription unit. Furthermore, enhancers have the unique ability to control genes that are far away (like thousands of bases away) from the enhancer's position. Nonetheless, enhancers are very tissue-specific in their activities — they only influence genes that are normally activated in that particular cell type.

  Like the proteins that turn transcription on, enhancers seem to have the ability to rearrange nucleosomes and pave the way for transcription to occur. The enhancer teams up with transcription factors to form a complex called the *enhanceosome*. The enhanceosome attracts chromatin-remodeling proteins to the team along with RNA polymerase to allow the enhancer to supervise transcription directly.

- **Silencers:** These are gene sequences that hook up with repressor proteins to slow or stop transcription. Like enhancers, silencers can be many thousands of bases away from the genes they control. Silencers work to keep the DNA tightly packaged and unavailable for transcription.

- **Insulators:** Sometimes called *boundary elements*, these sequences have a slightly different job. Insulators work to protect some genes from the effects of silencers and enhancers, confining the activity of those sequences to the right sets of genes. Usually, this protection means that the insulator must be positioned between the enhancer (or silencer) and the genes that are off limits to the enhancer's (or silencer's) activities.

Given that enhancers and silencers are often far away from the genes they control, you may be wondering how they're able to do their jobs. It is believed that the DNA must loop around to allow enhancers and silencers to come in proximity to the genes they influence. Figure 11-3 illustrates this looping action. The promoter region begins with the TATA box and extends to the beginning of the gene itself. Enhancers interact with the promoter region to regulate transcription.

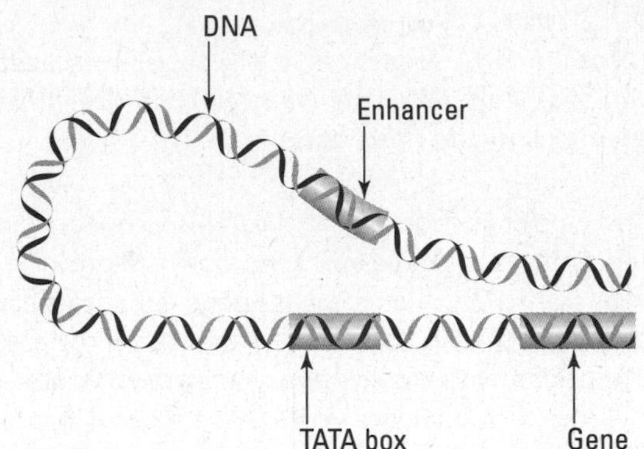

FIGURE 11-3: Enhancers loop around to turn on genes under their control.

Proteins controlling transcription

Transcription is also regulated by proteins that are transported into the nucleus and that bind to the promoter, enhancers, and silencers. These proteins are referred to as *transcription factors*, and they can either work to promote transcription or block transcription. Transcription factors function by binding to gene promoters and other regulatory sequences, such as enhancers or silencers. Each transcription factor contains a specific type of DNA-binding domain that recognizes a specific six to ten base pair sequence in the DNA.

- **General transcription factors:** This type of transcription factor is needed for the transcription of any gene and is found in all cell types. These proteins are part of the *transcription initiation complex*. In this complex, the transcription factors and *coactivator proteins* assist RNA polymerase, the enzyme responsible for transcription, locate and bind to the gene promoter (see Figure 11-4).
- **Regulatory transcription factors:** This type of transcription factor binds to specific regulatory sequences (such as enhancers) in order to help make sure the appropriate genes are turned on at the correct time and in the correct place (see Figure 11-4). Their function is more cell type- and gene-specific. Which regulatory transcription factors are present and active in a given cell depends on the specific cell type.

REMEMBER

To turn a gene on, both general and regulatory transcription factors bind to the promoter and enhancers of that gene. In doing so, transcription of that gene is activated. To turn a gene off, regulatory transcription factors bind to silencers, thereby preventing transcription.

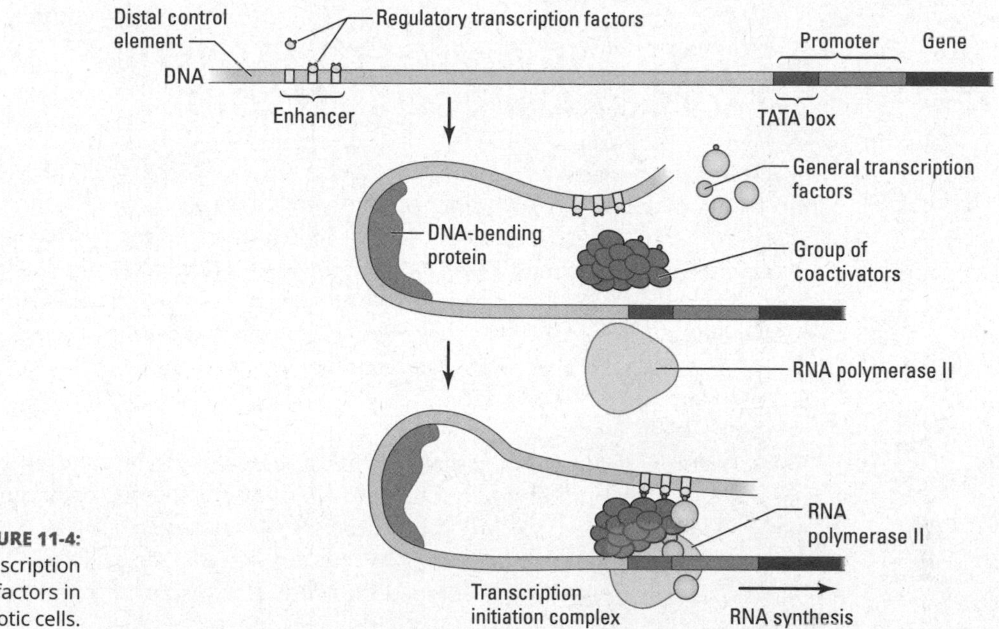

FIGURE 11-4: Transcription factors in eukaryotic cells.

Hormones controlling transcription

Hormones are complex chemicals that control gene expression. They're secreted by a wide range of tissues in the brain, gonads (ovaries and testes, which produce eggs and sperm), and other glands throughout the body. Hormones circulate in the bloodstream and can affect tissues far away from the hormones' production sites. In this way, they can affect genes in many different tissues simultaneously. Essentially, hormones act like a master switch for gene regulation all over the body. Take a look at the sidebar "A swing and a miss: The genetic effects of anabolic steroids" for more about the effects hormones can have on your body.

Some hormones are such large molecules that they often can't cross into the cells directly. These large hormone molecules rely on receptor proteins inside the cell to transmit their messages for them in a process called *signal transduction.* Other hormones, like steroids, are fat-soluble and small, so they easily pass directly into the cell to hook up with receptor proteins. Receptor proteins (and hormones small enough to enter the cell on their own) form a complex that moves into the cell nucleus to act as a transcription factor to turn specific genes on.

The genes that react to hormone signals are controlled by DNA sequences called *hormone response elements* (HREs). HREs sit close to the genes they regulate and bind with the hormone-receptor complex. Several HREs can influence the same gene — in fact, the more HREs present, the faster transcription takes place in that particular gene.

A SWING AND A MISS: THE GENETIC EFFECTS OF ANABOLIC STEROIDS

Anabolic-androgenic steroids are in the news a lot these days. These steroids are synthetic forms of testosterone, the hormone that controls male sex determination. The anabolic aspect refers to chemicals that increase muscle mass; the androgenic aspect refers to chemicals that control gonad functions such as sex drive and, in the case of men, sperm production. High-profile athletes, including some famous baseball players, have abused one or more of these drugs in an effort to improve performance. Reports also suggest that use of anabolic steroids is common among young athletes in high school and college.

Hormones like testosterone control gene expression. Research suggests that testosterone exerts its anabolic effects by depressing the activity of a tumor suppressor gene that produces the protein p27. When p27 is depressed in muscle tissue, the tissue's cells can divide more rapidly, resulting in the bulky physique prized by some athletes. Anabolic steroids apparently also accelerate the effects of the gene that causes male pattern baldness; thus, men carrying that allele and taking anabolic steroids become permanently bald faster and at a younger age than normal.

Defects in tumor-suppressor genes such as p27 are widely associated with cancer. Not only that, but some cancers depend on hormones to provide signals that tumor cells respond to (by multiplying). At least one study suggests that anabolic steroids are actually carcinogenic, meaning that their chemicals cause changes in the DNA that lead to cancer. Because illegally obtained steroids may also contain additional unwanted and potentially carcinogenic chemicals, these chemicals may also be introduced into the body while simultaneously depressing the activity of a tumor-suppressor gene.
It doesn't take a genius to realize that this is dangerous. Cancers associated with anabolic-androgenic steroid abuse include liver cancer, testicular cancer, leukemia, and prostate cancer.

Retroactive Control: Things That Happen after Transcription

After genes are transcribed into mRNA, their actions can still be controlled by events that occur later.

Nip and tuck: RNA splicing

As we discussed in Chapter 9, genes have sections called *exons* that actually code for protein products. In between the exons are *introns*, interruptions of non-coding DNA that may or may not do anything. When genes are transcribed, the whole thing is copied into a primary mRNA transcript (see Figure 9-8 in Chapter 9). This mRNA transcript then must be edited — meaning the introns are removed — in preparation for translation. When multiple exons and multiple introns are present in the unedited transcript, various combinations of exons can result from the editing process (see Figure 9-8). Specific exons can be included in the transcript or can be edited out, yielding new proteins when translation rolls around. This creative editing process allows genes to be expressed in new ways; one gene can code for more than one protein. This genetic flexibility is credited for the massive numbers of proteins you produce relative to the number of genes you have, as we discuss in Chapter 9.

TECHNICAL STUFF

One gene in which genetic flexibility is very apparent is *DSCAM.* Named for the human disorder it's associated with — Down Syndrome Cell Adhesion Molecule — *DSCAM* may play a role in causing the intellectual disabilities that accompany Down syndrome. (*DSCAM* in uppercase refers to the human gene; *Dscam* in lowercase refers to the fruit fly gene.) In fruit flies, *Dscam* is a large gene with 115 exons and at least 100 splicing sites. Altogether, *Dscam* is capable of coding for a whopping 30,016 different proteins. However, protein production from *Dscam* is tightly regulated; some of its products only show up during early stages of fly development. The human version of *DSCAM* is less showy in that it makes only a few proteins, but other genes in the human genome are likely to be as productive at making proteins as Dscam of fruit flies, making this a "fruitful" avenue of research. Humans have very few genes relative to the number of proteins we have in our bodies. Genes like *Dscam* may help geneticists understand how a few genes can work to produce many proteins.

With scientists wise to the nip and tuck game played by mRNA, the next step in deciphering this sort of gene regulation is figuring out how the trick is done and what controls it. Researchers know that a complex of proteins called a *spliceosome* carries out much of the work in cutting and pasting genes together (you can flip back to Chapter 9 for a review on splicing). How the spliceosome's activities are regulated is another matter altogether. Knowing how it all works will come in handy though, because some forms of cancer, most notably pancreatic cancer, can result from alternative splicing run amok.

Shut up! mRNA silencing

After transcription produces mRNA, genes may be regulated through *mRNA silencing.* mRNA silencing is basically interfering with the mRNA somehow so that it doesn't get translated. Scientists don't fully understand exactly how organisms like you and me use mRNA silencing, called *RNAi* (for *RNA interference*), to regulate genes. Geneticists know that most organisms use RNAi to stymie translation of unwanted mRNAs and that double-stranded RNA provides the signal for the initiation of RNAi, but the details are still a mystery. The discovery of RNAi has produced a revolution in the study of gene expression; see the sidebar "Interfering RNAs knock out genes" for more.

RNA silencing isn't just used to regulate the genes of an organism; sometimes it's used to protect an organism from the genes of viruses. When the organism's defenses detect a double-stranded virus RNA, an enzyme called *dicer* is produced. Dicer chops the double-stranded RNA into short bits (about 20 or 25 bases long). These short strands of RNA, now called *small interfering RNAs* (siRNAs), are then used as weapons against remaining viral RNAs. The siRNAs turn traitor, first pairing up with RNA-protein complexes produced by the host and then guiding those complexes to intact viral RNA. The viral RNAs are then summarily destroyed and degraded.

INTERFERING RNAs KNOCK OUT GENES

The world of RNAi (RNA interference; see the section "Shut up! mRNA silencing") is creating quite a splash in the understanding of how gene expression is controlled. The breakthrough moment came when two geneticists, Andrew Fire and Craig Mello, realized that by introducing certain double-stranded RNA molecules into roundworms, they could shut off genes at will. It turns out that scientists can put the RNAi into roundworm food and knock out gene function not only in the worm that eats the concoction but also in its offspring!

Since this discovery in 2003, geneticists have identified naturally occurring interfering RNAs in all sorts of organisms. The most well-known RNAi tend to be very short (only about 20 or so bases long) and hook up with special proteins, called *argonautes*, to regulate genes (mostly by silencing them). The argonaute proteins actually do the work, guided to the right target by the RNAi. RNAi finds its complementary mRNA (the product of the gene to be regulated), and the argonaute breaks down the mRNA, rendering it nonfunctional. New RNAi's are being discovered all the time, and their full importance in regulating genes is only just being realized. Longer, non-coding RNAs (over 200 bases long) are also produced during transcription; scientists are hard at work determining what functions those have.

The most promising applications for RNAi are in gene therapy (jump to Chapter 16 for that discussion). Using synthetic RNAi, geneticists have knocked out genes in all sorts of organisms, including chickens and mice. Work is also underway to knock out the function of genes in viruses and cancer cells.

mRNA expiration dates

After mRNAs are sliced, diced, capped, and tailed (see Chapter 9 for how mRNA gets dressed up), they're transported to the cell's cytoplasm. From that moment onward, mRNA is on a path to destruction because enzymes in the cytoplasm routinely chew up mRNAs as soon as they arrive. Thus, mRNAs have a relatively short lifespan, the length of which (and therefore the number of times mRNA can be translated into protein) is controlled by a number of factors. But the mRNA's poly-A tail (the long string of adenines tacked on to the 3′ end) seems to be one of the most important features in controlling how long mRNA lasts. Key aspects of the poly-A tail include:

- **Tail length:** The longer the tail, the more rounds of translation an mRNA can support. If a gene needs to be shut off rapidly, the poly-A tail is usually pretty short. With a short tail, when transcription comes to a halt, all the mRNA in the cytoplasm is quickly used up without replacement, thus halting protein production, too.
- **Untranslated sequences before the tail:** Many mRNAs with very short lives have sequences right before the poly-A tail that, even though they aren't translated, shorten the mRNA's lifespan.

Hormones present in the cell may also affect how quickly mRNAs disappear. In any event, the variation in mRNA expiration dates is enormous. Some mRNAs last a few minutes, meaning those genes are tightly regulated; other mRNAs hang around for months at a time.

Gene Control Lost in Translation

Translation of mRNA into amino acids is a critical step in gene expression. (See Chapter 10 for a review of the players and process of translation.) But sometimes genes are regulated during or even after translation.

Modifying where translation occurs

One way gene regulation is enforced is by corralling in mRNAs in certain parts of the cytoplasm. That way, proteins produced by translation are found only in certain parts of the cell, limiting their utility. Embryos use this strategy to direct their own development. Proteins are produced on different sides of the egg to create the front and back, so to speak, of the embryo.

Modifying when translation occurs

Just because an mRNA gets to the cytoplasm doesn't mean it automatically gets translated. Some gene expression is limited by certain conditions that block translation from occurring. For example, an unfertilized egg contains lots of mRNAs supplied by the female. Translation actually occurs in the unfertilized egg, but it's slow and selective. All that changes when a sperm comes along and fertilizes the egg: Preexisting mRNAs are slurped up by waiting ribosomes, which are signaled by the process of fertilization. New proteins are then rapidly produced from the maternal mRNAs.

Controlling gene expression by controlling translation can occur in several ways. One way is that the proteins needed for translation, such as the initiator proteins that interact with ribosomes, can be modified to increase or decrease how effectively translation occurs. For example, a protein known as eIF-2 (eukaryotic initiation factor 2) binds to one of the small rRNA subunits and is needed for translation to begin. When this protein is phosphorylated (that is, a phosphate group is attached to it), it is unable to bind to the small rRNA subunit, thereby preventing translation.

Another way that translation can be regulated is through a sequence in the mRNA. Basically, the mRNA carries a message that controls when and how it gets translated. All mRNAs carry short sequences on their 5′ ends that aren't translated (known as the 5′ untranslated region or 5′ UTR; see Figure 11-5), and these sequences can carry messages about the timing of translation. The untranslated sequences are recognized with the help of translation initiation factors that help assemble the ribosome at the start codon of the mRNA. Some cells produce mRNAs but delay translation until certain conditions are met. Some cells respond to levels of chemicals that the cell's exposed to. For example, the protein that binds to iron in the blood is created by translation only when iron is available, even though the mRNAs are being produced all the time. In other cases, the condition of the organism sends the message that controls the timing of translation. For example, insulin, the hormone that regulates blood sugar levels, controls translation, but when insulin's absent, the translation factors lock up the needed mRNAs and block translation from occurring. When insulin arrives on the scene, the translation factors release the mRNAs, and translation rolls on, unimpeded.

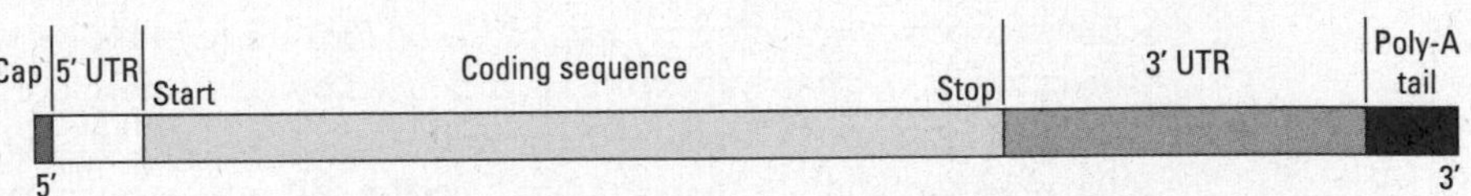

FIGURE 11-5: mRNAs have untranslated regions (UTRs) at the 5' and 3' ends.

Modifying the protein shape

The proteins produced by translation are the ultimate form of gene expression. Protein function, and thus gene expression, can be modified by adding certain components to the protein. This is referred to as post-translational modification. The chemical modification of proteins can be used to regulate protein activity and protein lifespan. Post-translational modifications can directly affect regions of the protein that are needed to function or it may change the way the protein is folded (also changing its functions).

A common modification of proteins is *phosphorylation* (the addition of a phosphoryl group – PO_3^- – to the protein). This is frequently a mechanism by which certain proteins can become activated or deactivated (like eIF-2 mentioned in the previous section). Other common modifications include *methylation* (addition of a methyl group – CH_3) and *acetylation* (addition of an acetyl group - CH_3CO). In addition, ubiquitination, or the attachment of the small molecular ubiquitin to a protein, occurs in all types of cells. One of the main functions of ubiquitination is to signal that the protein is to be degraded. Ubiquitination may also help with the localization of the protein within the cell or with interactions with other proteins.

Prokaryotic Gene Expression

The regulation of gene expression in prokaryotes differs from that of eukaryotes, primarily because of the way bacterial genes are organized and transcribed.

Bacterial gene organization

In eukaryotes, each gene is controlled by its own promoter. In bacteria, multiple genes located next to each other can be controlled by a single regulatory sequence (see Figure 11-6). The arrangement of multiple genes plus a single promoter controlling all of them is called an *operon*. The promoter is located at the beginning of the operon. Next, there is a regulatory sequence called an *operator*, where certain DNA-binding proteins bind and help regulate gene transcription. The genes in an operon typically encode structural proteins that all work together in a single process.

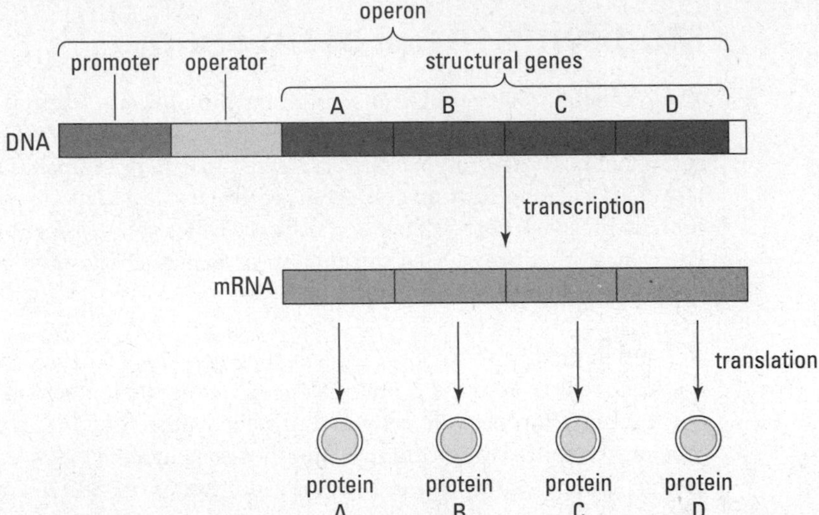

FIGURE 11-6: Organization of bacterial genes in an operon.

Bacterial gene expression

Gene expression in bacteria is regulated by DNA-binding proteins that can either help turn the genes on or turn the genes off. These DNA-binding proteins are, in turn, regulated by the binding of small molecules, such as sugars or amino acids. When the appropriate small molecule binds a regulatory protein, the protein changes form and is either activated or inactivated. When active, some regulatory proteins function to *induce* gene transcription, while others function to *repress* gene transcription.

Bacterial gene expression is generally initiated after the cell receives some type of signal from the environment. This signal takes the form of the small molecules that bind regulatory proteins. After a regulatory protein binds the appropriate small molecule, it either binds to the regulatory sequences in the DNA or is released from the DNA, depending on the role of the protein in gene expression. This, in turn, either activates transcription or prevents transcription of the genes in the operon.

3 Genetics and Your Health

IN THIS PART . . .

Understand the ways in which DNA can be altered and the consequences of those changes.

Find out what happens when chromosomes aren't divided up properly during cell division or when there are structural changes in those chromosomes.

Discover how genetic counselors read your family tree to better understand your family medical history and determine your risk of disease.

See how genetics are used to treat a variety of conditions, including genetic syndromes and cancer.

IN THIS CHAPTER

» Considering the different types and causes of DNA sequence changes

» Realizing the consequences of and repairs for DNA sequence changes

Chapter 12
When Things Go Wrong: Changes in DNA Sequence

Your DNA is remarkably similar to the DNA of every other human on earth. In fact, more than 99 percent of the human genome sequence is the same in any two individuals, whether they are related or not. The differences in DNA sequence are what make each of us unique. Variations in DNA sequence are responsible for all phenotypic variation, such as eye color and height in humans, the flavor of different kinds of apples, the differences among dog breeds, and the different types of bees. Unfortunately, changes in DNA sequence can also lead to disease.

Changes in DNA occur all the time, spontaneously and pretty much randomly. The effects of alterations in DNA sequence depend on where the changes occur (in a gene or in a non-coding region of the genome), how the change affects the protein product (if located in a gene), and whether the change occurs in a germ cell (an egg or sperm) or in a somatic cell (a non-sex cell). In this chapter, you discover what leads to changes in DNA sequence (whether good or bad), how DNA can repair itself in the face of these changes, and what the consequences are when repair attempts fail.

Heritable or Not Heritable?

Some changes in DNA sequence (referred to as *variants*) can be passed from parent to child, while others cannot. Whether or not a change in sequence can be inherited depends on the kind of cell in which the change occurs.

When talking about the *heritability* of a sequence variant (that is, the ability of the change to be passed from one generation to the next), there are two main categories to consider, and the distinction between the two is important to keep in mind:

- **Somatic variants:** These are sequence variants in body cells that are not eggs or sperm. Somatic cells, which divide only by mitosis (discussed in Chapter 2), constitute all the other cells of the body. Changes that occur in the DNA of somatic cells are *not* heritable — they can't be passed from parent to offspring. However, they can affect the person with the variant.
- **Germ-cell variants (also called germline variants):** These are sequence variants in the germ cells (also referred to as sex cells; that is, eggs and sperm). These are cells that lead to embryo formation. Unlike somatic variants, germ-cell variants can be passed from parent to child.

TECHNICAL STUFF

Some disorders have elements of both somatic and germ-cell sequence variants. Some cancers that run in families arise as a result of somatic variants in persons who are already susceptible to the disease because of variants they inherited from one or both parents. You can find out more about heritable cancers in Chapter 14.

Facing the Consequences of Sequence Variants

When a gene is changed and the sequence variant is passed along to the next generation, the new, altered version of the gene is considered a new allele. *Alleles* are simply alternative versions of a gene. For most genes, many alleles exist. The effect of various alleles can fall in to one of two categories:

- **Neutral change:** When the protein produced from the altered gene is a fully functional protein, the variant is considered neutral. See Chapter 10 to review how proteins are made.
- **Functional change:** A functional change occurs when the protein created results in a change in function of the protein. A *gain-of-function variant* results in a protein with enhanced activity or a completely new function. Genetic

conditions that result from gain-of-function variants are typically inherited in a dominant manner, meaning that you only need one copy of that variant in order to develop the condition. If a sequence variant causes the protein to stop functioning altogether or vastly affects normal function, it's considered a *loss-of-function variant.* Conditions that result from loss-of-function variants are generally inherited in a recessive manner, meaning that both copies of the gene involved need to be changed in order to have the condition (that is, you need two disease-causing alleles). When you have only one changed allele, the normal, unchanged allele is still producing a normal protein product — usually enough to compensate for the allele with the variant.

Sorting Out Terminology

Not only can variants be classified (very generally) based on whether or not they affect protein function, they can be classified based on the phenotypic effect (whether or not the change is disease-causing) and by the specific type of sequence change (the precise way in which the change affects the DNA sequence and the protein produced).

The phenotypic effect of a DNA sequence change

Historically, changes in DNA sequence have been classified in one of two ways: mutation or polymorphism. The term *mutation* is typically defined as a change in DNA sequence that leads to a functional change in the protein and/or has been associated with disease. Specific mutations are typically rare in the general population. *Polymorphism*, on the other hand, is generally reserved for changes in DNA sequence that do not cause a problem. These are considered *neutral* changes in the DNA. Polymorphisms are frequently common among the population.

TECHNICAL STUFF

More recently, the classification of *sequence variants* (any change in DNA sequence from the reference sequence for the human genome) has become more specific. Sequence variants that are known or likely to cause disease are referred to a *pathogenic* or *likely pathogenic* (pathogenic meaning disease-causing). This is the equivalent of a mutation. Sequence variants that are not expected to cause disease, because they do not affect protein structure or function, or because they are exceedingly common in the population, are referred to as *benign* or *likely benign*. This is the equivalent of a polymorphism.

The classification of whether a variant is pathogenic or benign is not trivial, especially since many people are using the results of DNA testing to make important health care and life decisions. Because of this, the classification of variants follows a very detailed and strict set of rules based on scientific evidence. No one wants to be told that the variant found in one of their genes is benign, only to find out later (after more data is available) that it is actually disease-causing and that they should have been monitored or treated for the disease in question.

With all the DNA sequencing that is now being performed (a process we describe in Chapter 8), the number of sequence variants that have been identified has grown drastically. Unfortunately, many sequence variants are unique to a person or family and there is no evidence showing whether it is a mutation or polymorphism. In other words, many sequence variants are being described for which the effect is unclear because of a lack of data. These are referred to as *variants of unknown significance*. Eventually, with more data, these may be reclassified as either pathogenic or benign.

The type of DNA sequence change

Another way that sequence variants are categorized is based on the type of change in the DNA and the effect of the change on the protein that is produced. The main types of variants (summarized in Table 12-1) include the following:

- **Missense variant:** Missense variants occur when there is a substitution of one base for another, leading to the substitution of one amino acid for another. Substitutions are sometimes called *point mutations.* This type of variant can also be broken down further into two categories:
  - **Transition variant:** When a purine base (adenine or guanine) is substituted for the other purine, or one pyrimidine (thymine or cytosine) is substituted for the other pyrimidine. (For a review of the chemistry of DNA, turn to Chapter 5). Transition mutations are the most common form of substitution errors.
  - **Transversion variant:** When a purine replaces a pyrimidine (or vice versa).
- **Silent variant:** Silent variants occur when there is a base substitution in the DNA but there is no change in the amino acid. Silent variants result from the redundancy of the genetic code. The code is redundant in the sense that multiple combinations of bases have identical meanings. Silent variants are often benign, with no effect on protein function. However, if a silent variant

occurs near the splice junction (where an exon and intron meet), it can affect splicing of the mRNA. In this case, the variant could have a drastic effect on the final protein and could be considered a functional variant.

- **Nonsense variant:** Nonsense variants occur when a substitution in the DNA sequence results in a codon for an amino acid being changed to a stop codon. The result of a stop codon too early in the gene is an early end to translation and a shortened protein product. In addition, if the change occurs early enough in the gene, the resulting mRNA will not be stable and will be degraded before translation can even occur. This is referred to as *nonsense-mediated decay* of mRNA.
- **Insertions and deletions of one or more bases:** When an extra base is added to a strand, the error is called an *insertion.* Dropping a base is considered a *deletion.* When the change happens within a gene, both insertions and deletions can lead to a change in the way the genetic code is read during translation, as discussed in Chapter 10. Translation involves reading the genetic code in three-letter batches, so when one or two bases are added or deleted, the reading frame is shifted. A *frameshift variant* results in a completely different interpretation of what the code says and produces an entirely different amino acid strand. In addition, frameshifts often lead to introduction of an early stop codon, which results in premature truncation of the protein. As you can imagine, these effects have disastrous consequences, because the expected gene product isn't produced. If three bases are added or deleted (or a multiple of three bases), the overall reading frame isn't affected. The result of a three-base insertion or deletion, called an *in-frame insertion or in-frame deletion,* is that one amino acid is either added (insertion) or lost (deletion). In-frame variants can be just as bad as frameshift variants, but sometimes they can be benign and have no significant effect on the protein.
- **Splice site variants:** When a sequence change affects how splicing of the gene occurs, it is referred to as a *splice site variant* (or *splicing variant*). Changes in DNA sequence at or near the junctions between exons (coding sequence of the gene) and introns (non-coding sequences located between exons) can lead to the skipping of exons or inclusion of intron sequences in the final mRNA. In either case, the structure of the protein will be altered because of changes to the mRNA translated.

TABLE 12-1 **Types and Effects of DNA Sequence Variants**

Type of Sequence Variant	Definition	Example: Original Sequence	Example: Altered DNA Sequence	Effect on Amino Acid Sequence and Protein
Missense variant	Substitution of a base leading to an amino acid substitution	*DNA*: ATG CTA TGC *Protein*: Met Leu Cys	*DNA*: ATG CAA TGC *Protein*: Met Gln Cys	Can affect protein structure and/or function.
Nonsense variant	Substitution of a base leading to a stop codon	*DNA*: ATG CTA TGC *Protein*: Met Leu Cys	*DNA*: ATG CTA TGA *Protein*: Met Leu STOP	Leads to production of a shortened protein or no protein at all if the mRNA is degraded.
Silent variant	Substitution of a base that does not change the amino acid	*DNA*: ATG CTA TGC *Protein*: Met Leu Cys	*DNA*: ATG CTC TGC *Protein*: Met Leu Cys	Does not change the amino acid but could affect splicing, depending on the location of the change.
Frameshift deletion	Deletion of a base (or bases) that results in a change in the reading frame	*DNA*: ATG CTA TGC AGG *Protein*: Met Leu Cys	*DNA*: ATG CAT GCA *Protein*: Met His Ala	Shifts the reading frame so that the amino acid sequence is completely different downstream of the change. Leads to a completely different protein and often the introduction a premature stop codon.
Frameshift insertion	Insertion of a base (or bases) that results in a change in the reading frame	*DNA*: ATG CTA TGC *Protein*: Met Leu Cys	*DNA*: ATG CTT ATG *Protein*: Met Leu Met	Shifts the reading frame so that the amino acid sequence is completely different downstream of the change. Leads to a completely different protein and often the introduction of a premature stop codon.

Type of Sequence Variant	Definition	Example: Original Sequence	Example: Altered DNA Sequence	Effect on Amino Acid Sequence and Protein
In-frame deletion	Deletion of bases (in multiples of 3) that does not change the reading frame	*DNA*: ATG CTA TGC AGG *Protein*: Met Leu Cys Arg	*DNA*: ATG TGC AGG *Protein*: Met Cys Arg	Results in the deletion of one amino acid (if a 3-base deletion). Could affect 2 neighboring amino acids, depending on which bases are deleted.
In-frame insertion	Insertion of bases (in multiples of 3) that does not change the reading frame	*DNA*: ATG CTA TGC *Protein*: Met Leu Cys	*DNA*: ATG CTA CTA TGC *Protein*: Met Leu Leu Cys	Results in the insertion of one amino acid (if a 3-base insertion). Could affect 2 neighboring amino acids, depending on where the bases are inserted.
Splice site variant	A change in the DNA sequence that affects splicing	*DNA**: ATG CTA TGC AAG gta agt *Protein*: Met Cys Leu Lys * Uppercase: exon; lowercase: intron	*DNA*: ATG CTA TGC AAG ATA AGT *Protein*: The effect of this type of variant depends on the neighboring sequences.	This variant could result in the skipping of an exon (with the exon being spliced out) or it could result in the inclusion of intron sequences into the coding sequence of the gene. Laboratory studies are often needed to determine the exact effect of a specific splice site change.

Abbreviations: *Met – Methionine; Leu – Leucine; Cys – cysteine; His – histidine; Ala – alanine; Arg – arginine; Lys – lysine*

What Causes Sequence Variants?

DNA sequence changes can occur for a whole suite of reasons. In general, the changes are either the result of random mistakes in DNA replication or exposure to outside agents such as chemicals or radiation. In the sections that follow, we delve into each of these causes.

Spontaneous sequence variants

REMEMBER

Spontaneous DNA sequence change occurs randomly and without any urging from some external cause. It's a natural, normal occurrence. Because the vast majority of your DNA doesn't code for anything, most spontaneous sequence variants go unnoticed. (You can check out Chapter 8 for more details about your non-coding "junk" DNA.) But when sequence changes occur within a gene, the function of the gene can be changed or disrupted. Those changes can then result in unwanted side effects such as cancer, which we address in Chapter 14.

Scientists are all about counting, sorting, and quantifying, and it's no different with sequence variants. Spontaneous sequence variants are measured in the following ways:

- **Frequency:** DNA sequence variants are sometimes measured by the frequency of occurrence. *Frequency* is the number of times some event occurs within a group of individuals. When you hear that one in some number of persons has a particular disease-causing allele, the number is a frequency. For example, one study estimates that the X-linked disease hemophilia has a frequency of 13 cases for every 100,000 males.
- **Rate:** Another way of looking at mutations is in the framework of a *rate,* like the number of sequence changes occurring per round of cell division, or the number of sequence changes per gamete or per generation. Sequence variant rates appear to vary a lot from organism to organism. Even within a species, the rates vary depending on which part of the genome you're examining. Some convincing studies show that sequence variant rate even varies by sex and that the rate of DNA sequence changes are higher in males than females (check out the sidebar "Dad's age matters, too" for more on this topic). Regardless of how it's viewed, spontaneous sequence changes occur at a steady but very low rate.

REMEMBER

Most spontaneous sequence variants occur because of mistakes made during replication (all the details of how DNA replicates itself are in Chapter 7). Here are the three main sources of error that can happen during replication:

- Mismatched bases are overlooked during proofreading.
- Strand slip-ups lead to deletions or insertions.
- Spontaneous but natural chemical changes cause bases to be misread during replication, resulting in substitutions or deletions.

Mismatches during replication

Usually, *DNA polymerase* catches and fixes mistakes made during replication. DNA polymerase has the job of reading the template, adding the appropriate complementary base to the new strand, and then proofreading the new base before moving to the next base on the template. DNA polymerase can snip out erroneous bases and replace them, but occasionally, a wrong base escapes detection. Such an error is possible because noncomplementary bases can form hydrogen bonds through what's called *wobble pairing.* As you can see in Figure 12-1, wobble pairing can occur:

- **Between thymine and guanine,** without any modifications to either base (because these noncomplementary bases can sometimes form bonds in odd spots).
- **Between cytosine and adenine,** only when adenine acquires an additional hydrogen atom (called *protonation*).

Thymine
Guanine
Cytosine
Adenine

FIGURE 12-1: Wobble pairing allows mismatched bases to form bonds.

If DNA repair crews don't catch the error and fix it (see the section "Evaluating Options for DNA Repair," later in this chapter), and the mismatched base remains in place, the mistake is perpetuated after the next round of replication, as shown in Figure 12-2. The mistaken base is read as part of the template strand, and its complement is added to the newly replicated strand opposite. Thus, the sequence variant is permanently added to the structure of the DNA in question.

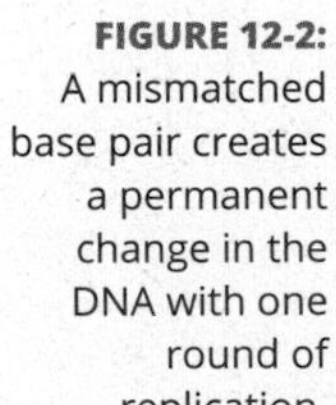

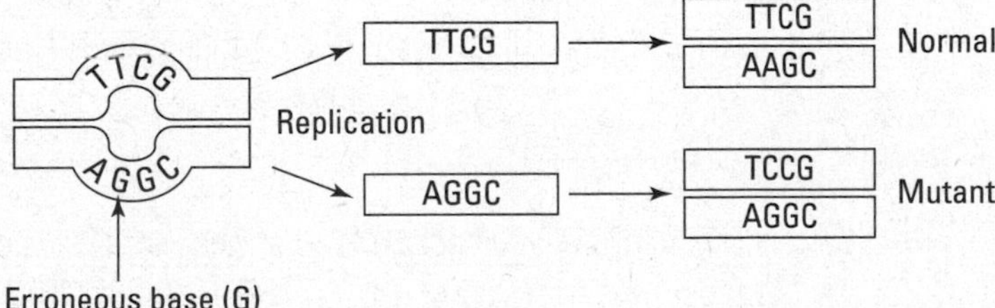

FIGURE 12-2: A mismatched base pair creates a permanent change in the DNA with one round of replication.

DAD'S AGE MATTERS, TOO

The relationship between maternal age and an increased incidence of chromosome problems, such as Down syndrome, is very well-known. *Nondisjunction events* — the failure of chromosomes to separate normally during meiosis — in developing eggs are thought to be a consequence of aging in women (whose egg cells are present from the beginning of life). This type of error occurs less frequently in men because they produce new sperm throughout their lifetime. However, older men are susceptible to germ-cell changes in DNA sequence that can cause heritable disorders in their children.

The reason that older men are more susceptible to spontaneous germ-cell mutations is the same reason that they're less likely to have nondisjunctions — males produce sperm throughout their life. With this continued sperm production comes continued DNA replication. As DNA ages, replication gets less accurate and repair mechanisms become faulty. Thus, older fathers have an increased risk (although it's still only slight) of fathering children with certain genetic disorders that are inherited in a dominant manner (meaning that you need only one copy of a mutation to have the condition). *Achondroplasia* (a form of dwarfism that's typified by shortened limbs, short stature, and a larger than average head size), *osteogenesis imperfecta type 1* (brittle bone disease) and *neurofibromatosis type 1* (a disease characterized by skin changes, benign growths in the colored part of the eye, and benign tumors of nerve cells) are all associated with an older age (>50 years) in fathers.

Strand slip-ups

During replication, both strands of DNA are copied more or less at the same time. Occasionally, a portion of one strand (either the template or the newly synthesized strand) can form a loop in a process called *strand slippage.* In Figure 12-3, you can see that strand slippage in the new strand results in an insertion, and slippage in the template strand results in a deletion.

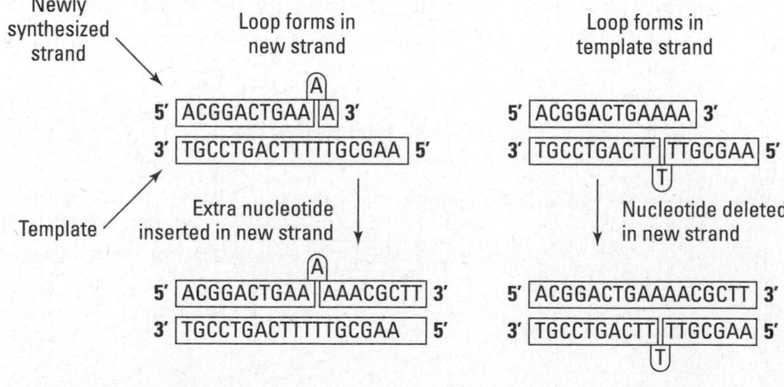

FIGURE 12-3: Strand slippage causes loops to form during replication, resulting in deletions and insertions.

REMEMBER

Strand slippage is associated with repeating bases. When one base is repeated more than five times in a row (AAAAAA, for example), or when any number of bases are repeated over and over (such as AGTAGTAGT), strand slippage during replication is far more likely to occur. In some cases, the mistakes produce lots of variation in non-coding DNA, and the variation is useful for determining individual identity; this is the basis for DNA fingerprinting, which we discuss in Chapter 18. When repeat sequences occur within genes, however, the addition of new repeats leads to an elongated protein and can sometimes lead to a stronger effect of the gene. This strengthening effect, called *anticipation,* occurs in genetic disorders such as Huntington disease. You can find out more about anticipation and Huntington disease in Chapter 4.

Another problem that repeated bases generate is unequal crossing-over. During meiosis, homologous chromosomes are supposed to align exactly so that exchanges of information are equal and don't disrupt genes (see Chapter 2 for a meiosis review). Unequal crossing-over occurs when the exchange between chromosomes results in the swapping of uneven amounts of material. Repeated sequences cause unequal crossovers because so many similar bases match. The identical bases can align in multiple ways that result in mismatches elsewhere along the chromosome. Unequal crossover events lead to large-scale chromosome changes, such as those we describe in Chapter 13. Chromosomes in cells affected by cancer are also vulnerable to crossing-over errors, as we describe in Chapter 14.

Spontaneous chemical changes

DNA can undergo spontaneous changes in its chemistry that result in both deletions and substitutions. DNA naturally loses purine bases at times in a process called *apurination.* Most often, a purine is lost when the bond between adenine and the sugar, deoxyribose, is broken. (See Chapter 5 for a reminder of what a nucleotide looks like.) When a purine is lost, replication treats the spot occupied by the orphaned sugar as if it never contained a base at all, resulting in a deletion.

Deamination is another chemical change that occurs naturally in DNA. It's what happens when an amino group (composed of a nitrogen atom and two hydrogens — NH_2) is lost from a base. Figure 12-4 shows the before and after stages of deamination. When cytosine loses its amino group, it's converted to uracil. Uracil normally isn't found in DNA at all because it's a component of RNA. If uracil appears in a DNA strand, replication replaces the uracil with a thymine, creating a substitution error. Until it's snipped out and replaced during repair (see "Evaluating Options for DNA Repair," later in this chapter), uracil acts as a template during replication and pairs with adenine. Ultimately, what was a C-G pair transitions into an A-T pair instead.

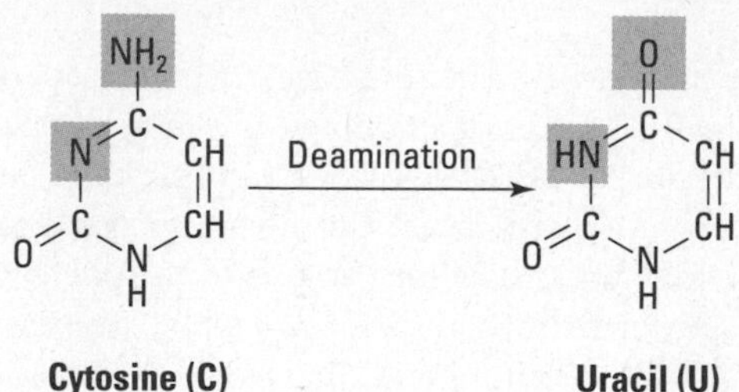

FIGURE 12-4: Deamination converts cytosine to uracil.

Induced sequence variants

REMEMBER

Induced sequence variants result from exposure to some outside agent such as chemicals or radiation. It probably comes as no surprise to you to find out that many chemicals can cause DNA mutations. *Carcinogens* (chemicals that cause cancers) aren't uncommon; the chemicals in cigarette smoke are probably the biggest offenders. In addition to certain chemicals, radiation (from X-rays to sunlight) can be highly mutagenic. A *mutagen* is any factor that causes an increase in the rate of DNA sequence changes. Mutagens may or may not have phenotypic effects — it depends on what part of the DNA is affected. The following sections cover two major categories of mutagens: chemicals and radiation. Each causes different damage to DNA.

Chemical mutagens

The ability of chemicals to cause permanent changes in the DNA of organisms was discovered by Charlotte Auerbach in the 1940s (see the sidebar "The chemistry of mutation" for the full story). There are many types of mutagenic chemicals; the following sections address four of the most common.

BASE ANALOGS

Base analogs are chemicals that are structurally very similar to the bases normally found in DNA. Base analogs can get incorporated into DNA during replication because of their structural similarity to normal bases. One base analog, 5-bromouracil, is almost identical to the base thymine. Most often, 5-bromouracil (also known as 5BU), which is pictured in Figure 12-5, gets incorporated as a substitute for thymine and, as such, is paired with adenine. The problem arises when DNA replicates again with 5-bromouracil as part of the template strand; 5BU is mistaken for a cytosine and gets incorrectly paired with guanine. The series of events looks like this: 5-bromouracil is incorporated where thymine used to be, so T-A becomes 5BU-A. After one round of replication, the pair is 5BU-G, because 5BU is prone to chemical changes that make it a mimic of cytosine, the base normally paired with guanine. After a second round of replication, the pair ends up as C-G, because 5BU isn't found in normal DNA. Thus, an A-T ends up as a C-G pair.

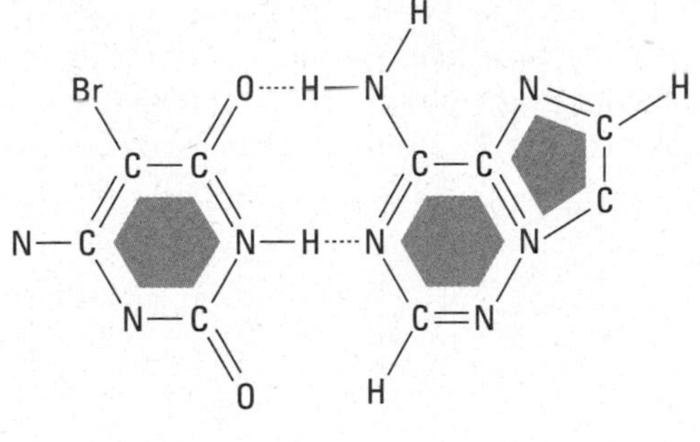

FIGURE 12-5: Base analogs, such as 5-bromouracil, are very similar to normal bases.

Another class of base analog chemicals that foul up normal base pairing is *deaminators.* Deamination is a normal process that causes spontaneous changes in DNA sequence; however, problems arise because deamination can speed up when cells are exposed to chemicals that selectively knock out amino groups converting cytosines to uracils.

ALKYLATING AGENTS

Like base analogs, *alkylating agents* induce mispairings between bases. Alkylating agents, such as the chemical weapon mustard gas, add chemical groups to the existing bases that make up DNA. As a consequence, the altered bases pair with the wrong complement, thus introducing the change in DNA sequence. Surprisingly, alkylating agents are often used to fight cancer as part of chemotherapy; therapeutic versions of alkylating agents may inhibit cancer growth by interfering with the replication of DNA in rapidly dividing cancer cells.

FREE RADICALS

Some forms of oxygen, called *free radicals,* are unusually reactive, meaning they react readily with other chemicals. These oxygens can damage DNA directly (by causing strand breaks) or can convert bases into new unwanted chemicals that, like most other chemical mutagens, then cause mispairing during replication. Free radicals of oxygen occur normally in your body as a product of metabolism, but most of the time, they don't cause any problems. Certain activities — such as cigarette smoking and high exposure to radiation, pollution, and weed killers — increase the number of free radicals in your system to dangerous levels.

INTERCALATING AGENTS

Many different kinds of chemicals wedge themselves between the stacks of bases that form the double helix itself, disrupting the shape of the double helix. Chemicals with flat ring structures, such as dyes, are prone to fitting themselves between bases in a process called *intercalation.* Figure 12-6 shows intercalating agents at work. Intercalating agents create bulges in the double helix that often result in insertions or deletions during replication, which in turn cause frameshift sequence variants.

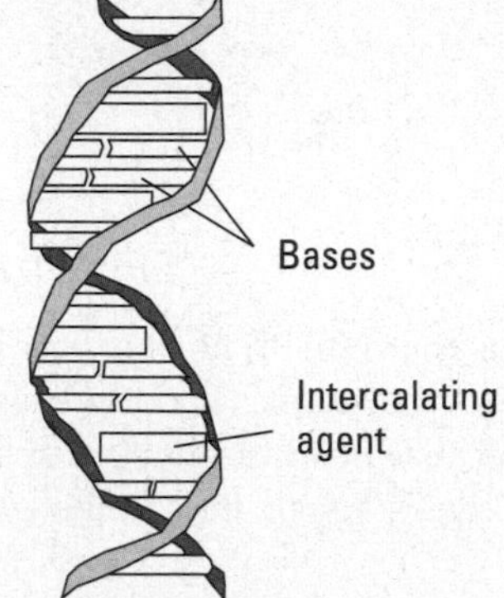

FIGURE 12-6: Intercalating agents fit between the stacks of bases to disfigure the double helix.

Radiation

Radiation damages DNA in a couple of ways. First, radiation can break the strands of the double helix by knocking out bonds between sugars and phosphates (see Chapter 5 for a review of how the strands are put together). If only one strand is broken, the damage is easily repaired. But when two strands are broken, large parts of the chromosome can be lost; these kinds of losses can affect cancer cells (see Chapter 14) and cause a variety of genetic syndromes (see Chapter 13).

Second, radiation causes DNA sequence changes through the formation of *dimers.* Dimers (*di-* meaning "two"; *mer* meaning "thing") are unwanted bonds between two bases stacked on top of each other (on the same side of the helix, rather than on opposite sides). They're most often formed when two thymines in a DNA sequence bind together, which you can see in Figure 12-7.

Thymine dimers can be repaired, but if damage is extensive, the cell dies (see Chapter 14 to learn how cells are programmed to die). When dimers aren't repaired, the machinery of DNA replication assumes that two thymines are present and puts in two adenines. Unfortunately, cytosine and thymine can also form dimers, so the default repair strategy sometimes introduces a sequence variant instead.

HISTORICAL STUFF

THE CHEMISTRY OF MUTATION

If ever anyone had an excuse to give up, it was Charlotte Auerbach. Born in Germany in 1899, Auerbach was part of a lively and highly educated Jewish family. In spite of her deep interest in biology, she became a teacher, convinced that higher education would be closed to her because of her religious heritage. As anti-Jewish sentiment in Germany grew, Auerbach lost her teaching job in 1933, when every Jewish secondary-school teacher in the country was fired. As a result, she emigrated to Britain, where she earned her PhD in genetics in 1935.

Charlotte Auerbach didn't enjoy the respect her degree and abilities deserved. She was treated as a lab technician and instructed to clean the cages of experimental animals. All that changed when she met Herman Muller in 1938. Like Auerbach, Muller was interested in how genes work; his approach to the problem was to induce DNA mutations using radiation and then examine the effects produced by the defective genes. Inspired by Muller, Auerbach began work on chemical mutagens. She focused her efforts on mustard gas, a horrifically effective chemical weapon used extensively during World War I. Her research involved heating liquid mustard gas and exposing fruit flies to the fumes. It's a wonder her experiments didn't kill her.

What Charlotte's experiments did do was show that mustard gas is an alkylating agent, a mutagen that causes substitution variants (substitution of one base for another). Shortly after the end of World War II, and after persevering through burns caused by hot mustard gas, Auerbach published her findings. At last, she received the recognition and respect her work warranted.

Charlotte Auerbach went on to have a long and highly successful career in genetics. She stopped working only after old age robbed her of her sight. She died in Edinburgh, Scotland, in 1994, at the age of 95.

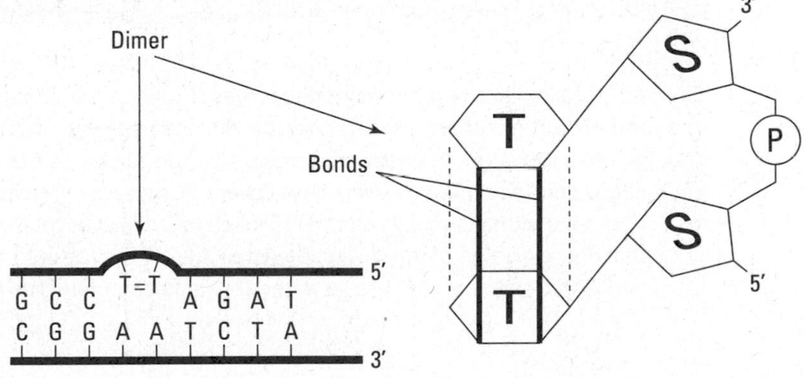

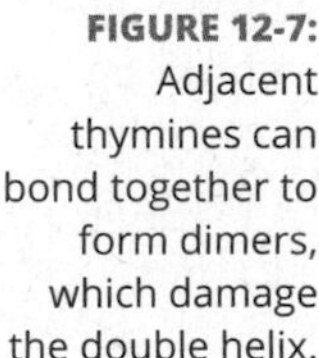

FIGURE 12-7: Adjacent thymines can bond together to form dimers, which damage the double helix.

Evaluating Options for DNA Repair

Mutations in your DNA can be repaired in four major ways:

- **Mismatch repair:** Incorrect bases are found, removed, and replaced with the correct, complementary base. Most of the time, DNA polymerase, the enzyme that helps make new DNA, immediately detects mismatched bases put in by mistake during replication. DNA polymerase can back up and correct the error without missing a beat. But if a mismatched base gets put in some other way (through strand slip-ups, for example), a set of enzymes that are constantly scrutinizing the double strand to detect bulges or constrictions signals a mismatched base pair. The mismatch repair enzymes can detect any differences between the template and the newly synthesized strand, so they clip out the wrong base and, using the template strand as a guide, insert the correct base.
- **Direct repair:** Bases that are modified in some way (such as when oxidation converts a base to some new form) are converted back to their original states. Direct repair enzymes look for bases that have been converted to some new chemical, usually by the addition of some unwanted group of atoms. Instead of using a cut-and-paste mechanism, the enzymes clip off the atoms that don't belong, converting the base back to its original form.
- **Base-excision repair:** Base-excisions and nucleotide-excisions (check out the next bullet) work in much the same way. *Base-excisions* occur when an unwanted base (such as uracil; see the section "Spontaneous chemical changes," earlier in this chapter) is found. Specialized enzymes recognize the damage, and the base is snipped out and replaced with the correct one.
- **Nucleotide-excision repair:** *Nucleotide-excision* means that the entire nucleotide (and sometimes several surrounding nucleotides as well) gets removed all at once. When intercalating agents or dimers distort the double helix, nucleotide-excision repair mechanisms step in to snip part of the strand, remove the damage, and synthesize fresh DNA to replace the damaged section.

  As with base excision, specialized enzymes recognize the damaged section of the DNA. The damaged section is removed, and newly synthesized DNA is laid down to replace it. In nucleotide-excision, the double helix is opened, much like it is during replication (which we cover in Chapter 7). The sugar-phosphate backbone of the damaged strand is broken in two places to allow removal of that entire portion of the strand. DNA polymerase synthesizes a new section, and DNA ligase seals the breaks in the strand to complete the repair process.

IN THIS CHAPTER

» Understanding how things can go wrong with chromosomes

» Learning about the more common chromosome disorders

» Learning how chromosomes are examined

Chapter 13
Chromosome Disorders: It's All a Numbers Game

The study of chromosomes is, in part, the study of cells. Geneticists who specialize in *cytogenetics*, the genetics of the cell, often examine chromosomes as the cell divides because that's when the chromosomes are easiest to see. Cell division is one of the most important activities that cells undergo; it's required for normal life, and a special sort of cell division prepares sex cells for the job of reproduction. Chromosomes are copied and divvied up during cell division. Getting the right number of chromosomes in each cell as it divides is critical. Most chromosome disorders, such as Down syndrome, occur because of mistakes during meiosis (the cell division that makes sex cells).

This chapter helps you understand how and why chromosome disorders occur. You see some of the ways geneticists study the chromosome content of cells. Knowing how many chromosomes are in each cell allows scientists to decode the mysteries of inheritance, especially when the number of chromosomes (called *ploidy*) gets complicated. Counting chromosomes also allows doctors to determine the origin of physical abnormalities caused by the presence of too many or too few chromosomes.

In this chapter, we are discussing chromosome problems that occur either in the egg or sperm; chromosome abnormalities that occur during a person's lifetime (which may contribute to diseases like cancer) will be discussed in Chapter 14.

TIP

If you skipped Chapter 2 or Chapter 6, you may want to review them before reading this chapter to get a handle on the basics of cell division and chromosome structure.

Chromosome Numbers: No More and No Less

Every species has a typical number of chromosomes revealed by its karyotype. For example, humans have 46 total chromosomes (humans are diploid, $2n$, and $n = 23$). Dogs are also diploid and have 78 total chromosomes, while house cats have $2n = 38$. *Euploidy* is when an organism has one or more complete sets of chromosomes. In a euploid organism, its total number of chromosomes is an exact multiple of its haploid number (n). When an organism has extra chromosomes or is missing chromosomes, so that the total number of chromosomes is not an exact multiple of its haploid number, the condition is referred to as *aneuploidy*. Aneuploidy is the most common type of chromosome abnormality.

Aneuploidy: Extra or missing chromosomes

Shortly after Thomas Hunt Morgan discovered that certain traits are linked to the X chromosome (see Chapter 15 for the whole story), his student Calvin Bridges discovered that chromosomes don't always play by the rules. The laws of Mendelian inheritance depend on the segregation of chromosomes — an event that takes place during the first phase of meiosis. But sometimes chromosomes don't segregate correctly; two or more copies of the same chromosome are sent to one *gamete* (sperm or egg), leaving another gamete without a copy of one chromosome.

Through his study of fruit flies, Bridges discovered the phenomenon of *nondisjunction,* the failure of chromosomes to segregate properly. Figure 13-1 shows nondisjunction at various stages of meiotic division. During the first round of meiosis, the homologous pairs of chromosomes should separate. If that doesn't happen, some eggs get two copies of one chromosome, and others have none.

If nondisjunction occurs in the second round of meiosis, some gametes will have only one copy of the chromosome in question, which is what normally happens, but others will have no copy of the chromosome or two copies of the chromosome. After fertilization, you could end up with a zygote that contains three copies of the chromosome in question (trisomy), one copy of the chromosome (monosomy), or two copies of the chromosome (normal diploid), depending on when the nondisjunction occurred.

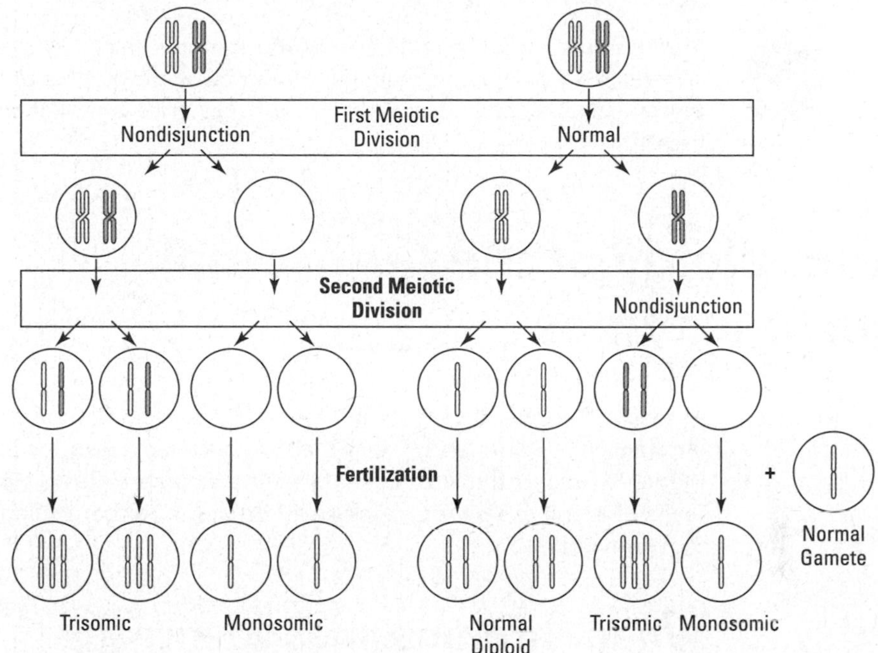

FIGURE 13-1: The results of nondisjunction during meiosis.

Many human chromosomal disorders arise from a sort of nondisjunction similar to that of fruit flies. Chromosomal abnormalities in the form of aneuploidy are very common among humans. Roughly 8 percent of all conceptions are aneuploid, and it's estimated that up to half of all miscarriages happen because of some form of chromosome disorder. Sex chromosome disorders are the most commonly observed type of aneuploidy in humans because X-chromosome inactivation allows individuals with more than two X chromosomes to compensate for the extra "doses" and survive the condition (see Chapter 6 for more about X inactivation).

Four different types of aneuploidy crop up in humans:

- **Nullisomy:** When a chromosome is missing altogether. Embryos that are nullisomic don't survive to be born.
- **Monosomy:** When one chromosome lacks its homolog.
- **Trisomy:** When one extra copy of a chromosome is present.
- **Tetrasomy:** When four total copies of a chromosome are present. Tetrasomy is extremely rare.

Chromosome conditions that are the result of missing or extra chromosomes are referred to by category of aneuploidy followed by the number of the affected chromosome. For example, trisomy 13 means that three copies of chromosome 13 are present.

Aneuploidy of the Autosomal Chromosomes

Most aneuploidies involving autosomal chromosomes (non-sex chromosomes; chromosomes 1 through 22) are lethal. Having too much or too little genetic information is frequently detrimental to the developing embryo. However, aneuploidies of several different autosomes are found in liveborn infants: chromosomes 13, 18, and 21.

When chromosomes go missing

Monosomy of autosomal (non-sex) chromosomes (when one chromosome lacks its homolog) in humans is very rare. The majority of embryos with monosomies don't survive to be born. The only autosomal monosomy reported in liveborn infants is monosomy 21, and the cases of this are exceedingly rare. Signs and symptoms of monosomy 21 are similar to those of trisomy 21 (Down syndrome; covered later in this section). However, infants with monosomy 21 rarely survive for longer than a few days or weeks. Monosomy 21 is the result of nondisjunction during meiosis.

Many monosomies are partial losses of chromosomes, meaning that only part of the chromosome is a missing chromosome (referred to as a partial monosomy). Most frequently this is the result of the chromosome breaking and part of it becoming attached to another chromosome. Movements of parts of chromosomes to other nonhomologous chromosomes are the result of *translocations.* We cover translocations in more detail in the section "Translocations," later in this chapter.

When too many chromosomes are left in

Trisomies (when one extra copy of a chromosome is present) are the most common sorts of chromosomal abnormalities in humans. The most common autosomal trisomy found in infants is Down syndrome, or trisomy 21. Other, less common trisomies include trisomy 18 (Edward syndrome) and trisomy 13 (Patau syndrome). All of these trisomies are usually the result of nondisjunction during meiosis. Trisomies of autosomal chromosomes other than 21, 18, and 13 always result in early miscarriage and are, therefore, not found in infants or children.

Down syndrome

Trisomy of chromosome 21, commonly called *Down syndrome*, affects between 1 in 600 to 1 in 800 infants. Children and adults with Down syndrome typically have distinct facial features, low muscle tone, and short stature. They also experience developmental delays and mild to moderate intellectual disability. In addition, certain types of heart defects are relatively frequent among those with an extra chromosome 21. One of the strengths frequently noted among those with Down syndrome is relatively advanced social skills compared to their peers. The average lifespan of someone with the condition is approximately 60 years (in the absence of serious heart conditions).

One of the most notable aspects of Down syndrome (and trisomies in general) is the increase in the chance of having a child with one of these conditions as a woman gets older (see Figure 13-2). Women between 18 and 25 have a very low risk of having a baby with trisomy 21 (roughly 1 in 1340 for women 21 years old at the time of delivery). The risk increases slightly but steadily for women between 25 and 35 (about 1 in 940 for women 30 years old at delivery) and then starts to increase more dramatically. By the time a woman is 40, the probability of having a child with Down syndrome is 1 in 85, and by the age of 45, the probability has increased to 1 in 35. The risk of having a child with Down syndrome increases as women age because of when and how their eggs develop.

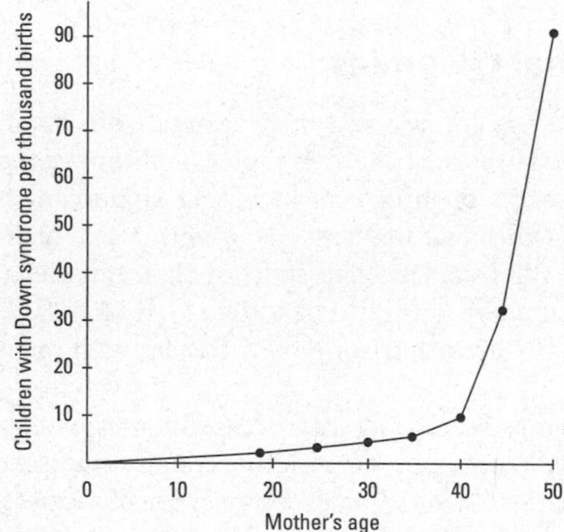

FIGURE 13-2: Risk of having a child with Down syndrome increases with maternal age.

The majority of Down syndrome cases seem to arise from nondisjunction during meiosis. In human females, meiosis begins in oogonia of the developing fetus (the cells that give rise to egg cells; see Chapter 2 for a review of gametogenesis in humans). All developing eggs go through the first round of prophase, including recombination. Meiosis in future egg cells then stops in a stage called *diplotene*, the stage of crossing-over, where homologous chromosomes are hooked together and are in the process of exchanging parts of their DNA. Meiosis doesn't start back up again until a particular developing egg is going through the process of ovulation. At that point, the egg completes the first round of meiosis (meiosis I, described in Chapter 2) and then halts again. When sperm and egg unite, the nucleus of the egg cell finishes meiosis just before the nuclei of the sperm and egg fuse to complete the process of fertilization. In contrast to females, meiosis in human males doesn't begin until puberty. Moreover, meiosis is continuous, without the pauses that occur in females.

Roughly 75 percent of the nondisjunctions responsible for Down syndrome occur during the first round of meiosis (meiosis I). Oddly, most of the chromosomes that fail to segregate seem also to have failed to undergo crossing-over, suggesting that the events leading up to nondisjunction begin early in life. Scientists have proposed a number of explanations for the cause of nondisjunction and its associated lack of crossing-over, but they haven't reached an agreement about what actually happens in the cell to prevent the chromosomes from segregating properly.

REMEMBER

Every pregnancy is an independent genetic event. So, although age is a factor in calculating risk of trisomy 21, Down syndrome with previous pregnancies doesn't necessarily increase a woman's risk of having another child affected by the disorder (unless the condition is the result of a translocation; this situation is discussed later in the chapter in the section "Robertsonian translocations").

Other autosomal trisomies

Trisomy 18, also called *Edward syndrome*, also results from nondisjunction. About 1 in 5000 to 1 in 8000 newborns has three copies of chromosome 18. The disorder is characterized by severe birth defects including significant heart defects and brain abnormalities. Other features associated with trisomy 18 include a small lower jaw relative to the face, clenched fingers, rigid muscles, and foot defects. Most infants with trisomy 18 don't live past their first birthdays. Like trisomy 21, the chance of having a baby with trisomy 18 increases with the mother's age.

Trisomy 13, also known as *Patau syndrome*, occurs in approximately 1 in 10,000 to 1 in 15,000 live births. Many embryos with this condition miscarry early in pregnancy. Babies born with trisomy 13 have a very short life expectancy — most die before the age of 12 months. However, some may survive until 2 or 3 years of age; very rarely, children with trisomy 13 have survived beyond the age of 5. Babies with trisomy 13 have extremely severe brain defects. In addition, abnormalities of the eyes, cleft lip, cleft palate, heart defects, and *polydactyly* (extra fingers and toes) are common among infants with an extra chromosome 13. Like other trisomies, the chance of having a child with trisomy 13 increases with maternal age.

Aneuploidy of the Sex Chromosomes

Aneuploidies of the sex chromosomes (the X and the Y chromosomes) are found more frequently than autosomal aneuploidies in infants and children. Because the body has a mechanism for balancing out the expression of most of the genes on the sex chromosomes (namely, X inactivation, which we discuss in Chapter 6), these conditions are much more survivable than those involving the autosomes. Some of these conditions are associated with maternal age (having an increased chance of having them as a mother gets older), but others are not. The features associated with sex chromosome aneuploidies also tend to be more subtle than those associated with autosomal aneuploidies. In some cases, a person may not even know that they have a sex chromosome aneuploidy until they undergo chromosome testing for something like infertility.

Extra Xs

Both males and females can have extra X chromosomes, each with different genetic and phenotypic consequences. When females have an extra X chromosome, two of which are inactivated in each cell, the condition is referred to as *triple X syndrome* (or *trisomy X*). Females with triple X syndrome tend to be taller than expected based on the heights of family members. Most women with the condition develop normally and experience normal puberty, menstruation, and fertility. They may have a lower IQ than expected for their family, but significant intellectual disability is uncommon. About one in every 1,000 girls has three copies of the X chromosome. Having more than three copies of the X chromosome is also possible but is much less common. With more copies of the X chromosome, the chance of significant intellectual disability and possibly additional medical problems increases.

Males with an extra X chromosome are affected with *Klinefelter syndrome.* Roughly one in every 500 boys has two X chromosomes and one Y chromosome (XXY). Like females, males with Klinefelter syndrome undergo X inactivation so that only one X chromosome is active in each cell. However, genes located in the pseudoautosomal regions (PARs, described in Chapter 6) of the X chromosome escape X inactivation. Therefore, these genes will be expressed in excess in males with the condition. Generally, males with Klinefelter are taller than average and have infertility (are unable to father children without assistive reproductive technologies). Men with Klinefelter often have reduced secondary sexual characteristics (such as less facial hair) and sometimes have some breast enlargement due to impaired production of testosterone. Intelligence tends to be within the normal range in males with an extra X chromosome, although they may experience certain types of learning disabilities. Like in females, the more copies of the X chromosome a male has, the more likely he is to experience intellectual disability and other medical problems. Like the previous aneuploidies discussed, the chance of having a son with an extra X chromosome does increase with the mother's age at the time of pregnancy.

Extra Ys

Approximately 1 in 1000 males have two Y chromosomes and one X chromosome (referred to as *XYY syndrome*). Most men with XYY syndrome have normal male development and features, but they're often taller and, as children, grow a bit faster than their peers with a normal number of chromosomes. In addition, infertility is not expected. Boys with XYY syndrome may have some learning difficulties or a low normal IQ, but significant intellectual disability is not associated with this condition. Unlike the other aneuploidies discussed, XYY syndrome is *not* associated with maternal age, since boys inherit their Y chromosome from their fathers.

Monosomy X

Monosomy X, which is also known as Turner syndrome, occurs when there is one X chromosome and no second sex chromosome. Individuals with just one X chromosome are female. Girls with Turner syndrome are shorter than average and do not typically undergo puberty as expected. They may also have certain physical features, such as a broad neck, unusually shaped ears, and swelling of the hands and feet. Heart defects, hearing loss, and changes in kidney structure are also common. As adults, women with Turner syndrome experience infertility. Turner syndrome affects about one in 2,500 girls and is *not* associated with the mother's age.

Exploring Variations in Chromosome Structure

The range of chromosomal disorders extends well beyond missing or extra chromosomes. A variety of structural changes in chromosomes can occur, which may or may not cause problems. If a structural change does not affect the total amount of DNA (that is, no DNA is missing or added) and does not interrupt a gene or genes, the chromosome change might not cause any problems. It could, however, lead to problems in the children of the individual (as illustrated later in the section "Translocations"). If a structural change does change the amount of DNA or does interrupt a gene or genes, it is very likely to cause a chromosome disorder. Changes in chromosome structure can affect any chromosome and can be almost any size. Consequently, there are essentially an infinite number of chromosome changes that are possible. Certain chromosome problems are much more common than others, however.

Large-scale chromosome changes are called *chromosomal rearrangements.* There are a variety of rearrangements that can occur. The most common kinds of chromosomal rearrangements, shown in Figure 13-3, include:

- **Duplication:** A duplication is when a segment of a chromosome is copied and re-inserted, such that the chromosome now contains two copies of the same segment.
- **Deletion:** A deletion is when a segment of a chromosome is lost.
- **Inversion:** An inversion is when a section of a chromosome gets turned around, reversing the order of DNA sequence.
- **Translocation:** A translocation is when segments of two different chromosomes are exchanged.

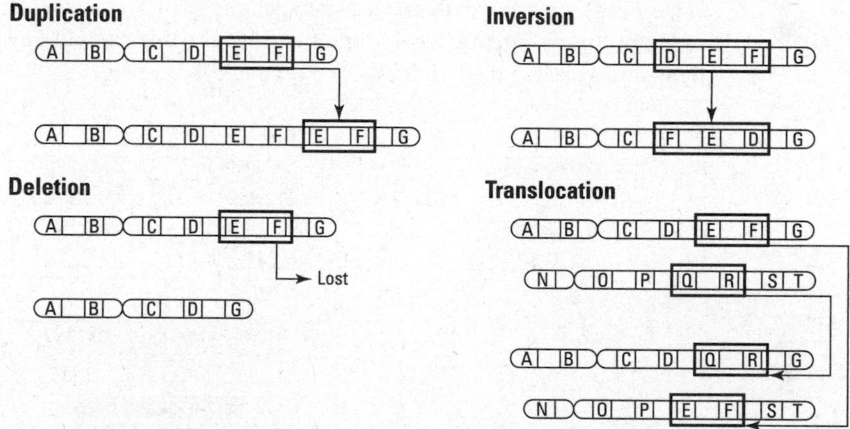

FIGURE 13-3: Common types of chromosomal rearrangements.

Duplications

Duplications, which occur when a section of a chromosome is duplicated and re-inserted into the chromosome, most often arise from unequal crossing-over. Disorders arising from duplications may be considered partial trisomies because a segment of one chromosome is present in triplicate (two copies on one chromosome and one copy on the homologous chromosome). Chromosome duplications can cause genetic disorders because having extra copies of a gene can lead to the production of too much of that gene's protein product. Organisms require just the right level of protein production during development, and sometimes having too much of a particular protein can negatively affect developmental processes, leading to developmental disorders and genetic syndromes.

Deletions

Deletions occur when a segment of DNA is lost from a chromosome. Deletion of a large section of a chromosome usually occurs in one of two ways:

- » The chromosome breaks during interphase of the cell cycle (see Chapter 2 for details about the cell cycle), and the broken piece is lost when the cell divides.
- » Parts of chromosomes are lost because of unequal crossing-over during meiosis.

Normally, when chromosomes start meiosis, they evenly align completely with no overhanging parts. If chromosomes align incorrectly, crossing-over can create a deletion in one chromosome and an insertion of extra DNA in the other, as shown in Figure 13-4. Unequal crossover events are more likely to occur where many repeats are present in the DNA sequence (see Chapter 8 for more on DNA sequences). Just like with duplications, changes in the levels of gene products can cause problems during development and lead to a variety of developmental problems and possibly birth defects.

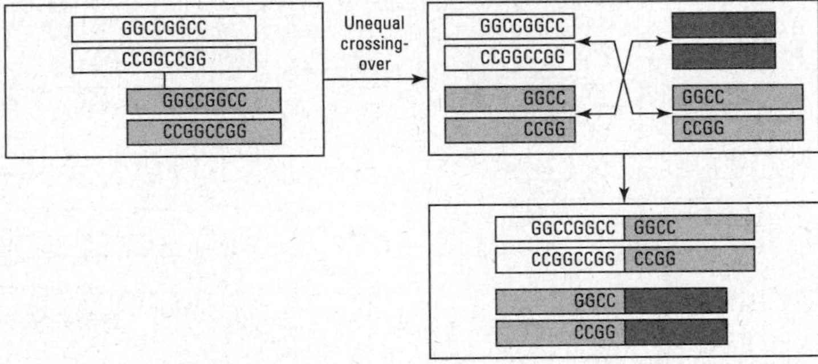

FIGURE 13-4: Unequal crossover events cause large-scale deletions of chromosomes.

Inversions

If a chromosome break occurs, sometimes DNA repair mechanisms (explained in Chapter 12) can repair the strands. If two breaks occur, part of the chromosome may be reversed before the breaks are repaired. When a large part of the chromosome is reversed and the order of the DNA sequence is changed, the event is called an *inversion*. When inversions involve the centromere (one break in each chromosome arm), they're called *pericentric*; inversions that don't include the centromere (two breaks in the same chromosome arm) are called *paracentric*. If an inversion disrupts a gene or alters regions that are involved in the regulation of gene expression, a genetic disorder can result. If nothing is disrupted, it is possible that the carrier of the inversion will have no problems as a result and may never even know they carry the inversion.

Translocations

Translocations involve the exchange of large portions of chromosomes — one segment of one chromosome switches places with a segment of another chromosome. They occur between nonhomologous chromosomes and come in several types:

- **Reciprocal translocation:** Reciprocal translocations involve an exchange of large segments between chromosomes. Reciprocal translocations, which can involve any chromosome, are found in approximately 1 in 670 to 1 in 1000 individuals.
- **Robertsonian translocation:** Robertsonian translocations involve an exchange between two acrocentric chromosomes (chromosomes with the centromere located near one end of the chromosomes), such that the long arms of the two chromosomes are attached together. Acrocentric chromosomes include chromosomes 13, 14, 15, 21, and 22. Robertsonian translocations occur in approximately 1 in 1000 newborn infants.

When two chromosomes are broken, they can exchange pieces and result in either a balanced rearrangement (with no gain or loss of DNA, just a rearrangement) or an unbalanced rearrangement (with a gain or loss, or both gains and losses, of one or more segments of DNA). In addition, the breaks can interrupt one or more genes, typically rendering the affected genes nonfunctional.

In some cases, a translocation event occurs spontaneously in one parent, who then passes the disrupted chromosomes on to his or her offspring. If the carrier parent has a balanced translocation and no genes are interrupted, he/she may have no symptoms of a chromosome disorder. However, because of the way in which these chromosomes align and segregate during meiosis, the children of a balanced carrier may have an unbalanced version of the translocation, resulting in a partial trisomy and/or a partial deletion. This could result in either a miscarriage or a child with symptoms of a chromosome disorder.

Reciprocal translocations

Most reciprocal translocations are unique among families. However, a recurring translocation involving chromosomes 11 and 22 has been found in many different families, resulting in cleft palate, heart defects, ear anomalies, and intellectual disabilities in those with an unbalanced version. Family members who have a balanced version of this translocation (with no missing or extra chromosomal segments) do not have these features; however, it appears that they may have a higher than average risk for developing breast cancer. They would also have an increased risk for miscarriages and for having a child with an unbalanced version and the related medical problems. Chromosome 11 seems particularly prone to breakage in an area that has many repeated sequences (where two bases, A and T, are repeated many times sequentially). Because both chromosome 11 and chromosome 22 contain similar repeat sequences, the repeats may allow crossover events to occur by mistake, resulting in a reciprocal translocation.

Robertsonian translocations

Robertsonian translocations are translocations in which the long arms of two acrocentric chromosomes are joined together. Acrocentric chromosomes are those with centromeres located toward one end of the chromosome, including chromosomes 13, 14, 15, 21, and 22. The short arms of acrocentric chromosomes contain only repetitive DNA; they have no genes. Therefore, the loss of the short arms is not expected to cause any problems. A balanced carrier of a Robertsonian translocation, therefore, generally has a total of 45 chromosomes per cell, with the long arms of the two acrocentric chromosomes involved in the translocation being attached together. When balanced, Robertsonian translocations are not expected to cause any medical problems. However, the children of someone with a balanced Robertsonian translocation have a chance of having an unbalanced translocation, resulting in a monosomy or trisomy of one of the chromosomes involved. Many of these unbalanced translocations result in early pregnancy loss. A translocation that results in trisomy 21 (Down syndrome) is the most common unbalanced Robertsonian translocation seen in liveborn infants. Down syndrome due to a translocation accounts for approximately 3 to 4% of Down syndrome cases. This is sometimes referred to as familial Down syndrome.

Familial Down syndrome is unrelated to maternal age. This disorder occurs as a result of the fusion of chromosome 21 to another acrocentric chromosome (often chromosome 14). In this case, long arms of chromosomes 21 and 14 are attached together. The leftover parts of chromosomes 14 and 21 (the short arms) also fuse together but are usually lost to cell division and aren't inherited. When a Robertsonian translocation occurs, affected persons can end up with several sorts of chromosome combinations in their gametes, as shown in Figure 13-5.

For the familial Down syndrome example, a balanced translocation carrier has one normal copy of chromosome 21, one normal copy of chromosome 14, and one fused translocation chromosome. Carriers aren't affected by Down syndrome because their fused chromosome acts as a second copy of the normal chromosomes. When a carrier's sex cells (eggs or sperm) undergo meiosis, some of their gametes have one translocated chromosome or get the normal complement that includes one copy of each chromosome. Fertilizations of gametes with a translocated chromosome and a normal chromosome 21 produce the phenotype of Down syndrome. Balanced carriers of a Robertsonian translocation involving chromosomes 21 and 14 have up to a 10 percent chance of having a child with trisomy 21 with each pregnancy. Carriers also have a higher chance than normal of miscarriage because of monosomy (of either 21 or 14) and trisomy 14.

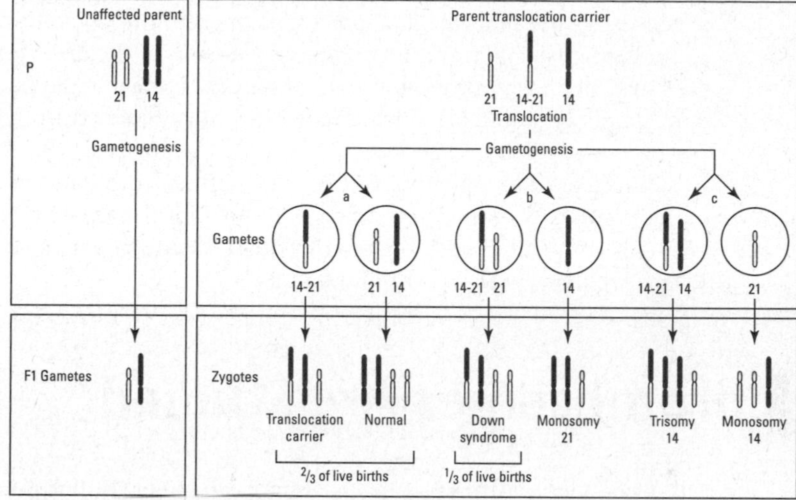

FIGURE 13-5: A translocation that leads to familial Down syndrome.

Other things that go awry with chromosomes

The types of chromosome abnormalities that are possible are quite varied and not limited to those we described in the previous sections. Two additional problems that you may read about are polyploidy and mosaicism.

Polyploidy

Polyploidy, the occurrence of more than two sets of chromosomes per cell, is extremely rare in humans. Two types of polyploidy that have been reported in liveborn infants are *triploidy* (three full chromosome sets per cell) and *tetraploidy* (four sets per cell). Most polyploid pregnancies result in miscarriage or stillbirth. All liveborn infants with either of these conditions have severe birth defects, and most don't survive longer than a few days.

Mosaicism

Mosaicism is a condition in which a chromosome abnormality is present in some cells of the body but not all. Early in embryo development, a change in the chromosomes (aneuploidy or structural rearrangement) can create two different cell lines. All cells that come from the cell with the abnormality will also have the abnormality. All cells that come from normal cells will not. The magnitude of the effects of mosaicism depends on when the error occurs. If the error happens very early, most of the individual's cells are affected. If it happens later, fewer cells will contain the abnormality and the effects may be lessened. The effect of the abnormality also depends on the precise change in the chromosomes and which types of cells carry the abnormality. Sometimes, mosaic chromosome abnormalities can be indistinguishable from the nonmosaic version of the disorder, with regards to the physical phenotype produced. In other cases, the phenotype can be so mild, it may remain undetectable unless found fortuitously when testing for other reasons.

In some cases, the abnormal cell line is confined to the placenta (referred to as *confined placental mosaicism*). Some embryos with placental mosaicism develop normally and suffer no ill effects. In other cases, this mosaicism can result in severe problems with growth in the fetus.

How Chromosomes Are Studied

There are a variety of ways that scientists can study chromosomes. What method they use depends on what they are looking for. Are they wanting to look at the total number of chromosomes? Are they looking for large-scale rearrangements of the chromosomes? Do they want to know if a person has a very small change in chromosome structure? Do they know what they are looking for and want to look at one specific chromosome, or do they want to look at the entire genome? The answers to these types of questions may dictate what testing should be done. And it may be that a combination of methods is used to find any chromosome changes that are present and could potentially be the cause a genetic disorder.

Big enough to see

One way a geneticist counts chromosomes is with the aid of microscopes and special dyes to see the chromosomes during *metaphase* — the one time in the cell cycle when the chromosomes take on a fat, easy-to-see, sausage shape. In order to examine chromosomes, a sample of cells is obtained. Almost any sort of dividing cell works as a sample, including root cells from plants, blood cells, or skin cells. These cells are then *cultured* — given the proper nutrients and conditions for growth — to stimulate cell division. Some cells are removed from the culture and treated to stop mitosis during metaphase, and dyes are added to make the chromosomes easy to see. Finally, the cells are inspected under a microscope. The chromosomes are sorted, counted, and examined for obvious abnormalities.

This process of chromosome examination, with the identification, pairing, and ordering of the chromosomes, is called *karyotyping.* A *karyotype* (shown in Chapter 6) reveals exactly how many chromosomes are present in a cell, along with some details about the chromosomes' structure. Scientists can only see these details by staining the chromosomes with special dyes. In the most common images of a karyotype, the chromosomes appear to be striped because of the stain used. When examining this type of karyotype, a geneticist looks at each individual chromosome. Every chromosome has a typical size and shape and a very specific banding pattern; the location of the centromere and the length of the chromosome arms (the parts on either side of the centromere) are what define each chromosome's physical appearance.

TECHNICAL STUFF

In some disorders, part of one of the chromosome arms is misplaced or missing. Therefore, geneticists often refer to the chromosome number along with the letter *p* or *q* to communicate which part of the chromosome is affected. The bands that are present after staining also have numbers. Therefore, you may see something like 22q11.2 when referring to a chromosome problem. This means that the alteration involves the long arm of chromosome 22 and that it is region 11.2 that is affected (see Figure 13-6). Basically, the q11.2 is like a map coordinate, telling you exactly where on chromosome 22 there is a change.

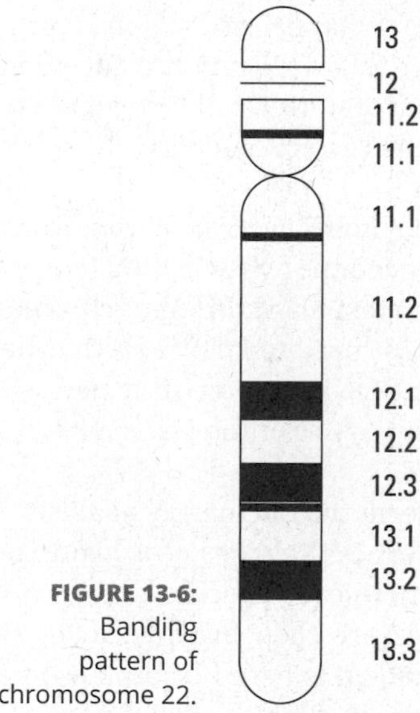

FIGURE 13-6: Banding pattern of chromosome 22.

Karyotyping is ideal for detecting *numerical* chromosome problems (aneuploidy or polyploidy) — it allows scientists to determine whether there are too many or too few chromosomes. In addition, because each chromosome has its own unique size and shape, as well as its own unique banding pattern when stained, they can determine specifically which chromosome is extra or missing. Because of those unique features, scientists can also use karyotyping to detect and identify chromosomal changes that involve huge sections of DNA, such as large deletions, duplications, inversions, and translocations. Unfortunately, karyotyping is unable to detect small changes in structure, such as deletions that are only a few thousand base pairs in length. With such a small deletion, the chromosome involved would look no different under the microscope than its normal homolog.

Too small for the naked eye

To identify small changes in chromosome structure, other techniques must be used. One method that may be used is *fluorescence in situ hybridization* (*FISH*). FISH is a pretty specific test. It uses a DNA probe (a small segment of DNA for which the sequence is known), which is fluorescently labeled. When in a single-stranded state (the two strands can be separated by raising the temperature), the probe can bind to its complementary sequence on the appropriate chromosome.

To see whether a specific region or gene is present in a person's chromosomes, the probe will be made up of a sequence located within that region or gene. After adding the fluorescent tag, the probe can be hybridized (bound) to the person's chromosomes on a slide and examined under a microscope. The number of signals from the fluorescent tags can then be counted to see if that gene or region is present in too many or too few copies (that is, if there is a duplication or deletion, respectively). The limitation of FISH analysis is that you need to know what you are looking for and have a probe that can identify the correct sequence.

A more recently developed method for identifying imbalances in chromosome number is *microarray analysis*, or array comparative genomic hybridization (aCGH). The advantage of microarray analysis is that you can test many different chromosome regions or genes at the same time, so you don't need to know exactly what you are looking for. The test is designed to identify *imbalances* in a person's DNA — that is, sequence that is present in too many copies or too few copies.

As shown in Figure 13-7, a *DNA chip* is used to perform the microarray analysis. A DNA chip is basically a slide with thousands of short DNA probes attached in an organized array, such that it is known exactly where each specific sequence is located on the array. Two different DNA samples are then hybridized to the array — a patient sample and a reference sample known to have no chromosomal imbalances (no deletions or duplications). One of these samples is labeled with a red fluorescent dye, and the other is labeled with a green fluorescent dye.

After the DNA samples are applied to the array and the DNA fragments in the samples stick to the probes, the array is analyzed by a computer. Anywhere that lacks an imbalance — where the patient and the reference sample have the same number of copies of DNA sequence — the signal will show up yellow. These sequences are present in the correct number.

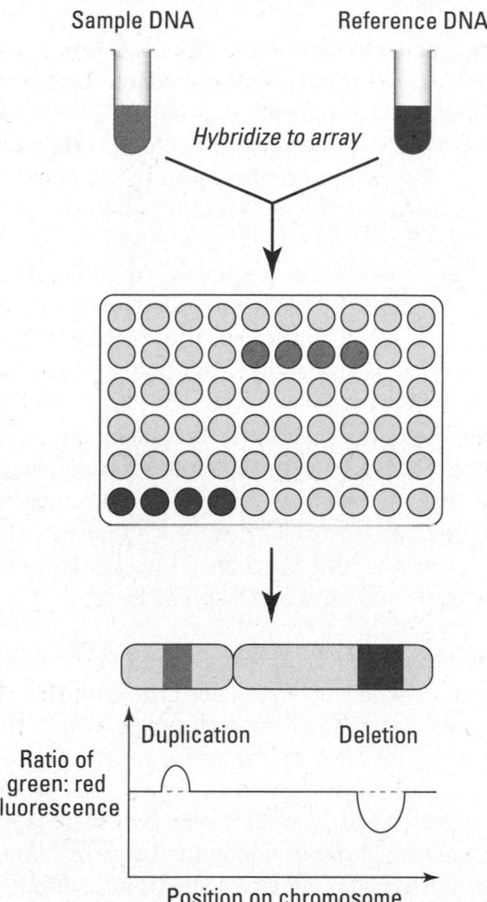

FIGURE 13-7: Microarray analysis used to test for chromosome imbalances.

If there is an imbalance (either a deletion or duplication, for example), the signal will be either red or green, depending on which sample has more copies of the sequence. The probes used for microarray analysis can be from throughout the whole genome. Therefore, the analysis can be used to test for deletions or duplications throughout the whole genome at one time. It is also an ideal test for identifying very small imbalances that could never be detected by karyotyping.

Non-Invasive Prenatal Testing for Aneuploidy

A relatively new type of chromosome study that is now available is referred to as *non-invasive prenatal testing* (*NIPT*, sometimes referred to as non-invasive prenatal screening). However, it differs from those described above in that it is only performed during pregnancy to assess the risk of a chromosome problem in the baby and it is a screening test that is *not* considered diagnostic. That means, it provides the likelihood of a chromosome problem, but it cannot tell for sure whether the baby has a chromosome disorder. Chromosome testing after amniocentesis or chorionic villus sampling (discussed in Chapter 15), or after birth, is necessary to establish the diagnosis of a specific condition.

NIPT is based on the fact that during pregnancy, fragments of DNA from the placenta can be found in the mother's bloodstream (along with fragments of her own DNA). This DNA is found outside of the cells present and is, therefore, referred to as *cell-free DNA* (cfDNA). The placental DNA should be the same as that of the developing baby, since they both come from the same fertilized egg. NIPT (was defined earlier in the section) is a screening test that analyzes the cfDNA found in a mother's blood. The test can be performed during pregnancy in order to determine the likelihood that the baby has a change in chromosome number (aneuploidy). It is considered non-invasive because it requires only a maternal blood sample, with no increased risk of pregnancy loss (as is seen with the invasive procedures amniocentesis and CVS; see Chapter 15).

To perform the test, a blood sample is taken from the mother any time after 9 to 10 weeks of gestation. DNA sequencing is performed on this sample using high-throughput next-generation sequencing techniques, which allow for sequencing of millions of fragments of DNA at the same time. The number of copies of sequences from each chromosome is then analyzed. Any chromosome that is represented too many or too few times can indicate that there is a higher chance that there is an extra or missing chromosome in the baby. In some cases, the test can also indicate whether there is an increased chance of a duplication or deletion in one of the chromosomes (a smaller segment that is either extra or missing).

IN THIS CHAPTER

» Defining what cancer is (and isn't)

» Understanding the genetic basis of cancer

» Looking at the different types of cancer

Chapter 14
Taking a Closer Look at the Genetics of Cancer

If you've had personal experience with cancer, you're not alone. We've lost family members, coworkers, students, and friends to this insidious disease — it's highly likely that you have, too. Second only to heart disease, cancer causes the deaths of around 609,000 persons a year in the United States alone, and it was estimated that nearly 1.7 million Americans would be diagnosed with cancer in 2019. Cancer is a genetic disorder that affects how cells grow and divide. Your likelihood of getting cancer is influenced by your genes (the genes you inherited from your parents) and your exposure to things in your environment, particularly certain chemicals and radiation. However, some cancers occur from random, spontaneous changes in the DNA — events that defy explanation and have no apparent cause. In this chapter, you find out what cancer is, the genetic basis of cancers, and some details about the most common types of heritable cancer.

TIP

If you skipped over the discussion of cells in Chapter 2, you may want to backtrack before delving into this chapter, because cell information helps you understand what you read here. All cancers arise from changes in the DNA; you can discover how and why sequence variants occur in Chapter 12.

Defining Cancer

Cancer is, in essence, cell division running out of control. As we explain in Chapter 2, the cell cycle is normally a carefully regulated process. Cells grow and divide on a schedule that's determined by the type of cell involved. Skin cells grow and divide continuously because replacing dead skin cells is a never-ending job. Some cells retire from the cell cycle: The cells in your brain and nervous system don't take part in the cell cycle; no growth and no cell division occur there during adulthood. Cancer cells, on the other hand, don't obey the rules and have their own, often frightening, agendas and schedules. Table 14-1 lists the probability of developing one of the six most common cancers in the United States.

TABLE 14-1 Lifetime Probability of Developing Cancer

Type of Cancer	Risk
Breast	1 in 8 women (12.5%)
Lung	1 in 15 men (6.9%) and 1 in 17 women (5.9%)
Prostate	1 in 9 men (11.6%)
Colon and rectum	1 in 22 men (4.5%) and 1 in 24 women (4.2%)
Skin (melanoma)	1 in 36 men (2.8%) and 1 in 58 women (1.7%)
Bladder	1 in 27 men (3.8%) and 1 in 89 women (1.1%)

REMEMBER

In the following sections, we outline the two basic categories of tumors — benign and malignant. *Benign tumors* grow out of control but don't invade surrounding tissues. *Malignant tumors* are invasive and have a disturbing tendency to travel and show up in new sites around the body. The term *cancer* refers to a malignant tumor.

Benign growths: Not always so harmless

In a benign tumor, the cells divide at an abnormally high rate but remain in the same location. Benign tumors tend to grow rather slowly, and they create trouble because of their size and location. In general, a *tumor* is any mass of abnormal cells. Benign tumors often cause problems because they take up space and can compress nearby organs. For example, a tumor that grows near a blood vessel can eventually cut off blood flow just by virtue of its bulk. Benign growths can sometimes also interfere with normal body function and even affect genes by altering hormone production (see Chapter 11 for more about how hormones control genes).

Generally, benign growths are characterized by their lack of invasiveness. A benign tumor is usually well-defined from surrounding tissue, pushes other tissues aside, and can be easily moved about. The cells of benign tumors usually bear a strong resemblance to the tissues they start from. For example, under a microscope, a cell from a benign skin tumor looks similar to a normal skin cell.

A different sort of benign cell growth is called a *dysplasia*, a cell with an abnormal appearance. Dysplasias aren't cancerous (that is, they don't divide out of control) but are worrisome because they have the potential to go through changes that lead to malignant cancers. When examined under the microscope, dysplasias often have enlarged cell nuclei and a "disorderly" appearance. In other words, they have irregular shapes and sizes relative to other cells of the same type. Tumor cells sometimes start as one cell type (benign) but, if left untreated, can give rise to more invasive types as time goes on.

Treatment of benign growths (including dysplasias) varies widely depending on the size of the tumor, its potential for growth, the location of the growth, and the probability that cell change may lead to malignancy (invasive tumors; see the following section). Some benign growths shrink and disappear on their own, and others require surgical removal.

Malignancies: Seriously scary results

Probably one of the most frightening words a doctor can utter is "malignant." *Malignancy* is characterized by cancer cells' rapid growth, invasion into neighboring tissues, and the tendency to metastasize. *Metastasis* occurs when cancer cells begin to grow in other parts of the body besides the original tumor site; cancers tend to metastasize to the bones, liver, lungs, and brain. Like benign growths, malignancies are tumors, but malignant tumors are poorly defined from the surrounding tissue — in other words, it's difficult to tell where the tumor ends and normal tissue begins. (See the section "Metastasis: Cancer on the move" later in this chapter for more info on the process.)

Malignant cells tend to look very different from the cells they arise from (Figure 14-1 shows the differences). The cells of malignant tumors often look more like tissues from embryos or stem cells than normal "mature" cells. Malignant cells tend to have large nuclei, and the cells themselves are usually larger than normal. The more abnormal the cells appear, the more likely it is that the tumor may be invasive and able to metastasize.

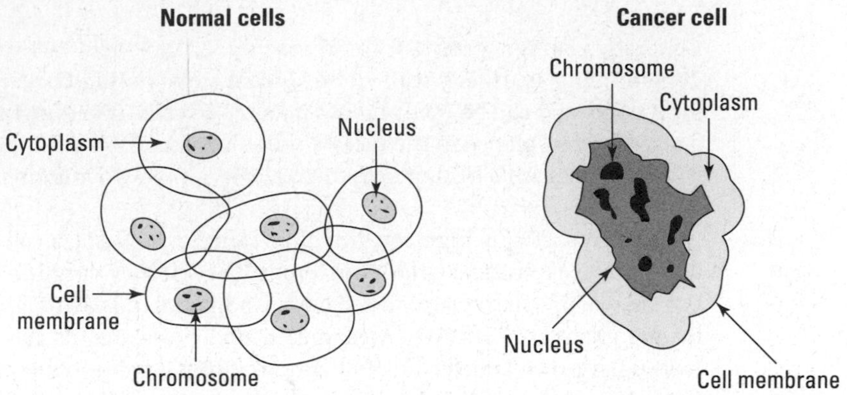

FIGURE 14-1: Normal cells and malignant cells look very different.

Malignancies fall into one of five categories based on the tissue type they arise from:

- **Carcinomas** are associated with skin, nervous system, gut, and respiratory tract tissue.
- **Sarcomas** are associated with connective tissue (such as muscle) and bone.
- **Leukemias** (related to sarcomas) are cancers of the blood.
- **Lymphomas** develop in glands that fight infection (lymph nodes and glands scattered throughout the body).
- **Myelomas** start in the bone marrow.

Cancer can occur in essentially any cell of the body. The human body has 200 or so different cell types, and doctors have identified more than 100 different forms of cancer.

Treatment of malignancy varies depending on the location of the tumor, the degree of invasion, the potential for metastasis, and a host of other factors. Treatment may include surgical removal of the tumor, surrounding tissues, and *lymph nodes* (little knots of immune tissue found in scattered locations around the body). *Chemotherapy* (administering of anticancer drugs) and radiation may also be used to combat the growth of invasive cancers. Some forms of gene therapy, which we address in Chapter 16, may also prove helpful.

Metastasis: Cancer on the move

Cells in your body stay in their normal places because of physical barriers to cell growth. One such barrier is called the *basal lamina*. The basal lamina (or basement membrane) is a thin sheet of proteins that's sandwiched between layers of cells.

Metastatic cells produce enzymes that destroy the basal lamina and other barriers between cell types. Essentially, metastatic cells eat their way out by literally digesting the membranes designed to keep cells from invading each other's space. Sometimes, these invasions allow metastatic cells to enter the bloodstream, which transports the cells to new sites where they can set up shop to begin a new cycle of growth and invasion.

Another consequence of breaking down the basal lamina is that the action allows tumors to set up their own blood supply in a process called *angiogenesis.* Angiogenesis is the formation of new blood vessels to supply the tumor cells with oxygen and nutrients. Tumors may even secrete their own growth factors to encourage the process of angiogenesis.

Recognizing Cancer as a DNA Disease

Normally, a host of genes regulate the cell cycle. Thus, at its root, cancer is a disease of the DNA. Mutations (disease-causing sequence variants) damage DNA, and mutations can ultimately lead to the phenotype (physical trait) of cancer. The good news is it takes more than one mutation to give a cell the potential to become cancerous. The transformation from normal cell to cancer cell is thought to require certain genetic changes. These mutations can happen in any order — it's not a 1-2-3 process.

- A mutation occurs that starts cells on an abnormally high rate of cell division.
- A mutation in one (or more) rapidly dividing cells confers the ability to invade surrounding tissue.
- Additional mutations accumulate to confer more invasive properties or the ability to metastasize.

REMEMBER

Most cancers arise from two or more mutations that occur in the DNA of *one* cell. Tumors result from many cell divisions. The original cell containing the mutations divides, and that cell's "offspring" divide over and over to form a tumor (see Figure 14-2).

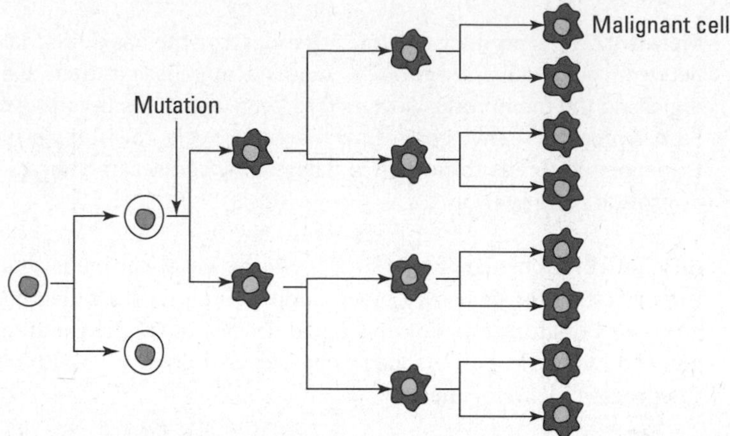

FIGURE 14-2: Tumors start out from mutations in the DNA of one cell.

Exploring the cell cycle and cancer

The cell cycle and division (called *mitosis*, which we cover in Chapter 2) is tightly regulated in normal cells. Cells must pass through checkpoints, or stages of the cell cycle, in order to proceed to the next stage. If DNA synthesis isn't complete or damage to the DNA hasn't been repaired, the checkpoints prevent the cell from moving into another stage of division. These checkpoints protect the integrity of the cell and the DNA inside it. Figure 14-3 shows the cell cycle and the checkpoints that occur throughout.

REMEMBER

Chapter 2 explains two checkpoints of the cell cycle. Four major conditions — basically, quality control points — must be met for cells to divide:

- DNA must be undamaged (no mismatches, bulges, or strand breaks like those described in Chapter 12) for the cell to pass from G_1 of interphase into S phase (DNA synthesis).
- All the chromosomes must complete replication for the cell to pass out of S phase.
- DNA must be undamaged to start prophase of mitosis.
- Spindles required to separate chromosomes must form properly for mitosis to be completed.

If any of these conditions aren't met, the cell is "arrested" and not allowed to continue to the next phase of division. Many genes and the proteins they produce are responsible for making sure that cells meet all the necessary conditions for cell division.

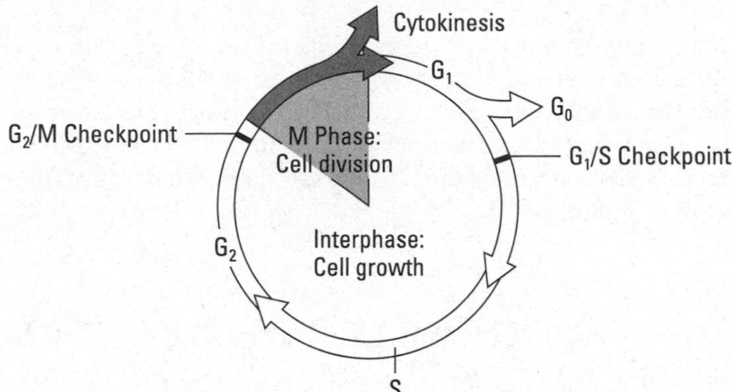

FIGURE 14-3: Quality control points in the cell cycle protect your cells from mutations that can cause cancer.

When it comes to cancer and how things go wrong with the cell cycle, two types of genes are especially important:

- **Proto-oncogenes,** which stimulate the cell to grow and divide, basically acting to push the cell through the checkpoints.
- **Tumor-suppressor genes,** which act to stop cell growth and tell cells when their normal life spans have ended.

Basically, two things go on in the cell: One set of genes (and their products) acts like an accelerator to tell cells to grow and divide, and a second set of genes puts on the brakes, telling cells when to stop growing, when not to divide, and even when to die.

REMEMBER

The mutations that cause cancer to turn proto-oncogenes into either *oncogenes* (turning the accelerator permanently "on") or damage tumor-suppressor genes (removing the brakes).

Genes gone wrong: Oncogenes

You can think of oncogenes as "on" genes because that's essentially what these genes do: They keep cell division turned on. Many genes, when altered, can become oncogenes. All oncogenes have several things in common:

- Their mutations usually represent a gain of function (described in Chapter 12).
- They're dominant in their actions. Only one copy of the gene needs to be altered in order to have an effect.
- Their effects cause excessive numbers of cells to be produced.

Oncogenes were the first genes identified to play a role in cancer. In 1910, Peyton Rous identified a virus that caused cancer in chickens. It took 60 years for scientists to identify the gene carried by the virus, the first known oncogene. It turns out that many viruses can cause cancer in animals and humans; for more on how these viruses do their dirty work, see the sidebar "Exploring the link between viruses and cancer."

EXPLORING THE LINK BETWEEN VIRUSES AND CANCER

It's becoming clearer that viruses play a significant role in the appearance of cancer in humans. Second only to the risk factor of cigarette smoking, viruses may be responsible for approximately 15 percent of all malignancies.

One class of viruses implicated in cancer is *retroviruses.* One familiar retrovirus that makes significant assaults on human health is HIV (Human Immunodeficiency Virus), which causes AIDS (Acquired Immunodeficiency Syndrome). And if you have a cat, you may be familiar with feline leukemia, which is also caused by a retrovirus (humans are immune to this cat virus). Most retroviruses use RNA as their genetic material. Viruses aren't really alive, so to replicate their genes, they have to hijack a living cell. Retroviruses use the host cell's machinery to synthesize DNA copies of their RNA chromosomes. The viral DNA then gets inserted into the host cell's chromosome where the virus genes can be active and wreak havoc with the cell and, in turn, the entire organism. Retroviruses that cause cancer copy their oncogenes into the host cell. The oncogenes and additional mutations in infected cells can then result in the development of cancer.

If you've ever had a wart, then you're already acquainted with the harmless version of a virus whose relatives can cause cancer. Human papilloma virus (HPV) causes genital warts and is linked to cervical cancer in women. Infection with the HPV associated with cervical cancer usually starts with *dysplasia* (the formation of abnormal but noncancerous cells). It usually takes many years for cervical cancer to develop, which it does only rarely. Nevertheless, it is expected that more than 13,000 women in the United States alone will be diagnosed with cervical cancer in 2019. Early screening for cervical cancer, in the form of Pap smears, has improved detection and saved the lives of countless women. In addition, the development and use of an HPV vaccine in adolescents and young adults is leading to a dramatic decrease in the incidence of HPV-related malignancies.

You have at least 70 naturally occurring proto-oncogenes in your DNA. Normally, these genes carry out regulatory jobs necessary for normal functioning. It's only when these genes gain mutations and become oncogenes that they switch from good genes to cancer-causers. Proto-oncogenes can become an oncogene in several ways. One way a proto-oncogene can become an oncogene is by acquiring a DNA sequence change that leads to an increase in the protein's functions (that is, gain of function mutations; described in Chapter 12). Another mechanism by which proto-oncogenes become oncogenic is gene amplification. Cancer cells tend to have multiple copies of oncogenes because the genes somehow duplicate themselves in a process called *amplification*. This duplication allows those genes to have much stronger effects than they normally would. Finally, chromosome translocations that move a proto-oncogene to a region where it is more highly expressed can result in the gene becoming an oncogene.

HISTORICAL STUFF

The first oncogene identified in humans resides on chromosome 11. The scientists responsible for its discovery were looking for the gene responsible for bladder cancer. They took cancerous cells from the bladder and isolated their DNA; then they introduced small parts of the cancer cell's DNA into bacteria and allowed the bacteria to infect normal cells growing in test tubes. The scientists were looking for the part of the DNA present in cancer cells that would transform the normal cells into cancerous ones. The gene they found, now called *HRAS1*, was very similar to a virus oncogene that had been found in rats. The mutation that makes *HRAS1* into an oncogene affects only three bases of the genetic code (called a *codon*; see Chapter 10). This tiny change causes *HRAS1* to constantly send the signal "divide" to affected cells.

Since the discovery of *HRAS1*, a whole group of oncogenes has been found; they're known collectively as the *RAS* genes. All the *RAS* genes work much the same way and, when mutated, turn the cell cycle permanently "on." In spite of their dominant activities, a single mutated oncogene usually isn't enough to cause cancer all by itself. That's because tumor-suppressor genes (see the next section) are still acting to put on the brakes and keep cell growth from getting out of control.

Oncogenes aren't usually implicated in inherited forms of cancer. Most oncogenes show up as somatic mutations (mutations that occur in the non-sex cells of the body), which can't be passed on from parent to child.

The good guys: Tumor-suppressor genes

Tumor-suppressor genes are the cell cycle's brakes. Normally, these genes work to slow or stop cell growth and to put a halt to the cell cycle. When these genes fail, cells can divide out of control, meaning that mutations in tumor-suppressor genes are loss-of-function mutations (covered in Chapter 12). Loss-of-function mutations are generally recessive, meaning that both copies of the gene need to be altered in order to cause any problems. So for a tumor suppressor gene to lose its function as the cell cycle's brakes, both copies of the gene (both the one from mom and the one from dad) need to have mutations. Typically, only one of these mutations is inherited; the other occurs during the person's lifetime (a somatic mutation).

HISTORICAL STUFF

The first gene recognized as a tumor suppressor is associated with a cancer of the eye, called *retinoblastoma*. Retinoblastoma often runs in families and shows up in very young children. In 1971, geneticist Alfred Knudson suggested that one mutated allele of the gene was being passed from parent to child and that a mutation event in the child was required for the cancer to occur (referred to as the *Two-Hit Hypothesis*; see Figure 14-4). Therefore, it was much more likely for the cancer to develop in someone who already had one mutation (the first "hit") at birth. It should be noted, however, that we now know that more genetic changes are likely necessary for the development of a malignancy, although those changes are much more likely once the tumor suppressor gene has been inactivated.

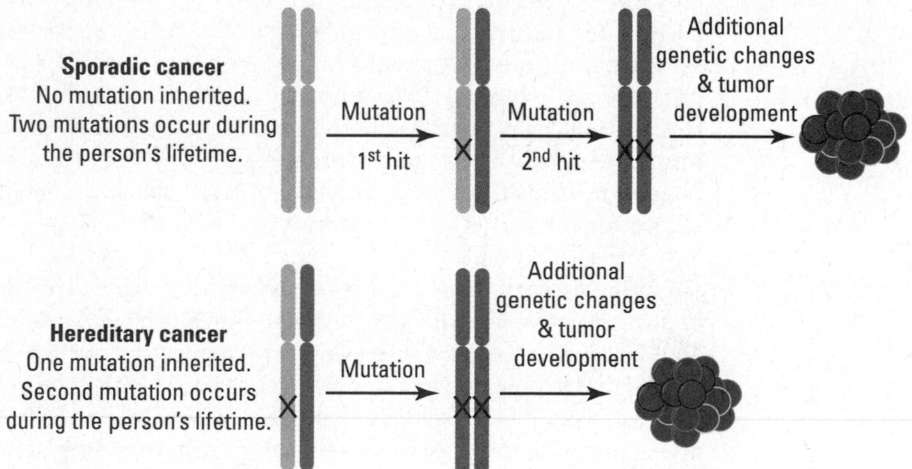

FIGURE 14-4: The Two-Hit Hypothesis.

The gene responsible, called *RB1*, was mapped to chromosome 13 and is implicated in other forms of cancer such as breast, bladder, and bone (osteosarcoma). *RB1* turns out to be a very important gene. If both copies are mutated in embryos, the mutations are lethal, suggesting that normal *RB1* function is required for survival. *RB1* regulates the cell cycle by interacting with transcription factors (transcription factors are discussed in greater detail in Chapter 11). These particular transcription factors control the expression of genes that push the cell through the checkpoint at the end of G_1, just before DNA synthesis. When the proteins that *RB1* codes for (called *pRB*) are attached to the transcription factors, the genes that turn on the cell cycle aren't allowed to function. Normally, pRB and the transcription factors go through periods of being attached and coming apart, turning the cell cycle on and off. If both copies of *RB1* have mutations, then this important brake system goes missing. As a result, affected cells move through the cell cycle faster than normal and divide without stopping. *RB1* not only interacts with transcription factors to control the cell cycle; it's also thought to play a role in replication, DNA repair, and *apoptosis* (programmed cell death).

One of the most important tumor-suppressor genes identified to date is *TP53*, found on chromosome 17. This gene codes for the cell-cycle regulating protein p53. Mutations that lead to loss of p53 function are implicated in a wide variety of cancers. The most important of p53's roles may be in regulating when cells die:

- When DNA has been damaged, the cell cycle is stopped to allow repairs to be carried out.
- If repair isn't possible, the cell receives the signal to die (apoptosis).

If you've ever had a bad sunburn, then you have firsthand experience with apoptosis. Apoptosis, also known by the gloomy moniker "programmed cell death," occurs when the DNA of a cell is too damaged to be repaired. Rather than allow the damage to go through replication and become cemented into the DNA as mutation, the cell voluntarily dies. In the case of your severe sunburn, the DNA of the exposed skin cells was damaged by the sun's radiation. In many cases, the DNA strands were broken, probably in many different places. Those skin cells killed themselves off, resulting in the unpleasant skin peeling that you suffered. When your DNA gets damaged from too much sun exposure or because of any other mutagen (see Chapter 12 for examples), a protein called p21 stops the cell cycle. Encoded by a gene on the X chromosome, p21 is produced when the cell is stressed. The presence of p21 stops the cell from dividing and allows repair mechanisms to heal the damaged DNA. If the damage is beyond repair, the cell may skip p21 altogether. Instead, the tumor-suppressor protein p53 signals the cell to kill itself.

TECHNICAL STUFF

When the cell gets the message that says, "Die!" a gene called *BAX* swings into action. *BAX* sends the cell off to its destruction by signaling the mitochondria — those energy powerhouses of the cell — which release a wrecking-crew of proteins that go about breaking up the chromosomes and killing the cell from the inside. When your cells die due to injury (like a burn or infection), the process is a messy one: The cells explode, causing surrounding cells to react in the form of inflammation. Not so in apoptosis. The cells killed by the actions of apoptosis are neatly packaged so that surrounding tissues don't react. Cells that specialize in garbage collection and disposal, called *phagocytes* (meaning cells that eat), do the rest.

Drugs used to fight cancer often try to take advantage of the apoptosis pathway to cell death. The drugs turn on the signals for apoptosis to trick the cancer cells into killing themselves. Radiation therapy, also used to treat cancer by introducing double-strand breaks (see Chapter 12 for more on this kind of DNA damage), relies on the cells knowing when to die. Unfortunately, some of the mutations that create cancer in the first place make cancer cells resistant to apoptosis. In other words, in addition to growing and dividing without restriction, cancer cells don't know when to die.

Demystifying chromosome abnormalities

Large-scale chromosome changes — the kinds that are visible when karyotyping (chromosome examination; you can flip back to Chapter 13 for details) is done — are associated with some cancers. These chromosome changes (like losses of chromosomes) often occur after cancer develops and occur because the DNA in cancer cells is unstable and prone to breakage. Normally, damaged DNA is detected by proteins that keep tabs on the cell cycle. When breaks are found, either the cell cycle is stopped and repairs are initiated or the cell dies. Because the root of cancer is the loss of genetic quality control functions provided by proto-oncogenes and tumor-suppressor genes, it's no surprise that breaks in the cancer-cell DNA lead to losses and rearrangements of big chunks of chromosomes as the cell cycle rolls on without interruption. One of the biggest problems with all this genetic instability in cancer cells is that a tumor is likely to have several different genotypes among its many cells, which makes treatment difficult. Chemotherapy that's effective at treating cells with one sort of mutation may not be useful for another.

Three types of damage — deletions, inversions, and translocations — have been shown to interrupt tumor-suppressor genes, rendering them nonfunctional. Translocations and inversions may change the positions of certain genes so that the gene gets regulated in a new way (see Chapter 11 for more about how gene expression is regulated by location). Chronic myeloid leukemia, for example, is caused by a translocation event between chromosomes 9 and 22. This form of leukemia is a cancer of the blood that affects the bone marrow.

TECHNICAL STUFF

Translocations generally result from double-stranded DNA breaks. In the case of chronic myeloid leukemia, the translocation event makes chromosome 22 unusually short. This shortened version of the chromosome is called the *Philadelphia chromosome* because geneticists working in that city discovered it. The translocation event causes two genes (*BCR* and *ABL*), one from each chromosome, to become fused together (see Figure 14-5). The new gene product acts as a powerful oncogene, leading to out-of-control cell division and eventually leukemia.

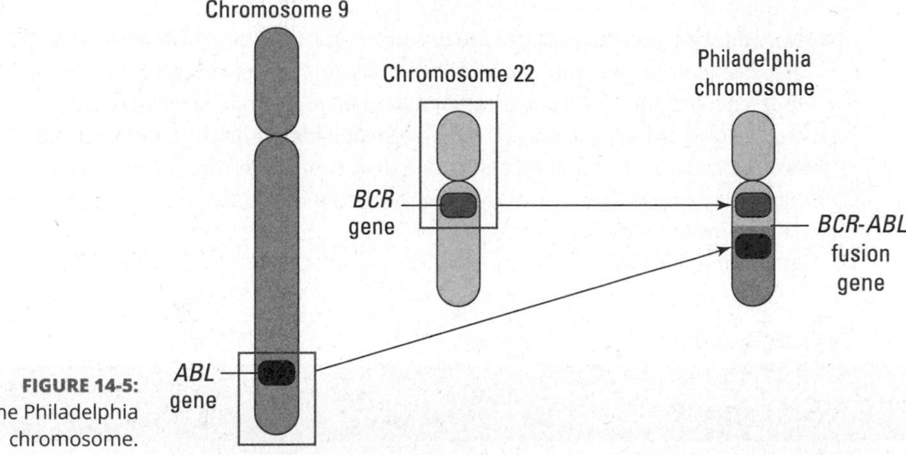

FIGURE 14-5: The Philadelphia chromosome.

Certain cancers seem prone to losing particular chromosomes altogether, resulting in monosomies (similar to those described in Chapter 13). For instance, one copy of chromosome 10 often goes missing in the cells of glioblastomas, a deadly form of brain cancer. Cancer cells are also prone to nondisjunction leading to trisomies in tumor cells. It appears that mutations in the p53 gene (*TP53*) are linked to these localized changes in chromosome number. Basically, mutations that render p53 nonfunctional can lead to genome instability in a cell, which then acquires chromosome abnormalities that further promote the development of a malignancy.

Breaking Down the Types of Cancers

Earlier in this chapter, we discussed how cancers can be classified based on the tissue in which the tumor started (such as a leukemia or sarcoma). Well, a cancer can also be classified based on its cause and its heritability (how likely it is to recur in other members of the same family). There are three main groups into which a cancer may fall (each described below): sporadic, familial, or hereditary. Table 14-2 further breaks down the vital information on each type.

- **Sporadic cancers:** Most cancers are sporadic — not the result of inherited gene mutations. *Sporadic cancers* typically develop later in life and result from environmental exposures and the accumulation of mutations that occur during a person's lifetime. They occur by chance.
- **Familial cancers:** Other cancers are familial in nature. Individuals with *familial cancers* tend to have a family history of the same type of cancer. However, there is no clear pattern of inheritance. They likely result from a combination of genetic and environmental factors, which are shared among family members.

» **Hereditary cancers:** The last type of cancers is hereditary. *Hereditary cancers* are the result of gene mutations that are passed from parent to child. Those that inherited the disease-causing mutation have a significantly increased chance of developing cancer. They are also at a higher risk of multiple tumors, which frequently occur at an earlier age than sporadic tumors. Those with hereditary cancer generally have a family history of the same type of cancer or related malignancies.

TABLE 14-2 **Types of Cancer**

Type of Cancer	Percent of Cases	Family History Characteristics	Age at Disease Onset	Genetic Basis
Sporadic	70-80%	No family history of the same type of cancer; risk of cancer to others in the family not increased above the general population risk for the specific type of cancer	Cancer typically develops later in life (after age 50)	The result of gene mutations that are acquired during the person's lifetime (not inherited)
Familial	15-20%	Some family history of the same type of cancer, with no clear inheritance pattern; risk of cancer to others in the family may be slightly increased above the general population risk for the specific type of cancer	Cancer typically develops later in life (after age 50)	Likely the result of shared genetic background and environmental factors; no single gene mutation running through the family
Hereditary	5-10%	Strong family history of the same type of cancer (or related malignancies); inherited in an autosomal dominant manner (with reduced penetrance); risk of cancer to family members with the same gene mutation is significantly elevated relative to the general population risk for the specific type of cancer	Cancer typically develops earlier than average (before age 50, and often before age 40)	Typically due to an autosomal dominant mutation in a hereditary cancer syndrome gene; carriers have a 50% chance of passing the mutation on to each child

REMEMBER

Both familial and hereditary cancers run in families. However, no one ever inherits cancer; what's inherited is the predisposition to certain sorts of cancer. What this means is that certain cancers tend to run in families because one or more mutations are being passed on from parent to child. Most geneticists agree that additional mutations are required to trigger the actual disease. Just because you have a family history of a particular cancer doesn't mean you'll get it. The opposite is also true: Just because you don't have a family history doesn't mean you won't get cancer.

Examples of hereditary cancer syndromes include hereditary breast and ovarian cancer syndrome and Lynch syndrome. These conditions are inherited in an autosomal dominant manner. Any individual who inherits one copy of the disease-causing mutation will have a significantly increased risk of developing cancer.

TIP

For more information on all types of cancers, visit the American Cancer Society website at `www.cancer.org` and the National Cancer Institute website at `www.cancer.gov`.

Hereditary Breast Cancer

Breast cancer is the most common malignancy among American women (refer to Table 14-1). Generally, the first symptom of breast cancer is a lump in the breast tissue. The lump may be painless or sore, hard (like a firm knot) or soft; the edges of the lump may not be easy to detect, but in some cases they're very easy to feel. Other symptoms include swelling, changes in the skin of the breast, nipple pain or unexpected discharge, and a swelling in the armpit. Different sorts of breast cancer are distinguished by the part of the breast that develops the tumor. Regardless of the type of breast cancer, the number one risk factor appears to be a family history of the disease. However, only 5 to 10 percent of breast cancer cases are hereditary.

Researchers have identified two breast cancer genes: *BRCA1* and *BRCA2* (for BReast CAncer genes 1 and 2). These genes account for slightly less than 25 percent of inherited breast cancers, however. Mutations in the gene for p53, along with numerous other genes, are also associated with hereditary forms of breast cancer (see "The good guys: Tumor-suppressor genes" earlier in this chapter for more on p53). Breast cancers associated with mutations of *BRCA1* and/or *BRCA2* are inherited as *autosomal dominant disorders* (genetic disorders resulting from one bad copy of a gene; see Chapter 15 for more on inheritance patterns). Women who inherit a mutation in one of these genes have a significantly increased risk of both breast and ovarian cancer. Men who inherit a mutation have a significantly increased risk of both male breast cancer and prostate cancer.

When it comes to the risk of breast cancer in individuals with *BRCA1* or *BRCA2* mutations, penetrance is roughly 70 percent, meaning 70 percent of the women inheriting a mutation in one of the breast cancer genes will develop cancer. (This penetrance value is based on a life span of 85 years, by the way, so women living 85 years have a 70 percent chance of expressing the phenotype of cancer.) This is in comparison to the general population risk of 12 percent for a woman to develop breast cancer at some point in her life.

TECHNICAL STUFF

Both *BRCA* genes are tumor-suppressor genes. *BRCA1* encodes a protein that functions in the repair of damaged DNA. BRCA1 interacts with other tumor suppressor genes, cell cycle proteins, and DNA repair proteins. As for the BRCA2 protein, it also plays a role in DNA repair, especially of double-strand breaks. In addition, it has been proposed that this protein functions in the process of cytokinesis during cell division.

REMEMBER

Early detection of breast cancer is the best defense against the disease. Genetic tests are available to confirm the presence of mutations that are associated with the development of breast cancer. In those found to carry disease-causing mutations, increased surveillance is recommended and prophylactic treatment (such as surgical removal of the breasts and/or ovaries before any sign of disease) may significantly reduce the risk of cancer. After breast cancer is diagnosed, treatment options vary based on the kind of cancer. Breast cancer is considered very treatable, and the prognosis for recovery is very good for most patients.

Hereditary colorectal cancer

One cancer that's considered highly treatable when detected early is colon cancer. Your colon (the large intestine) is the bulky tube that carries waste products to your rectum for defecation. Over 100,000 people are likely to be diagnosed with colon cancer each year.

Almost all colon cancers start as benign growths called *polyps.* These polyps are tiny wart-like protrusions on the wall of the colon. If colon polyps are left untreated, a *RAS* oncogene often becomes active in the cells of one or more of the polyps, causing the affected polyps to increase in size (see the "Genes gone wrong: Oncogenes" section earlier in the chapter for more on how oncogenes work). When the tumors get big enough, they change status and are called *adenomas.* Adenomas are benign tumors but are susceptible to mutation, often of the tumor-suppressor gene that encodes p53. When p53 is lost through mutation, the adenoma becomes a *carcinoma* — a malignant and invasive tumor.

Lynch syndrome, historically referred to as hereditary nonpolyposis colorectal cancer (HNPCC), is a hereditary cancer syndrome in which affected individuals have a significantly increased risk of colon or rectal cancer (in men and women) and endometrial cancer (in women). The condition is also associated with an increased risk of other types of malignancies, including cancer of the stomach, small intestine, urinary tract, bile duct, and liver. Affected women also have an increased risk of ovarian cancer. It is estimated that approximately 3 percent of colorectal and endometrial cancers are the result of Lynch syndrome. Lynch syndrome results from mutations in genes that play a role in the repair of DNA damage, including the *MLH1*, *MSH2*, *MSH6*, and *PMS2* genes. The condition is inherited in an autosomal dominant manner.

REMEMBER

Early detection and treatment are critical to prevent colon polyps from becoming cancerous. If large numbers of polyps develop, the likelihood that at least one will become malignant is very high. The good news is that the changes in the colon usually accumulate slowly, over the course of several years. Individuals with Lynch syndrome do not develop more polyps than those in the general population, but their polyps tend to occur earlier and are more likely to become cancerous.

A combination of tumor testing (testing for the presence of certain proteins in a tumor sample) and genetic testing may be used to diagnose Lynch syndrome and to identify the disease-causing mutation. Once the disease-causing mutation has been identified in a family, then at-risk family members (siblings and children of a mutation carrier) may be tested in order to determine their risk of cancer. Those that are found to carry the mutation will have up to an 80 percent chance of developing colorectal cancer at some time in their life. Increased surveillance for tumors is recommended for all mutation carriers, and options for prophylactic surgery are available (such as removal of the large intestine, ovaries, and/or uterus).

IN THIS CHAPTER

» Understanding what genetic counselors do

» Examining family trees for different kinds of inheritance

» Using inheritance patterns to determine the chance of inheriting a trait or genetic disorder

» Exploring options for genetic testing

Chapter 15 Genetic Counseling, Risk Assessment, and Genetic Testing

If you're thinking of starting a family or adding to your brood, you may be wondering what your little ones will look like. Will they get your eyes or your dad's hairline? If you know your family's medical history, you may also have significant worries about genetic conditions running through your family. You may worry about your own health, too, as you contemplate news stories dealing with cancer, heart disease, and diabetes, for example. All these concerns revolve around genetics and the inheritance of a predisposition to a particular disease or the inheritance of the disorder itself.

Genetic counselors are specially and rigorously trained to help people learn about the genetic aspects of their family medical histories. This chapter explains the process of genetic counseling, including how counselors generate family trees and calculate probability of inheritance and how genetic testing is done when genetic disorders are possible or anticipated.

Getting to Know Genetic Counselors

Everyone has a family. You have a mother and a father, grandparents, and perhaps children of your own. You may not think of them, but you also have many ancestors — people you've never met — whose genes you carry and may pass down to descendants in the centuries to come.

Genetic counselors help people examine their families' genetic histories and uncover inherited conditions. They work as part of a healthcare team, alongside physicians and nurses. Genetic counselors usually hold a master's degree in genetic counseling and have an extensive background in genetics (and can solve genetics problems in a snap) so that they can spot patterns that signal an inherited disorder. For more on genetic counselors and other career paths in genetics, see Chapter 1.

Genetics counselors perform a number of functions, including:

- » Obtaining and analyzing medical information about the patient and their family.
- » Constructing and interpreting family trees, also called *pedigrees,* to assess the likelihood that various inherited conditions will be (or have been) passed on to a particular generation.
- » Counseling individuals about options for genetic testing and what the potential results would mean for the patient and their family.
- » Educating patients and their families about genetic conditions running through the family, explaining recurrence risks (the chance a given person will inherit the causative gene change(s)), and discussing the implications for the patient and their family.
- » Helping individuals make important decisions regarding genetic testing.

Physicians refer a wide range of people to genetic counselors for evaluation and potentially genetic testing, including:

- » Individuals who have a family history of (or who have been diagnosed with) an inherited genetic disorder.
- » Children who are suspected of having a genetic disorder based on a history of developmental delays, intellectual disability, birth defects, and/or other atypical features.
- » Older women (typically over the age of 35) who are pregnant or planning a pregnancy.

» Women who have had an abnormal screening test, such as an ultrasound, during a pregnancy.

» Pregnant women who are concerned about exposure to substances known to cause birth defects (such as radiation, viruses, drugs, and chemicals).

» Couples who have experienced multiple miscarriages, a stillbirth, or problems with infertility.

» People with a family history of later onset conditions (such as Huntington disease [described in Chapter 4] or certain cancers such as breast, ovarian, or colon cancer) who may be considering genetic testing to determine their risk of getting the disease.

TIP

We cover many of the scientific reasons for the inheritance of genetic disorders elsewhere in this book. Sequence variants within genes are the cause of many genetic disorders (including cystic fibrosis, Tay-Sachs disease, and sickle cell anemia); the types of sequence variants that may cause disease are described in detail in Chapter 12. We also explain chromosomal disorders such as Down syndrome in Chapter 13, and discuss the causes and genetic mechanics of cancer in Chapter 14.

Building and Analyzing a Family Tree

One of the key steps in genetic counseling is drawing a family tree (also known as a *pedigree*). The tree usually starts with the person for whom the tree is initiated; this person is called the *proband*. The proband can be a newly diagnosed child, a woman planning a pregnancy, or an otherwise healthy person who's curious about risk for inherited disease. Often, the proband is simply the person who meets with the genetic counselor and provides the information used to plot out the family tree. The proband's position in the family tree is always indicated by an arrow, and they may or may not be affected by an inherited disorder.

Genetic counselors use a variety of symbols on family trees to indicate personal traits and characteristics. For instance, certain symbols convey each person's sex, whether a person is a gene carrier, whether a person is deceased, and whether a person's history is unknown. The manner in which symbols are connected show relationships among family members, such as which offspring belong to which parents, whether someone is adopted, and whether someone is a twin. Check out Figure 15-1 for a detailed key to the symbols typically used in pedigree analysis.

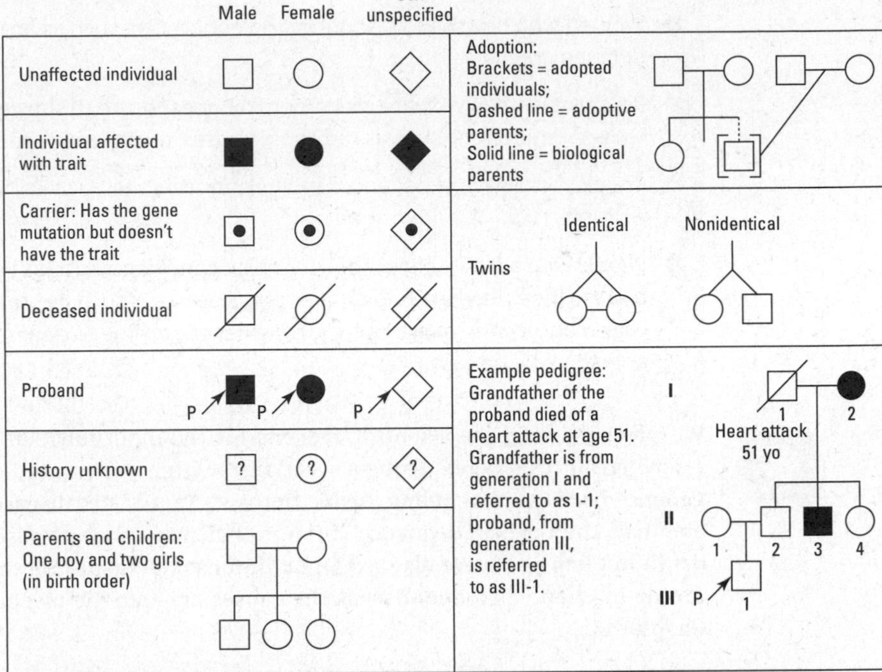

FIGURE 15-1: Symbols commonly used in the pedigree analysis.

In a typical pedigree, the age or date of birth of each person is noted on the tree. If deceased, the person's age at the time of death and the cause of death are listed. Some genetic traits are more common in certain regions of the world, so it's also useful to include all kinds of other details about family history on the pedigree, such as what countries people immigrated from or how they're related. Pedigrees generally include three generations, and every member of the family should be listed, along with any medical information known about that person, including the age at which certain medical disorders occurred. In the example in Figure 15-1, the grandfather of the proband died of a heart attack at age 51. Including this information creates a record of all disorders so that the counselor is more likely to detect every inherited disease present in the family. (Medical information doesn't appear in Figure 15-1, but it's normally a part of a pedigree.)

Medical problems often listed on pedigrees include the following:

- Known genetic conditions;
- Birth defects;
- Developmental delays and intellectual disability;

- Miscarriages or stillbirths;
- Psychiatric illness;
- Cancer;
- Other significant medical conditions that require regular treatment (such as kidney disease).

Humans have only a few children relative to other creatures, and humans start producing offspring after a rather long childhood. Geneticists rarely see neat offspring ratios (such as four siblings with three affected and one unaffected) in humans that correspond to those observed in animals (see Chapters 3 and 4 for more on common offspring ratios). Therefore, genetic counselors must look for very subtle signs to detect particular patterns of inheritance in humans.

When the genetic counselor knows what kind of disorder or trait is involved, the counselor can determine the likelihood that particular person will possess the trait or pass it on to their children. Genetic counselors use the following terms to describe the individuals in a pedigree:

- **Affected:** Any person diagnosed with or having symptoms of a given disorder.
- **Heterozygote:** Any person possessing one disease-causing allele in the gene associated with a given disorder. For recessive disorders, an unaffected heterozygote is called a *carrier.*
- **Homozygote:** Any person possessing two copies of a disease-causing allele for a disorder. This person can also be described as *homozygous.*
- **Compound Heterozygote:** Any person possessing two different disease-causing allele for a disorder (they are heterozygous for each mutation).

The way in which many human genetic disorders are passed down to later generations — the *mode of inheritance (or inheritance pattern)* — is well established. After a genetic counselor determines which family members are affected or are likely to be carriers, it's relatively straightforward to determine the probability of another person being a carrier or inheriting the disorder.

In the following sections, we explore the modes of inheritance for human genetic disorders, how genetic counselors identify these patterns, and how you (and your counselor) can figure out the probability of passing these traits on to offspring.

Autosomal Inheritance: No Differences Among the Sexes

Genes located on the autosomes (chromosomes 1 through 22, the non-sex chromosomes) are found in pairs. It doesn't matter whether you are male or female — one copy is inherited from your mother, and one copy is inherited from your father. Consequently, the chance of inheriting DNA sequence changes found on the autosomes is the same among male and female offspring.

Autosomal dominant traits and disorders

For a *dominant* trait or disorder, only one allele known to cause the trait or disorder is needed in order to show the related phenotype (physical features). *Autosomal dominant* means that the gene (and the causative allele) is carried on an autosome (a chromosome other than the X or Y chromosomes). Genes located on the autosomes come in pairs, one inherited from each parent. In human pedigrees, autosomal dominant traits have some typical characteristics:

- Affected children have an affected parent.
- Both males and females are affected with equal frequency.
- If neither parent is affected, usually no child is affected.
- The trait doesn't typically skip generations.

Figure 15-2 shows the pedigree of a family with an autosomal dominant trait. In the figure, affected persons are shaded, and you can clearly see how only affected parents have affected children. The trait can be passed to a child from either the mother or the father. With each pregnancy, an affected parent has a 50 percent chance of passing on the causative allele and having an affected child.

Some examples of autosomal dominant disorders are:

- **Achondroplasia:** A form of dwarfism;
- **Huntington disease:** A progressive and fatal disease affecting the brain and nervous system;
- **Marfan syndrome:** A disorder affecting the skeletal system, heart, and eyes.

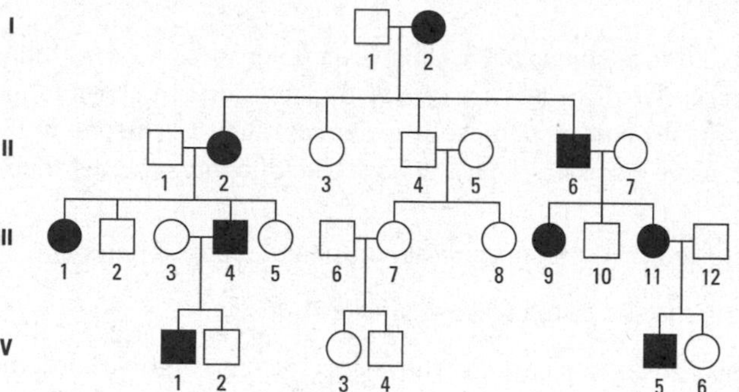

FIGURE 15-2: A typical family tree with an autosomal dominant inheritance pattern.

The normal pattern of autosomal dominant inheritance has three exceptions:

- **Reduced penetrance:** *Penetrance* is the percentage of individuals having a particular genotype (specifically a disease-causing allele) that actually display the physical characteristics associated with the gene mutation (or express the phenotype, scientifically speaking). Many autosomal dominant disorders have complete penetrance, meaning that every person inheriting the gene mutation shows features of the condition. But some disorders have *reduced penetrance,* meaning only a certain percentage of individuals inheriting the gene mutation show the phenotype. When an autosomal dominant disorder shows reduced penetrance, the phenotype can skip generations. See Chapter 4 for more details on reduced penetrance.
- **New mutations:** New mutations are those that occur in the egg or the sperm of an individual, as opposed to being inherited from a parent. In the case of new mutations that are autosomal dominant, the trait appears for the first time in one individual of a single generation and can appear in every generation thereafter. See Chapter 12 to find out more details about mutations — how they occur and how they're passed on.
- **Variable expressivity:** Expressivity is the degree to which a trait is expressed. Some conditions may be milder in some individuals and more severe in others. The condition could even go undiagnosed in certain generations because the symptoms are so mild they are never noted. See Chapter 4 to find out more about expressivity.

Autosomal recessive traits and disorders

For *recessive traits or disorders*, both alleles of the associated gene need to have the necessary change(s) in order to show the related phenotype. If a person carries two copies of the *same* disease-causing allele, it's then said that the individual is a *homozygote*. If a person carries two *different* disease-causing alleles in the same gene, they are referred to as *compound heterozygotes*. Like autosomal dominant disorders, autosomal recessive disorders are caused by mutations in genes found on chromosomes other than sex chromosomes. In pedigrees, such as the one in Figure 15-3, autosomal recessive disorders typically have the following characteristics:

- Affected children are generally born to unaffected parents.
- Both males and females are affected equally.
- Children born to parents who share a common ethnic background are more likely to be affected than those of parents with different backgrounds.
- The disorder or trait typically shows up in a single generation (often in siblings), or it skips one or more generations.

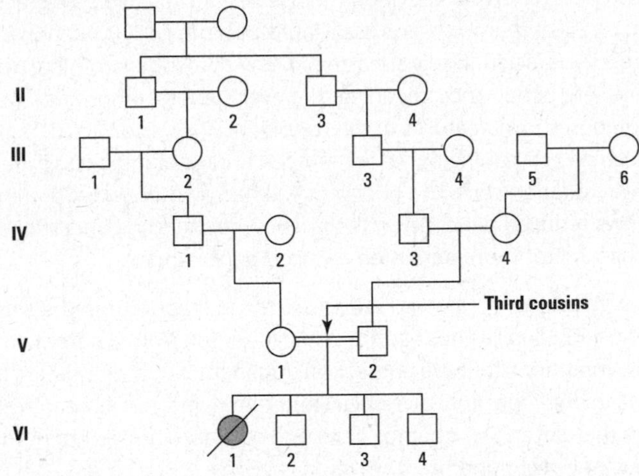

FIGURE 15-3: An example of an autosomal recessive disorder in a family tree.

The probability of inheriting an autosomal recessive disorder varies depending on the parents' genotypes (see Chapter 3 for all the details on how the odds of inheritance are calculated):

- **When both parents are heterozygous carriers,** there is a 25 percent chance of both parents passing on a disease-causing allele and having an affected child. The chance is the same for each and every pregnancy.
- **When one parent is a carrier and the other isn't,** each child has a 50 percent chance of being a carrier. No child will be affected because they also carry a normal allele that they inherited from the noncarrier parent.

» **When one parent is a carrier and the other is affected,** there will be a 50 percent chance of having an affected child with each pregnancy. The affected parent will always pass on a disease-causing allele, while the carrier can pass on either a normal allele or a disease-causing allele. All unaffected children from the union will be heterozygous carriers.

» **When one parent is affected and the other is unaffected (and not a carrier),** all children born to the couple will be heterozygous carriers. They will all get a disease-causing allele from the affected parent and a normal allele from the unaffected noncarrier parent. No children will be affected.

TECHNICAL STUFF

Cystic fibrosis (CF) is an autosomal recessive disorder that causes severe lung and digestive problems in affected persons. As with all autosomal recessive disorders, if both members of a couple are carriers for cystic fibrosis, they have a 25 percent chance of having an affected child with each pregnancy they have. That's because both the man and the woman are heterozygous for an allele that causes cystic fibrosis, and each has a 50 percent chance of contributing the CF allele to each child. To get this figure, you calculate the probability of *both* members of the couple contributing CF alleles in one fertilization event by multiplying the probability of each event happening independently. The probability the father contributes his CF allele is 50 percent, or 0.5; the probability the mother contributes her CF allele is also 50 percent, or 0.5. The probability that both contribute their allele is $0.5 \times 0.5 = 0.25$, or 25 percent. For more details on how to calculate probabilities of inheritance, see Chapters 3 and 4.

Some autosomal recessive disorders are more common among people of certain ethnic groups, because people belonging to those groups tend to marry within the group. After many generations, everyone within the group shares common ancestry. When cousins or other close relatives marry (such as illustrated in the sixth generation of the family shown in Figure 15-3), such relationships are referred to as *consanguineous* (meaning "same blood"). The closer the individuals are (such as first cousins versus third cousins), the higher their risk of having a child with a recessive disorder. When populations are founded by rather small groups of people, those groups often have higher rates of certain genetic disorders than the general population; for more details, see the sidebar "Genetic disorders in small populations." In these cases, autosomal recessive disorders may no longer skip generations, because so many individuals in the population are heterozygous carriers of the disease-causing allele.

GENETIC DISORDERS IN SMALL POPULATIONS

The Pennsylvania Amish don't have electricity in their homes, don't drive cars, and don't use email or cellphones. They live simply in the modern world as a religious way of life. Because many Amish families are descended from a small number of Amish ancestors, certain genetic disorders are common.

Amish families come by horse and buggy to the Clinic for Special Children in Strasburg, Pennsylvania, for evaluation and genetic testing. By partnering with an ultra-high-tech company, the clinic provides rapid, inexpensive genetic testing. Among the clinic's findings is the fact that at least 21 children from the Old Order Amish of southeastern Pennsylvania were born with and died from a devastating form of sudden infant death syndrome (SIDS). A collaboration between the clinic and a genetic testing company identified the gene that causes this condition and now all of the Amish newborns in the community are tested for this condition.

Another finding among the Amish population is that an extremely rare metabolic condition known as glutaric aciduria type 1 is found in 1 in 400 births, much higher than in the general population. With improved diagnosis and treatment, physicians at the clinic have been able to prevent the severe complications that can occur in the children with this condition.

Another metabolic disorder known as maple syrup urine disease, which is fatal if untreated, is also found at a very high frequency in this Amish community — found in about 1 in 100 individuals in the Amish, compared to 1 in 180,000 individuals elsewhere in the world. Because of the diagnosis and treatment now available for this condition at the clinic, the deaths due to this condition have been essentially eliminated.

Found on Sex Chromosomes: Sex-linked Inheritance

Sex has a lot to do with which genes are expressed and how. *Sex-linked genes* are ones that are actually located on the sex chromosomes themselves. Some traits are truly X-linked (such as hemophilia) or Y-linked (such as hairy ears). Other traits are expressed differently in males and females even though the genes that control the traits are located on non-sex chromosomes (autosomes). This section explains how sex influences (and sometimes controls) the phenotypes of various genetic conditions.

X-linked recessive traits

X-linked traits and disorders are the result of sequence variants in genes located on the X chromosome. As we discuss in Chapter 6, females have two X chromosomes and males have an X and a Y. Since males have only one copy of the X chromosome, they don't have a second X to offset the expression of any disease-causing alleles on their X chromosome. For all X-linked recessive traits, the gene acts like a dominant gene when it's in the *hemizygous* (one copy) state. Thus, similar to autosomal dominant disorders, X-linked recessive disorders express the trait fully in males, even though they have only one disease-causing allele. Females rarely show X-linked recessive disorders, because it is rare for them to have two disease-causing alleles and when they have just one, the other normal allele is typically enough to prevent symptoms of the condition. In pedigrees, X-linked recessive disorders have the following characteristics:

- Affected sons are typically born to unaffected mothers.
- Far more males than females are affected.
- The trait is *never* passed from father to son.
- The disorder can skip one or more generations.

Unaffected parents can have unaffected daughters and one or more affected sons. Women who are carriers frequently have brothers with the disorder, but if families are small, a carrier may have no immediate family members who have the condition. Sons of affected fathers are never affected, but daughters of affected fathers are always carriers, because daughters must inherit one of their X chromosomes from their fathers. In this case, that X chromosome will always carry the allele for the disorder. The pedigree in Figure 15-4 is a classic example of a well-researched family possessing many carriers for the X-linked disorder hemophilia, a serious condition that prevents normal clotting of the blood. For more on the royal families whose history is pictured in Figure 15-4, see the sidebar "A royal pain in the genes."

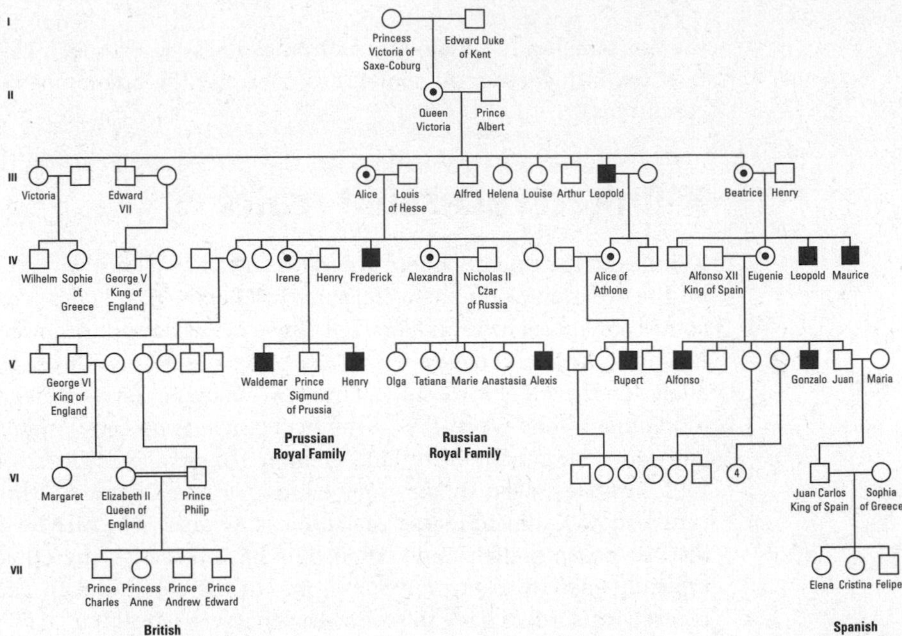

FIGURE 15-4: The X-linked recessive disorder hemophilia works its way through the pedigree of the royal families of Europe and Russia.

The probability of inheritance of X-linked disorders depends on sex. Female carriers have a 50 percent likelihood of passing the gene on to each child. Males determine the sex of their offspring, making the chance of any particular child being a boy 50 percent. Therefore, the likelihood of a carrier mom having an affected son is 25 percent (chance of having a son = 0.5; chance of a son inheriting the affected X = 0.5; therefore, $0.5 \times 0.5 = 0.25$, or 25 percent).

Rarely, female carriers of a disease-causing allele for an X-linked recessive disorder will manifest symptoms of the condition. The reason this happens is because of *skewed X inactivation*. X inactivation is when most of the genes on one of a female's two X chromosomes are turned off in each cell, in order to compensate for the fact that males only have one X chromosome (that is, to balance things out). Which X is inactivated in each cell is typically random. However, in some cases, things can be skewed such that the same X chromosome gets inactivated in all or most of the cells. If the inactive X contains the normal allele, then the allele with the mutation would be on the active X and would be the only copy of the gene expressed. When this happens, the female carrier can show signs of the condition. The severity of the condition would depend on how many cells are expressing the disease-causing allele.

A ROYAL PAIN IN THE GENES

You can find one of the most famous examples of an X-linked family pedigree in the royal families of Europe and Russia. Queen Victoria of England had one son affected with hemophilia. It's not clear if Queen Victoria inherited the allele from a parent or if she had a new gene mutation. In any event, two of her daughters were carriers, and she had one affected son, Leopold. Queen Victoria's granddaughter Alexandra was also a carrier. Alexandra married Nicholas Romanov, who became the czar of Russia, and together they had five children: four daughters and one son. The son, Alexis, suffered from hemophilia.

The role Alexis's disease played in his family's ultimate fate is debatable. Clearly, however, one of the men who influenced the downfall of Russia's royal family was linked to the family as Alexis's "doctor." Gregory Rasputin was a self-proclaimed faith healer; in photographs, he appears wild-eyed and deeply intense. He's generally perceived to have been a fraud, but at the time, he had a reputation for miraculous healings, including helping little Alexis recover from a bleeding crisis. Despite Rasputin's talent for healing, Alexis didn't live to see adulthood. Shortly after the Russian Revolution broke out, the entire Russian royal family was murdered. (Rasputin himself had been murdered some two years earlier.)

In a bizarre final twist to the Romanov tale, a road repair crew discovered the family's bodies in 1979. Oddly, two of the family members were missing. Eleven people were supposedly killed by firing squad on the night of July 16, 1918: the Russian royal family (Alexandra, Nicholas, and their five children) along with three servants and the family doctor. However, the bodies of Alexis and his little sister, Anastasia, have never been found. Using DNA fingerprinting, researchers confirmed the identities of Alexandra and her children by matching their mitochondrial DNA to that of one of Queen Victoria's living descendants, Prince Philip of England. (To find out more about the forensic uses of DNA, see Chapter 18.)

X-linked dominant traits

Like autosomal dominant disorders, X-linked dominant traits don't skip generations. Every person who inherits the allele expresses the disorder. The family tree in Figure 15-5 shows many of the hallmarks of X-linked dominant disorders:

- Affected mothers can have both affected sons and daughters.
- Both males and females can be affected.
- All daughters of affected fathers are affected.
- The trait doesn't typically skip generations.

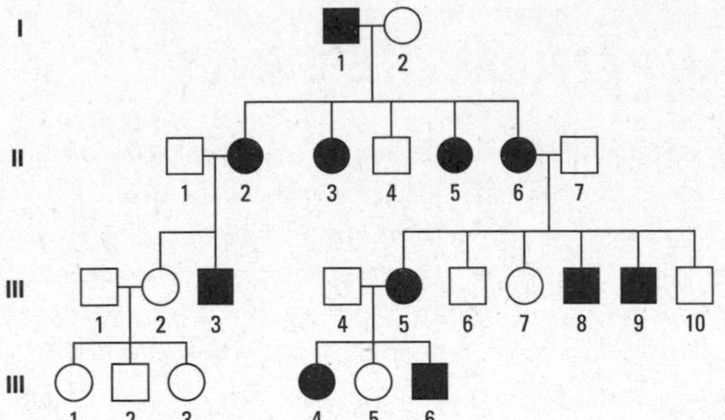

FIGURE 15-5: A family tree showing inheritance of an X-linked dominant trait.

X-linked dominant traits show up more often in females than males because females can inherit an affected X from either parent. In addition, some disorders are lethal in males who are *hemizygous* (having only one copy of the chromosome, not two; see Chapter 6). Affected females have a 50 percent chance of having an affected child of either sex. Males never pass their affected X to sons; therefore, sons of affected fathers and unaffected mothers have *no* chance of being affected, in contrast to daughters, who are always affected. The probability of an affected man having an affected child is 50 percent (that is, equal to the likelihood of having a daughter).

Y-linked traits

The Y chromosome is passed strictly from father to son. By definition, Y-chromosome traits are considered hemizygous. Y-chromosome traits are expressed as if they were dominant because there's only one copy of the allele per male, with no other allele to offset the effect of the gene. Y-linked traits are easy to recognize when seen in a pedigree, such as Figure 15-6, because they have the following characteristics:

- Affected men pass the trait to all their sons.
- No women are ever affected.
- The trait doesn't skip generations.

Because the Y chromosome is tiny and has relatively few genes, Y-linked traits are very rare. Most of the genes involved control male-only traits such as sperm production and testis formation. If you're female and your dad has hairy ears, you can relax — hairy ears is also considered a Y-linked trait.

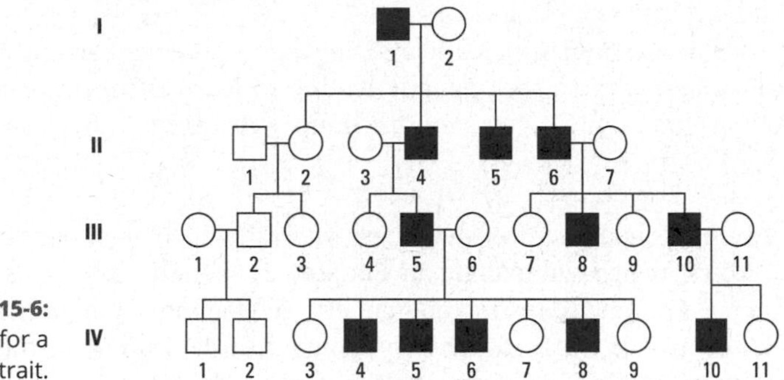

FIGURE 15-6: Pedigree for a Y-linked trait.

Sex-limited traits

Sex-limited traits are inherited in the normal autosomal fashion but are never expressed in one sex, regardless of whether the gene is heterozygous or homozygous. Such traits are said to have 100 percent penetrance in one sex and zero penetrance in the other. (*Penetrance* is the probability that an individual having a dominant allele will show its effects; see Chapter 4 for more.) Traits such as color differences between male and female birds are sex-limited; both males and females inherit the genes for color, but the genes are expressed only in one sex (usually the male). In mammals, both males and females possess the genes necessary for milk production, but only females express these genes, which are controlled by hormone levels in the female's body. See Chapter 11 for more about how gene expression is controlled.

Sex-influenced traits

Sex-influenced traits are coded by genes on autosomes, but the phenotype depends on the sex of the individual carrying the affected gene. Sex-influenced traits come down to the issue of penetrance: The traits are more penetrant in males than females. Horns, hair, and other traits that make male organisms look different from females are usually sex-influenced traits. In humans, male-pattern baldness is a sex-influenced trait.

Testing for Genetic Disorders

It is expected that every person in the world carries one or more alleles that cause genetic disease. Most of us never know which alleles or how many we carry (most are recessive and most of us haven't had our genomes sequenced). If you have a family member who's affected with a rare genetic disorder, particularly an autosomal dominant disorder with incomplete penetrance or delayed onset, you may be vitally concerned about which allele(s) you carry. Persons currently unaffected with certain disorders can seek genetic testing to learn if they're carriers. Most tests involve a blood sample, but some are done with a simple cheek swab or saliva sample.

With the advent of many new technologies (many of which grew out of the Human Genome Project, which we explain in Chapter 8), genetic testing is easier and cheaper than ever. Genetic testing and genetic counseling often go hand in hand. The genetic counselor works to identify which disorders occur in the family (or could occur in the family) and testing then examines the DNA directly to determine whether the disease-causing gene mutation or a chromosome problem is present. Genetic testing has many ethical implications, as we cover in Chapter 20.

The types of genetic tests available are quite varied. There are tests to look specifically at a person's DNA sequence, tests to look at the number of chromosomes in each cell, and tests to look for changes in chromosome structure. There are tests that look for any change present and tests that look specifically for a change known to be present in the family. The types of genetic testing that are available include:

- **Diagnostic testing:** Genetic testing performed in an affected individual in order to establish a specific diagnosis.
- **Prenatal diagnosis:** Genetic testing performed during pregnancy in order to diagnose the baby with a genetic condition before birth.
- **Carrier testing:** Genetic testing performed in a healthy individual in order determine whether they are a carrier who have a chance of having a child with recessive genetic condition.
- **Predictive and susceptibility testing:** Genetic testing performed in a healthy individual in order to determine whether they will develop a later-onset condition (predictive) or whether they have an increased chance of developing a particular disease (susceptibility).
- **Preimplantation genetic diagnosis (PGD):** Genetic testing performed in an early embryo created by *in vitro* fertilization in order to identify only those embryos *without* a specific genetic change, which would then be used for implantation.
- **Pharmacogenetic testing:** Genetic testing performed in order to determine what is the most appropriate medication for the patient and what dose would be best.

Chapters 17 and 22 cover ancestry and direct-to-consumer genetic testing, both of which are hot topics in the field of genetics.

Diagnostic testing

Genetic testing is frequently performed in order to establish a diagnosis. A genetic disorder may be suspected in individuals (frequently infants and children) with developmental delays, intellectual disabilities, unusual physical features, and/or birth defects. When a person comes to a genetics clinic with features of a genetic syndrome, the physician typically performs a variety of tests and evaluations to determine what the person has. The patient may have medical tests like X-rays and other imaging tests, blood tests, and other laboratory studies. The results of these tests, findings from a physical examination, and a review of the individual's personal and family history may then be used to determine what genetic testing may be most appropriate for the individual. Tests that are frequently used include chromosome studies (described in Chapter 13) and gene tests for specific genetic disorders.

Prenatal diagnosis

Prenatal testing is performed during pregnancy. It is used to determine whether a fetus carries a gene mutation or a chromosome problem. For a definitive diagnosis of a genetic disorder, testing requires cells from the affected person. Two common prenatal procedures used to obtain fetal tissue for testing are *chorionic villus sampling* (CVS) and *amniocentesis.* Both procedures require ultrasound to guide the instruments used to obtain the samples. In addition, both procedures are invasive and are associated with a slight increase in the risk of miscarriage. The key differences between the two procedures include:

- **CVS** is usually done late in the first trimester of pregnancy (weeks 10 to 12). A catheter is inserted vaginally and guided to the outer layer of the placenta, called the *chorion.* Gentle suction is used to collect a small sample of chorionic tissue. The placental tissue arises from the fetus, not the mother, so the collected cells should give an accurate picture of the fetus's chromosome number and genetic profile.
- **Amniocentesis** is usually done early in the second trimester of pregnancy (weeks 15 and beyond). Amniocentesis is used to obtain a sample of the amniotic fluid that surrounds the growing fetus, because amniotic fluid contains fetal cells (skin cells that have sloughed off) that can be examined for prenatal testing. The fluid is drawn directly from the amniotic sac using a needle inserted through the abdomen. Because fetal cells in the fluid are at a very low concentration, the cells must be grown in a lab to provide enough tissue for testing.

After a prenatal sample is obtained, it can be used to look at the baby's chromosomes (see Chapter 13 for information about the chromosome tests that can be used). A prenatal sample can also be used to test the baby for a gene mutation known to be running in the family, or for the cause of certain genetic disorders that may be suspected based on ultrasound findings. Prenatal testing is designed to allow time for couples to make pregnancy-related decisions and to prepare for what comes after delivery of an infant with special medical concerns.

Carrier testing

Carriers are individuals who are heterozygous for a mutation (have one allele with a mutation) known to cause a recessive condition. The individuals are unaffected because both copies of the gene need to have a mutation in order to have the condition. The presence of the second, normal copy is enough for the person to be unaffected. However, if both members of a couple are carriers for the same disorder, they will have a 25 percent chance of having an affected child with each pregnancy (chance that the mom will pass on the mutation = 0.5; chance that the dad will pass on the mutation = 0.5; therefore, 0.5 × 0.5 = 0.25, or 25 percent). Therefore, carrier testing is used to identify carriers who have an increased chance of having a child with the condition in question. Often, what testing a person is offered depends on the ethnicity of the individual, since certain conditions are more frequent in certain populations. However, it is now common practice to offer a wider range of tests, such as carrier tests for more than 100 conditions at a time. If both members of a couple are found to be carriers for the same condition, they can then be offered prenatal diagnosis for any pregnancies they may have.

Predictive and susceptibility testing

Genetic tests are also used to predict the development of certain conditions later in life. These tests are generally performed in individuals who have a family history of a genetic disorder that develops after birth, often in adulthood. If the disease-causing mutation is first identified in the family member who has the condition, then at-risk family members can be tested before they ever show symptoms of the condition.

Predictive testing is testing performed for conditions that *will* develop eventually in any person carrying the disease-causing mutation(s). For example, presymptomatic testing is often performed for Huntington disease, a progressive neurological condition that is inherited in an autosomal dominant manner (described in more detail in Chapter 4).

Susceptibility testing is testing performed for conditions that lead to an *increased risk* for disease, such as cancer. Susceptibility testing is performed for hereditary cancer syndromes, such as hereditary breast and ovarian cancer caused by mutations in the breast cancer genes *BRCA1* or *BRCA2*. As explained in Chapter 14, individuals who carry disease-causing mutations in one of these genes have a significantly increased risk for developing breast and ovarian cancer (and other malignancies). They have an increased susceptibility, but it does not mean that they will develop cancer for certain. Predictive and susceptibility testing offer individuals information that they can use for planning their futures, whether it be deciding whether or not to have children, or deciding on early disease surveillance and prophylactic treatment (treatment before any symptoms arise, such as prophylactic surgery for those at risk of cancer).

Preimplantation genetic diagnosis

Preimplantation genetic diagnosis (PGD, also referred to as preimplantation genetic testing or preimplantation genetic screening) involves testing an embryo very early in development. PGD is used alongside *in vitro* fertilization, in order to identify embryos with specific genetic conditions for which the children of the couple are at risk. The idea behind PGD is that only those embryos that do not carry the specific genetic change would be implanted into the mother's uterus, thereby decreasing the chance of having an affected child and increasing the chances of a successful pregnancy.

Several types of testing are available that can be used during PGD. For couples who have had genetic testing and are known to have an increased risk of a specific genetic disorder (such as cystic fibrosis or Tay-Sachs disease), the embryo can be tested for the specific gene mutation(s) carried by the parent(s). For couples in which one is known to carry a chromosome rearrangement (such as a balanced translocation; see Chapter 13), the embryo can be tested for a structural change in the chromosomes. Finally, for couples who have an increased risk of having a child with aneuploidy (too many or too few chromosomes; see Chapter 13), the embryo can be tested for changes in chromosome number.

In order to perform PGD, embryos are first created using *in vitro* fertilization (IVF). For IVF, sperm from the father is used to fertilize eggs from the mother. The embryos are created in the laboratory and grown for five to six days. At that point, there are a sufficient number of cells for a biopsy to be performed. The embryo is then frozen while the genetic testing is performed using the cells that were removed. After the results of testing are obtained, the couple and medical team review the results and decide if there are any embryos they would like to implant. Embryos with normal genetic testing results and with a typical physical appearance will have the highest chance of implanting and resulting in a healthy baby.

Pharmacogenetic testing

Pharmacogenetic testing involves examining genes known to be involved in drug metabolism. Because of differences in the sequence of some of the genes, different people respond differently to certain medications. Genes may be responsible for why different doses are needed for different people, and for why some people experience side effects while others do not. Consequently, pharmacogenetic testing may performed before a medication is prescribed, in order to determine whether it is the most appropriate medication and what dose would be best. Pharmacogenetic tests are available for a wide variety of drugs, including pain medications, psychiatric medications, and chemotherapeutic agents (used to treat cancer). See Chapter 16 to learn more about pharmacogenetics.

IN THIS CHAPTER

» Delivering healthy genes to treat or cure disease

» Discovering precision medicine

» Examining the benefits of pharmacogenetics

Chapter 16
Treating Genetic Disorders and Using Genetics to Tailor Treatment

The completion of the Human Genome Project (described in Chapter 8), along with the sequencing of nonhuman genomes, spawned an incredible revolution in the understanding of genetics. Simultaneously, geneticists raced to develop medicines to treat and cure diseases caused by genes gone awry.

Gene therapy, treatment that gets at the direct cause of genetic disorders, is sometimes touted as the magic bullet, the cure-all for inherited diseases and cancer. Gene therapy may even provide a way to block the genes of pathogens such as viruses, providing reliable treatments for illnesses that currently have none. Unfortunately, the shining promise of gene therapy has been hampered by a host of challenges, including finding the right way to supply the medicine to patients without causing worse problems than the ones being treated.

The advances in our understanding of genetics as it relates to disease and its treatment have also led to great advances in *precision medicine*. Precision medicine, also referred to as personalized medicine, involves tailoring medical treatment to each individual based on their genetics.

In this chapter, we examine the progress and perils of gene therapy. In addition, we discuss precision medicine and how it is being used in the management of a wide variety of conditions.

Alleviating Genetic Disease through Gene Therapy

Take a glance through Part 3 of this book for proof that your health and genetics are inextricably linked. Mutations cause disorders that are passed from generation to generation, and mutations acquired during your lifetime can have unwanted consequences such as cancer. Your own genes aren't the only ones that cause complications — the genes carried by bacteria, parasites, and viruses lend a hand in spreading disease and dismay worldwide.

So wouldn't it be great if you could just turn off those pesky bad genes? Just think: A mutation causes a loss of function in a tumor suppressor gene (described in Chapter 14), and you get a shot to turn that gene back on. A virus giving you trouble? Just take a pill that blocks the function of viral genes.

Some geneticists see the implementation of these genetic solutions to health problems as only a matter of time. Therefore, the development of gene therapy has focused on several courses of action:

- Supplying genes to provide desired functions that have been lost or are missing
- Blocking genes from producing unwanted products
- Delivering a new gene that can make a helpful protein to treat a specific disease

Inserting Healthy Genes into the Picture

Finding the right delivery system is a necessary step in mastering gene therapy, but to nab genes and put them to work as therapists, geneticists must also find the right ones. Prior to the sequencing of the human genome, finding the correct disease-causing gene was a major obstacle in the road to implementing gene therapy. Imagine you're handed a person's photograph and told to find them in New York City — no name, no address, no phone number. The task of finding that person includes figuring out their identity (maybe by finding out who their friends are), figuring out what they do for a living, narrowing your search to the borough they live in, and identifying their street, block, and finally, their address. This wild-goose chase is almost exactly like the gargantuan task of finding genes.

Your DNA has roughly 22,000 genes tucked away among about 3 billion base pairs of DNA. (See Chapter 5 to find out how DNA is sized up in base pairs.) Because most genes are pretty small, relatively speaking (often less than 5,000 base pairs long), finding just one gene in the midst of all the genetic clutter may sound like a nearly impossible task. Until recently, the main tool geneticists had in the search for genes was the observation of inheritance patterns (such as those shown in Chapter 15) and the subsequent comparisons of how different genetic markers were inherited. Geneticists use this method, called *linkage analysis*, to construct gene maps (locate where genes are on the chromosomes). Also, candidate genes were examined to see if disease-causing mutation(s) could be identified. This required large families with multiple affected individuals. It was also very labor-intensive, time-consuming, and costly. With the advent of automated DNA sequencing and the completion of the Human Genome Project (described in Chapter 8), however, the search for names and addresses of genes has reached a whole new level.

Currently, the gene-hunting safari depends on access to the reference genome sequence that has been completed. It also depends on vast computer databases that can be used to search DNA sequences, protein information, and information on disease phenotypes. Available databases, which the scientific community can easily access, also allow investigators to search professional journals to keep up with new discoveries by other scientists. Researchers are also constantly adding new pieces of the puzzle — such as newly identified proteins, newly discovered mutations, and novel phenotypes — to storehouses of data.

TIP

You can take a peek into the genetic data warehouse by visiting the Online Mendelian Inheritance in Man website (`https://ncbi.nlm.nih.gov/omim`) or the main website for the National Center for Biotechnology Information (NCBI; `https://ncbi.nlm.nih.gov`). From there, you can explore everything from DNA to protein data compiled by scientists from around the world.

REMEMBER

Recombinant DNA technology is the catchall phrase that covers most of the methods geneticists use to examine DNA in the lab. The word *recombinant* is used because DNA from the organism being studied is often popped into a virus or bacteria (that is, it's recombined with DNA from a different source) to allow further study. Scientists also use recombinant DNA for a vast number of other applications, including creating genetically engineered organisms and cloning. In the case of gene therapy, illustrated in Figure 16-1, molecular genetic and recombinant DNA technologies are used to:

- Identify the gene (or genes) involved in a particular disorder or disease.
- Cut the desired gene out of the surrounding DNA.
- Pop the gene into a vector (delivery vehicle) for transfer into the cells where treatment is needed.

For more information on genetically engineered organisms, see Chapter 19. For a description of how cloning works, see the bonus chapter on this book's web page on `www.dummies.com`.

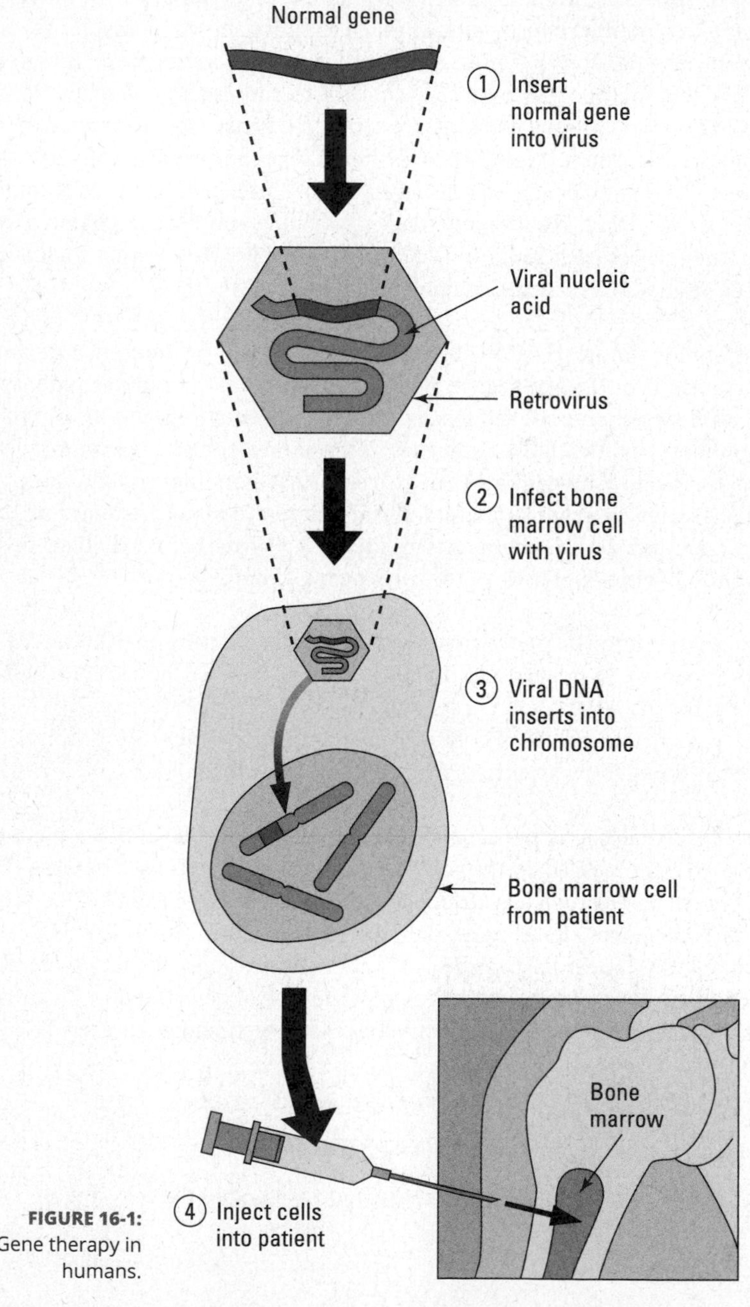

FIGURE 16-1: Gene therapy in humans.

Finding Vehicles to Get Genes to Work

The first step in successful gene therapy is designing the right delivery system to introduce a new gene or shut down an unwanted one. The delivery system for gene therapy is called a *vector.* A perfect vector:

- Must be harmless so that the recipient's immune system doesn't reject or fight the vector.
- Must be easy to manufacture in large quantities. Just one treatment may require over 10 billion copies of the vector, because you need one delivery vehicle for every cell in the affected organ.
- Must be targeted for a specific tissue. Gene expression is tissue-specific (see Chapter 11 for details), so the vector needs to be tissue-specific as well.
- Must be capable of integrating its genetic payload into each cell of the target organ so that new copies of each cell generated later by mitosis contain the gene therapy payload.

REMEMBER

Currently, viruses are the favored vector. Most gene therapies aim to put a new gene into the patient's genome (as shown in Figure 16-1), and this gene-sharing action is almost precisely what viruses do naturally.

When a virus latches onto a cell that isn't protected from the virus, the virus hijacks all that cell's activities for the sole purpose of making more viruses. Viruses reproduce this way because they have no moving parts of their own to accomplish reproduction. Part of the virus's attack strategy involves integrating virus DNA into the host genome in order to execute viral gene expression. The problem is that when a virus is good at attacking a cell, it causes an infection that the patient's immune system fights. So the trick to using a virus as a vector is taming it.

Making a virus harmless in order to use it as a vector usually involves deleting most of its genes. These deletions effectively rob the virus of almost all its DNA, leaving only a few bits. These remaining pieces are primarily the parts the virus normally uses to get its DNA into the host. Using DNA manipulation techniques like those we describe in the "Inserting Healthy Genes into the Picture" section of this chapter, the scientist inserts a healthy gene sequence into the virus to replace the deleted parts of the viral genome. But a helper is needed to move the payload from the virus to the recipient cell, so the scientist sets up another virus particle with some of the deleted genes from the vector. This second virus, called a *helper,* makes sure that the vector DNA replicates properly.

Geneticists conducting gene therapy have several viruses to choose from as possible delivery vehicles (vectors). These viruses fall into one of two classes (see Table 16-1):

- Those that integrate their DNA directly into the host's genome.
- Those that climb into the cell nucleus to become permanent but separate residents (called *episomes*).

Within these two categories, three types of viruses — oncoretroviruses, lentiviruses, and adenoviruses — are popular choices for gene therapy.

TABLE 16-1 Common Viral Vectors for Gene Therapy

Vector Source	Integrating or Episome	Advantage(s)	Disadvantage(s)
Oncoretrovirus	Integrating	Potential for an immune response is low	Can only treat cells that are actively dividing; can cause cancer in some cases
Lentivirus	Integrating	Potential for an immune response is low; can be used to treat cells that are not dividing	Can integrate into a gene causing a loss-of-function mutation
Adenovirus	Episome	Can be used to treat cells that are not dividing; tend to infect most treated cells	Potential for an immune response is higher; episome may not be replicated and transmitted to all daughter cells

Viruses that join right in

Two popular viruses for gene therapy integrate their DNA directly into the host's genome. *Oncoretroviruses* and *lentiviruses* are retroviruses that transfer their genes into the host genome; when the retrovirus genes are in place, they're replicated right along with the other host DNA. Retroviruses use RNA instead of DNA to code their genes and use a process called *reverse transcription* to convert their RNA into DNA, which is then inserted into a host cell's genome.

Oncoretroviruses, the first vectors developed for gene therapy, get their name from *oncogenes*, which turn the cell cycle on permanently — one of the precursors to development of full-blown cancer (described in detail in Chapter 14). Most of the oncoretrovirus vectors in use for gene therapy trace their history back to a virus that causes leukemia in monkeys (it's called *Moloney murine leukemia virus*, or MLV). MLV has proven an effective vector, but it's not without problems; MLV's propensity to cause cancer has been difficult to keep in check. Oncoretroviruses work well as vectors only if they're used to treat cells that are actively dividing.

Lentiviruses (a subtype of retroviruses), on the other hand, can be used to treat cells that aren't dividing. You're probably already familiar with a famous lentivirus: HIV. Vectors for gene therapy were developed directly from the HIV virus itself. Although the gutted virus vectors contain only 5 percent of their original DNA, rendering them harmless, lentiviruses have the potential to regain the deleted genes if they come in contact with untamed HIV virus particles (that is, the ones that have infected people with AIDS). Lentiviruses are also a bit dicey because they tend to put genes right in the middle of host genes, leading to loss-of-function mutations (see Chapter 12 to review the various types of mutations).

Viruses that are a little standoffish

Adenoviruses are excellent vectors because they pop their genes into cells regardless of whether cell division is occurring. In gene therapy, adenoviruses have been both promising and problematic. On the one hand, they're really good at getting into host cells. On the other hand, they tend to excite a strong immune response — the patient's body senses the virus as a foreign particle and fights it. To combat the immune reaction, researchers have worked to delete the genes that make adenoviruses easy for the host to recognize.

REMEMBER

Adenoviruses don't put their DNA directly into the host genome. Instead, they exist separately as episomes, so they aren't as likely as lentiviruses to cause mutations. The drawback is that the episomes aren't always replicated and passed on to daughter cells when the host cell divides. Nonetheless, researchers have used adenovirus vectors with notable success — and failure (described in the following section).

Progress on the Gene Therapy Front

As the Human Genome Project started fulfilling the dreams of geneticists worldwide, realizing gene therapy's promises seemed very much in reach. In fact, the first trials conducted in 1990 were a resounding success.

In those first attempts at gene therapy, two patients suffering from the same immunodeficiency disorder received infusions of cells carrying genes coding for their missing enzymes. The disorder was a form of severe combined immunodeficiency (SCID) that results from the loss of one enzyme: adenosine deaminase (ADA). SCID is so severe that affected persons must live in completely sterilized environments with no contact with the outside world, because even the slightest infection is likely to prove deadly. Because only one gene is involved, SCID is a natural candidate for treatment with gene therapy. Retroviruses armed with a healthy *ADA* gene were infused into the two affected children with dramatic results: Both children were essentially cured of the disease and now lead normal lives.

Implementation of other gene therapies has met with mixed results. At least 17 children have been treated for an X-linked version of SCID. These children also received a retrovirus loaded with a healthy gene and were apparently cured. However, four of the children have since been diagnosed with leukemia (a cancer of the blood). The virus that delivered the gene also plopped its DNA right into a proto-oncogene, switching it on.

HISTORICAL STUFF

The most famous failure of gene therapy occurred in 1999, when 18-year-old Jesse Gelsinger volunteered for a study aimed at curing a genetic disorder called ornithine transcarbamylase (OTC) deficiency. With this disorder, Jesse occasionally suffered a huge buildup of ammonia in his body because his liver lacked enough of the OTC enzyme to process all the nitrogen waste products in his blood. Jesse's disease was controlled medically — with drugs and diet — but other affected children often die of the disease. Researchers used an adenovirus to deliver a normal *OTC* gene directly into Jesse's liver. (See the earlier section "Viruses that are a little standoffish" for the scoop on adenoviruses.) The virus escaped into Jesse's bloodstream and accumulated in his other organs. His body went into high gear to fight what seemed like a massive infection, and four days after receiving the treatment that was meant to cure him, Jesse died. Oddly, another volunteer in the same experimental trial received the same dose of virus that Jesse did and suffered no ill effects at all.

More recently, gene therapy success has been achieved in monkeys. In 2009, researchers announced a successful trial of gene therapy for colorblindness in monkeys. The monkeys, which had a form of red-green colorblindness similar to the sort that humans get, were given viruses bearing a functional form of the missing gene. A few weeks later, the monkeys were able to see colors that they were unable to see prior to the therapy. Also in 2009, scientists reported that by adding three genes to the brains of monkeys suffering from a form of Parkinson disease, the animals showed a decrease in the involuntary movements that accompany the disease.

Though recent results seem optimistic, the struggle to find appropriate vectors continues. The future of gene therapy is complicated by discoveries that most genetic disorders involve several genes on different chromosomes. Not only that, but many different genes can cause a given disease (diabetes, for example, is associated with genes on at least five different chromosomes), making it difficult to know which gene to target. Finally, some genes are so large, such as the gene for Duchenne muscular dystrophy, that typical vectors can't carry them.

Utilizing Genetic Information for Precision Medicine

Precision medicine (also referred to as *personalized medicine*) is an approach to managing a person's health based on their genetics, lifestyle, and environment. While the term personalized medicine suggests that treatments could be designed specifically for each patient, the term precision medicine may more accurately describe this type of approach. This practice does not involve the development of new medication for each patient. Instead, the management of patients in general is more tailored to the individual. For example, a medication prescribed for one person may not be given to another because of differences in their genetics and other aspects of their life. In this section, we are going to introduce you to emerging trends in precision medicine. However, while lifestyle and environment do play key roles in this field, we are going to focus specifically on the role of genetics.

Pharmacogenetics (and pharmacogenomics)

Throughout the history of medicine, treatments have been based solely on a person's diagnosis. You may be prescribed a certain medication based on your symptoms — a medication that works for most people. Then, if the medication doesn't work or you start experiencing significant side effects, your doctor may switch you to another medication. And if that one doesn't work for you, you're back to square one. With precision medicine, the idea is that we can utilize our genetic information in order to determine which medication would work best for us, what dose we will need for it to be effective, and the likelihood that we will experience side effects, without all the trial and error.

Pharmacogenetics is a key part of precision medicine. Pharmacogenetics is the study of how changes in genes affect the metabolism of medications. Pharmacogenetic testing entails testing genes involved in drug metabolism. It has been shown that variants in certain genes can lead to changes in how a person's body breaks down certain medications, at least partly explaining the differences in response that different people have. The term pharmacogenomics is often used interchangeably with pharmacogenetics. However, *pharmacogenetics* really deals with specific gene-changing interactions, while *pharmacogenomics* deals more with the overall influence of the genome on response to medications. Pharmacogenomics may also be used to describe the effect of differences in gene *expression* on the safety and effectiveness of medical therapies (as opposed to differences in gene *sequence*).

HISTORICAL STUFF

In 1957, Arno Motulsky first reported on the heritability of adverse reactions to specific medications. Motulsky noted that serious side effects resulting from the use of primaquine (which is used to treat malaria) and of suxamethonium chloride (a muscle relaxant) appeared to run in families. In addition, he reported that these poor reactions appeared to be the result of deficits in the activity of specific enzymes involved in the breakdown of those medications.

Cytochrome P450 and drug metabolism

Cytochrome P450 enzymes are a large family of proteins that function in the breakdown of a wide variety of chemicals. Cytochrome P450 enzymes function primarily in cells in the liver, but they are also found in other cells throughout the body. There are numerous cytochrome P450 genes, each of which is named with CYP (short for cytochrome P450, followed by a number indicating the cytochrome P450 family number, a letter indicating the subfamily, and another number indicating the member within the subfamily. For example, *CYP2D6* is the gene that encodes the cytochrome P450 family 2, subfamily D, member 6 protein. Currently, cytochrome P450 proteins account for up to 80 percent of enzymes that function in the breakdown of medications.

An important example of the effects of gene variants on drug metabolism is illustrated by the role of CYP2D6 in the breakdown of codeine, a commonly prescribed opioid pain medication. Once in the body, codeine is broken down into its active form morphine. Morphine is what is actually responsible for the pain relief experienced by those taking codeine. The majority of the general population metabolizes codeine in the expected way (referred to as *extensive metabolizers*) (see Table 16-2). In these individuals, a standard dose should be safe and effective. Approximately 1 to 2 percent of individuals are what are referred to as *ultrarapid metabolizers*. The fast metabolism of codeine in these individuals leads to higher levels of morphine in a short amount of time, and an increased risk for drug toxicity. The remaining portion of the general population are considered either *poor metabolizers* (who may experience insufficient pain relief with a standard dose and need an alternative medication) or *intermediate metabolizers* (who may need a higher dose for adequate pain relief).

TABLE 16-2 **Metabolism of Medications by the CYP2D6 Enzyme**

	Ultrarapid Metabolizer	Extensive Metabolizer	Intermediate Metabolizer	Poor Metabolizer
Effect on drug metabolism	Increased enzyme activity leading to the rapid breakdown of the medication into its active form.	Normal enzyme activity and normal breakdown of the medication into its active form.	Slightly decreased enzyme activity and slightly slower breakdown of the medication into its active form.	Significantly decreased enzyme activity and impaired breakdown of the medication into its active form.
Effect on the patient	Responds to lower dose; increased risk of toxicity and adverse side effects.	Standard dose is typically safe and effective.	Standard dose may be ineffective; higher dose may be required.	Standard dose is ineffective; may require higher dose or alternative medication.

TECHNICAL STUFF

A number of alleles (different version of a gene due to changes in DNA sequence) have been reported in the *CYP2D6* gene. The combinations of alleles a person has (their genotype) determines how they will metabolize certain medications. Some combinations of alleles result in the person being an extensive metabolizer of medications that are broken down by this enzyme (including codeine). Other combinations result in the person being either an ultrarapid metabolizer or a poor metabolizer of drugs processed by CYP2D6. Genetic testing for variations in the *CYP2D6* gene is available and can tell what type of metabolizer you are based on the two alleles you carry (one allele inherited from your mother and one allele inherited from your father). If you carry two alleles associated with normal drug metabolism, you will be an extensive metabolizer and should be prescribed standard doses of the relevant medication. If you carry two variant alleles that are known to result in significantly *increased* CYP2D6 activity, you will be an ultrarapid metabolizer and should be prescribed either lower doses (and be monitored closely for adverse reactions) or an alternative medication.

The potential consequences of not knowing how a person metabolizes certain medications is illustrated in a case reported in 2006. In this case, a woman who had just given birth was prescribed codeine for pain relief. The new mother was breastfeeding her child. Unbeknownst to her healthcare providers, she was an ultrarapid metabolizer of codeine. Consequently, she had high levels of morphine in her system, which were passed on to her infant through her breast milk and resulted in the death of the baby due to morphine toxicity. Unfortunately, this is not the only case of death due to a morphine overdose in the infant of a breastfeeding mother being treated with an opioid pain medication.

Decreasing the risk of side effects of treatment

Another example of how pharmacogenetics is used to minimize the chance of adverse side effects is in the use of abacavir for the treatment of HIV (human immunodeficiency virus) infection. While abacavir is helpful in many patients with HIV, it unfortunately results in severe and potentially life-threatening side effects in some individuals, including difficulty breathing, vomiting, fever, and rash. After comparing sequence variants between those who do experience these side effects with those who do not, it was discovered that individuals who carry a specific variant allele in the human leukocyte antigen-B (*HLA-B*) gene have a significantly increased risk for problems when taking abacavir. The protein encoded by the *HLA-B* gene functions in the immune system, helping it to recognize and respond to foreign pathogens. Now, the FDA label for abacavir recommends that all patients that are to be treated with this medication be tested for the variant in the *HLA-B* gene, and that carriers of the variant should consider an alternative medication.

Increasing the effectiveness of treatment

Tailoring treatments to maximize efficiency is illustrated very well in the field of cancer therapy. One example is the development and use of the medication trastuzumab for breast cancers that overexpress a particular receptor on the surface of tumor cells. Trastuzumab was approved by the FDA in 1998 for the treatment of metastatic breast cancer that tested positive for the human epidermal growth factor receptor 2 (HER2). This would be a prime example of pharmacogenomics in practice.

The tumors of women with metastatic breast cancer are tested for the overexpression of the *HER2* gene by looking for the HER2 protein on tumor cells or looking for extra copies of the gene itself. In some tumors, the *HER2* gene undergoes gene amplification (described in Chapter 14), which results in multiple copies of the gene and increased expression. Approximately 20 percent of breast tumors have increased levels of HER2, which contributes to increased cell division and growth of the tumor.

TECHNICAL STUFF

To perform the test, a sample of the tumor is obtained by biopsy or surgery. This sample can then be used to look for extra copies of the *HER2* gene by examining the chromosomes with fluorescence in situ hybridization (which is explained in Chapter 13). Alternatively, the sample could be used to look for excess HER2 protein on the actual tumor cells using immunohistochemistry (IHC). With IHC, slides of tumor cells are stained with a labeled antibody designed to identify cells with the HER2 protein. The label on the antibody allows one to see exactly how much protein is present in the tumor sample and compare it to what is found in normal breast tissue.

Patients found to overexpress *HER2* may then be treated with trastuzumab. Trastuzumab is considered a targeted therapy; it is an antibody to the HER2 protein. In normal cells, a growth factor (specifically, epidermal growth factor) binds to the HER2 receptor, ultimately resulting in the activation of certain genes and the stimulation of cell division (see Chapter 11 to learn about turning genes on or off). In tumor cells that have too much HER2, the signal to turn genes on is always active, resulting in uncontrolled cell proliferation. Trastuzumab can bind to the receptor, blocking the epidermal growth factor and preventing the tumor cells from dividing. In addition, this antibody can help the body's own immune system attack and destroy cancer cells. Patients that test negative for HER2 will not respond to this therapy and may therefore be treated with a different medicine or therapeutic approach.

4 Genetics and Your World

IN THIS PART . . .

See how you can trace human history using genetics and how human activities affect the genetics of a population.

Understand how genetics is used to solve crimes and determine family relationships.

Find out how genetic technologies can be used to move genes from one organism to another or to change a DNA sequence in an organism.

Discover how genetics knowledge can provide more choices, complicate decision-making, and raise a variety of ethical issues.

IN THIS CHAPTER

» Relating the genetics of individuals to the genetics of groups

» Describing genetic diversity

» Understanding the genetics of evolution

Chapter 17
Tracing Human History and the Future of the Planet

It's impossible to overestimate the influence of genetics on our planet. Every living thing depends on DNA for its life, and all living things, including humans, share DNA sequences. The amazing similarities between your DNA and the DNA of other living things suggest that all living things trace their history back to a single source. In a very real sense, all creatures great and small are related somehow.

You can examine the genetic underpinnings of life in all sorts of ways. One powerful method for understanding the patterns hidden in your DNA is to compare the DNA of many individuals as a group. This specialty, called *population genetics,* is a powerful tool. Geneticists use this tool to study not only human populations but also animal populations to understand how to protect endangered species, for example. By comparing DNA sequences of various species, scientists also infer how natural selection acts to create evolutionary change. In this chapter, you find out how scientists analyze genetics of populations and species to understand where we came from and where we're going.

Genetic Variation Is Everywhere

The next time you find yourself channel surfing on the TV, pause a moment on one of the channels devoted to science or animals. The diversity of life on earth is truly amazing. In fact, scientists still haven't discovered all the species living on our planet; the vast rain forests of South America, the deep-sea vents of the ocean, and even volcanoes hold undiscovered species.

The interconnectedness of all living things, from a scientific perspective, can't be overstated. The sum total of all the life on earth is referred to as *biodiversity.* Biodiversity is self-sustaining and is life itself. Together, the living things of this planet provide oxygen for you (and everything else) to breathe, carbon dioxide to keep plants alive and regulate the temperature and weather, rainwater for you and your food supply, nutrient cycling to nourish every single creature on earth, and countless other functions.

Underlying the world's biodiversity is *genetic variation.* When you look around at the people you know, you see enormous variation in height, hair and eye color, skin tone, body shape — you name it. That phenotypic (physical) variation implies that each person differs genetically, too. Likewise, the individuals in all populations of other sexually reproducing organisms vary in phenotype (physical traits) and genotype (genetic information) as well. Scientists describe the genetic variation in *populations* (defined as groups of interbreeding organisms that exist together in both time and space) in two ways:

- **Allele frequencies:** How often do various alleles (alternate versions of a particular section of DNA) show up in a population?
- **Genotype frequencies:** What proportion of a population has a certain genotype (the two specific alleles at a given locus)?

Allele frequencies and genotype frequencies are both ways of measuring the contents of the gene pool. The *gene pool* refers to all the possible alleles of all the various genes that, collectively, all the individuals of any particular species have. Genes get passed around in the form of alleles that are carried from parent to child as the result of sexual reproduction. (Of course, genes can be passed around without sex — viruses leave their genes all over the place.)

Allele frequencies

Alleles are various versions of a particular section of DNA (like alleles for eye color; see Chapter 3 for a review of terms used in genetics). Most genes have many different alleles. Geneticists use DNA sequencing (which we explain in Chapter 8) to examine genes and determine how many alleles may exist. To count alleles, they examine the DNA of many different individuals and look for differences among base pairs — the As, Gs, Ts, and Cs — that comprise DNA. For the purposes of population genetics, scientists also look for individual differences in *non-coding DNA* (previously referred to as *junk DNA*).

Some alleles are very common, and others are rare. To identify and describe patterns of commonness and rarity, population geneticists calculate allele frequencies. What geneticists want to know is what proportion of a population has a particular allele. This information can be vitally important for human health. For example, geneticists have discovered that some people carry an allele that makes them immune to HIV infection, the virus that causes AIDS.

TIP

An allele's frequency — how often the allele shows up in a population — is pretty easy to calculate: Simply divide the number of copies of a particular allele by the number of copies of all the alleles represented in the population for that particular gene.

In a two-allele system (as illustrated by Mendel's peas, covered in Chapter 3), a lowercase letter p is usually used to represent one allele frequency, and q is used for the other. In this two-allele system, $p + q$ must equal 1. If you know the number of *homozygotes* (individuals having two identical copies of a particular allele) and *heterozygotes* (individuals having one copy of the allele), you can set the problem up using these two equations: $p + q = 1$ or $q = 1 - p$.

For example, say you want to know the frequency for the dominant allele (R) for round peas in a population of plants like the ones Mendel studied. You know that there are 60 RR plants, 50 Rr plants, and 20 rr plants. To determine the allele frequency for R (referred to as p), you multiply the number of RR plants by 2 (because each plant has two R alleles) and add that value to the number of Rr plants: $60 \times 2 + 50 = 120 + 50 = 170$. Divide the sum, 170, by two times the total plants in the population (because each plant has two alleles), or $2(60 + 50 + 20) = 260$. The result is 0.65, meaning that 65 percent of the alleles in the population of peas have the allele R. To get the frequency of the allele r (that is, q), simply subtract 0.65 from 1 (which is 0.35, or 35 percent of alleles).

REMEMBER

The situation gets pretty complicated, mathematically speaking, when several alleles are present, but the take-home message of allele frequency is still the same: All allele frequencies are the proportion of the population carrying at least one copy of the allele. And all the allele frequencies in a given population must add to 1 (which can be expressed as 100 percent, if you prefer).

Genotype frequencies

Most organisms have two copies of every gene (that is, they're diploid). Because the two copies don't necessarily have to be identical, individuals can be either heterozygous or homozygous for any given allele. Like alleles, genotypes can vary in frequency. Genotypic frequencies tell you what proportion of individuals in a population are homozygous and, by extension, what proportion are heterozygous. Depending on how many alleles are present in a population, many different genotypes can exist. Regardless, the sum total of all the genotype frequencies for a particular locus (location on a particular chromosome) must equal 1 (just like with allele frequency).

To calculate a genotype frequency, you need to know the total number of individuals who have a particular genotype. For example, suppose you're dealing with a two-allele locus in a population of 100 individuals. A total of 25 individuals are homozygous recessive (aa), 30 are heterozygous (Aa), and 45 are homozygous dominant (AA). The frequency of the three genotypes (assuming there are only two alleles, A and a) is shown in the following, where the total population is represented by *N*.

$$\text{Frequency of AA} = \frac{\text{Number of AA individuals}}{\text{N}}$$

$$\text{Frequency of Aa} = \frac{\text{Number of Aa individuals}}{\text{N}}$$

$$\text{Frequency of aa} = \frac{\text{Number of aa individuals}}{\text{N}}$$

REMEMBER

Allele frequency and genotype frequency are very closely related concepts because genotypes are derived from combinations of alleles. It's easy to see from Mendelian inheritance (described in Chapter 3) and pedigree analysis (described in Chapter 15) that if an allele is very common, homozygosity is going to be high. It turns out that the relationship between allele frequency and homozygosity is quite predictable. Most of the time, you can use allele frequencies to estimate genotypic frequencies using a genetic relationship called the *Hardy-Weinberg law* of population genetics, which we explain in the next section.

Breaking Down the Hardy-Weinberg Law of Population Genetics

HISTORICAL STUFF

Godfrey Hardy and Wilhelm Weinberg never met, yet their names are forever linked in the annals of genetics. In 1908, both men, completely independent of each other, came up with the equation that describes how genotypic frequencies are related to allele frequencies. Their set of simple and elegant equations accurately describes the genetics of populations for most organisms. What Hardy and Weinberg realized was that in a two-allele system, all things being equal, homozygosity and heterozygosity balance out. Figure 17-1 shows how the *Hardy-Weinberg equilibrium*, as this genetic balancing act is known, looks in a graph.

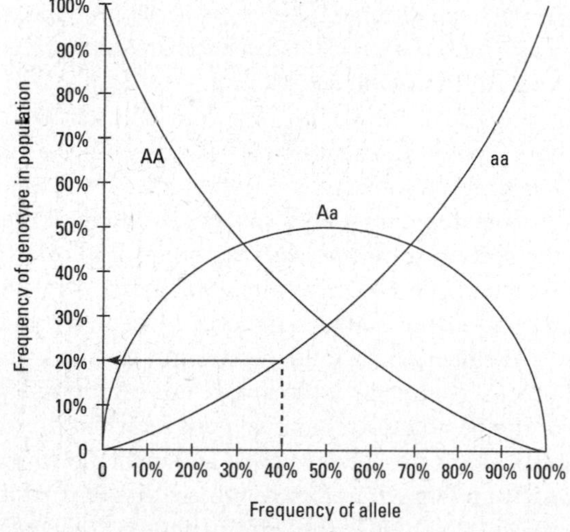

FIGURE 17-1: The Hardy-Weinberg graph describes the relationship between allele and genotype frequencies.

Relating alleles to genotypes

REMEMBER

An *equilibrium* occurs when something is in a state of balance. Genetically, an equilibrium means that certain values remain unchanged over the course of time. The Hardy-Weinberg law says that allele and genotype frequencies will remain unchanged, generation after generation, as long as certain conditions are met. In order for a population's genetics to follow Hardy-Weinberg's relationships:

» **The organism must reproduce sexually and be diploid.** Sex provides the opportunity to achieve different combinations of alleles, and the whole affair (pardon the pun) depends on having alleles in pairs (but many alleles can be used; you're not limited to two at a time).

- **The allele frequencies must be the same in both sexes.** If alleles depend entirely on maleness or femaleness, the relationships don't fall into place, because not all offspring have an equal chance to inherit the alleles — alleles on the Y chromosome violate Hardy-Weinberg rules.
- **The loci must segregate independently.** Independent segregation of loci is the heart of Mendelian genetics, and Hardy-Weinberg is directly derived from Mendel's laws.
- **Mating must be random with respect to genotype.** Matings between individuals need to be random, meaning that organisms don't sort themselves out based on the genotype in question.

Hardy-Weinberg makes other assumptions about the populations it describes, but the relationship is pretty tolerant of violations of those expectations. Not so with the four aforementioned conditions. When one of the four major assumptions of Hardy-Weinberg isn't met, the relationship between allele frequency and genotype frequency usually starts to fall apart.

The Hardy-Weinberg equilibrium relationship is often illustrated graphically, and the Hardy-Weinberg graph is fairly easy to interpret. On the left side of the graph in Figure 17-1 is the genotypic frequency (as a percentage of the total population going from 0 at the bottom to 100 percent at the top). Across the bottom of the graph is the frequency of the recessive allele, *a* (going 0 to 100 percent, left to right). To find the relationship between genotype frequency and allele frequency according to Hardy-Weinberg, just follow a straight line up from the bottom and then read the value off the left side of the graph. For example, if you want to know what proportion of the population is homozygous *aa* when the allele frequency of *a* is 40 percent, start at the 40 percent mark along the bottom of the graph and follow a path straight up (shown in Figure 17-1 with a dashed line) until you get to the line marked *aa* (which describes the genotype frequency for *aa*). Take a horizontal path (indicated by the arrow) to the left and read the genotypic frequency. In this example, 20 percent of the population is expected to be *aa* when 40 percent of the population carries the *a* allele.

It makes sense that when the allele *a* is rare, *aa* homozygotes are rare, too. As allele *a* becomes more common, the frequency of homozygotes slowly increases. The frequency of the homozygous dominant genotype, *AA*, behaves the same way but as a mirror image of *aa* (the genotype frequency for *aa*) because of the relationship of the alleles *A* to *a* in terms of their own frequency: $p + q$ must equal 1. If p is large (the proportion of alleles that is the *A* allele), q must be small (the proportion of alleles that is the *a* allele), and vice versa.

Check out the humped line in the middle of Figure 17-1. This is the frequency of heterozygotes, *Aa*. The highest proportion of the population that can be heterozygous is 50 percent. You may guess that's the case just by playing around with monohybrid crosses like those we describe in Chapter 3. No matter what combination of matings you try (*AA* with *aa*, *Aa* with *Aa*, *Aa* with *aa*, and so on), the largest proportion of *Aa* offspring you can ever get is 50 percent. Thus, when 50 percent of the population is heterozygous, the Hardy-Weinberg equilibrium predicts that 25 percent of the population will be homozygous for the *A* allele, and 25 percent will be homozygous for the *a* allele. This situation occurs only when *p* is equal to *q* — in other words, when both *p* and *q* equal 50 percent.

Many loci obey the rules of the Hardy-Weinberg law despite the fact that the assumptions required for the relationship aren't met. One of the major assumptions that's often violated among humans is random mating. People tend to marry each other based on their similarities, such as religious background, skin color, and ethnic characteristics. For example, people of similar socioeconomic backgrounds tend to marry each other more often than chance would predict. Nevertheless, many human genes are still in Hardy-Weinberg equilibrium. That's because matings may be dependent on some characteristics but are still independent with respect to the genes. The gene that confers immunity to HIV is a good example of a locus in humans that obeys the Hardy-Weinberg law despite the fact that its frequency was shaped by a deadly disease.

Violating the law

Populations can wind up out of Hardy-Weinberg equilibrium in several ways. One of the most common departures from Hardy-Weinberg occurs as a result of *inbreeding*. Put simply, inbreeding happens when closely related individuals mate and produce offspring. Purebred dog owners are often faced with this problem because certain male dogs sire many puppies, and a generation or two later, descendants of the same male are mated to each other. (In fact, selective inbreeding is what created the various dog breeds to begin with.)

Inbreeding tends to foul up Hardy-Weinberg because some alleles start to show up more and more often than others. In addition, homozygotes get more common, meaning fewer and fewer heterozygotes are produced. Ultimately, the appearance of recessive phenotypes becomes more likely. For example, the appearance of hereditary problems in some breeds of animals, such as deafness among Dalmatian dogs, is a result of generations of inbreeding.

The high incidence of particular genetic disorders among certain groups of people, such as Amish communities (see Chapter 15), is also a result of inbreeding. Even if people in a group aren't all that closely related anymore, if a small number of people started the group, everyone in the group is related somehow. (Relatedness shows up genetically even within large human populations; see the upcoming section, "Mapping the Gene Pool," to find out how.)

Loss of heterozygosity is thought to signal a population in peril. Populations with low levels of heterozygosity are more vulnerable to disease and stress, and that vulnerability increases the probability of extinction. Much of what's known about loss of heterozygosity and resulting problems with health of individuals — a situation ironically called *inbreeding depression* — comes from observations of captive animals, like those in zoos. Many animals in zoos are descended from captive populations, populations that had very few founders to begin with. For example, all captive snow leopards are reportedly descended from a mere seven animals.

GENETICS AND THE MODERN ARK

As human populations grow and expand, natural populations of plants and animals start getting squeezed out of the picture. One of the greatest challenges of modern biology is figuring out a way to secure the fate of worldwide biodiversity. Preserving biodiversity often takes two routes: establishment of protected areas and captive breeding.

Protected areas such as parks set aside areas of land or sea to protect all the creatures (animals and plants) that reside within its borders. Some of the finest examples of such efforts are found among America's national parks. However, while protecting special areas helps preserve biodiversity, it also allows populations to become isolated. With isolation, smaller populations start to inbreed, resulting in genetic disease and vulnerability to extinction. Sometimes, it's necessary for conservation geneticists to step in and lend a hand to rescue these isolated populations from genetic peril. For example, greater prairie chickens were common in the Midwest at one time. By 1990, their populations were tiny and isolated. Isolation contributed to inbreeding, causing their eggs to fail to hatch. In order to help rebuild a healthy population, biologists brought in more birds from populations elsewhere to increase genetic diversity. The strategy worked — the prairie chickens' eggs now hatch with healthy chicks that are hoped to bring the population back from the brink of extinction.

Captive breeding efforts by zoos, wildlife parks, and botanical gardens are also credited with preserving species. Twenty-five animal species that are completely extinct in the wild still survive in zoos thanks to captive breeding programs. Most programs are designed to provide not only insurance against extinction but also breeding stock for eventual reintroduction into the wild. Unfortunately, zoo populations often descend from very small founder populations, causing considerable problems with inbreeding. Inbreeding leads to fertility problems and the death of offspring shortly after birth. In the last 20 years, zoos and similar facilities have worked to combat inbreeding by keeping track of pedigrees (like the ones that appear in Chapter 15) and swapping animals around to minimize sexual contact between related animals.

Not just captive animals are at risk. As habitats for animals become more and more altered by human activity, natural populations get chopped up, isolated, and dwindle in size. Conservation geneticists work to understand how human activities affect natural populations of birds and animals. See the sidebar "Genetics and the modern ark" for more about how zoos and conservation geneticists work to protect animals from inbreeding depression and rescue species from extinction.

Mapping the Gene Pool

When the exchange of alleles, or *gene flow*, between groups is limited, populations take on unique genetic signatures. In general, unique alleles are created through changes in DNA sequence (described in Chapter 12). If groups of organisms are geographically separated and rarely exchange mates, altered alleles become common within populations. What this amounts to is that some alleles are found in only certain groups, giving each group a unique genetic identity. (After some time, these alleles usually conform to a Hardy-Weinberg equilibrium within each population; see the previous section, "Breaking Down the Hardy-Weinberg Law of Population Genetics," for details.) Geneticists identify genetic signatures of unique alleles by looking for distinct patterns within genes and certain sections of non-coding DNA.

REMEMBER

Altered alleles that show up outside the population they're usually associated with suggest that one or more individuals have moved or dispersed between populations. Geneticists use these genetic hints to trace the movements of animals, plants, and even people around the world. In the sections that follow, we cover some of the latest efforts to do just that.

One big happy family

With the contributions of the Human Genome Project (covered in Chapter 8), population geneticists have a treasure trove of information to sift through. Using new technologies, researchers are learning more than ever before about what makes various human populations distinct. One such effort is the study of single base pair changes, called SNPs (single nucleotide polymorphisms; pronounced "*Snips*"), in the DNA. Most of the tiny changes studied are found in non-coding DNA and have no effect on phenotype, but collectively they vary enough from one population to another to allow geneticists to discern each population's genetic signature.

After geneticists understood how much diversity exists among populations, they worked to create genetic maps that relate SNP alleles to geographic locations. Essentially, all humans tend to divide up genetically into the three continents of

Africa, Asia, and Europe. This isn't too surprising — humans have been in North and South America for only 10,000 years or so. When the genetic uniqueness of the Old World's people was described, geneticists examined populations in North America and other immigrant populations to see if genetics could predict where people came from. For example, genetic analyses of a group of immigrants in Los Angeles accurately determined which continent these people originally lived on. With more and more data, the genetic maps have become even more specific and can point people to specific countries where their ancestors once lived. This is the basis of ancestry tests that are now available to the general public, which we discuss in the next section, "Ancestry testing."

Because humans love to travel, geneticists have also compared rates of movements between men and women. Common wisdom suggests that, historically, men tended to move around more than women did (think Christopher Columbus or Leif Ericson). However, DNA evidence suggests that men aren't as prone to wander as previously believed. Geneticists compared mitochondrial DNA (passed from mother to child) with Y chromosome DNA (passed from father to son). It seems that women have migrated from one continent to another eight times more frequently than males. The tradition of women leaving their own families to join their husbands may have contributed to the pattern, but another possible explanation exists: A pattern of *polygyny,* men fathering children by more than one woman.

Ancestry testing

One huge area that has come from our increased knowledge of population genetics is that of ancestry testing (also known as *genetic genealogy*). Ancestry testing is now available through a variety of companies, and millions of individuals have already been tested, in hopes of finding out more about their families and where their ancestors came from.

Ancestry testing involves examining differences in DNA sequence in order to determine a person's ancestry. Specific variants tend to be shared among individuals of shared ancestry, since the variants are passed on from one generation to the next. Ancestry tests involve the analysis of a large number of SNPs (which are described in the preceding section). Most tests involve the examination of more than 600,000 different SNPs located throughout the genome. SNPs are variations in DNA sequence, most often found in non-coding regions of the genome. For example, one person may have an adenine at a specific location in the genome, while others may have a cytosine at that same location. SNPs can be unique to an individual or can be common among certain populations. Most of the SNPs examined are located on the autosomes. However, some of the DNA variants that are examined are found on the Y chromosome and are used to trace paternal lineage (since only males have Y chromosomes and all males get their Y chromosome from their fathers). Variants in the mitochondrial DNA (mtDNA), which is separate from the nuclear DNA, are also examined. As was discussed in Chapter 5, mtDNA is always inherited from the mother, which means that variants in the mtDNA can be used to trace the maternal lineage in a family. Both SNPs and mtDNA are discussed more in Chapter 18.

Once a person's DNA has been tested, the results are then compared with those of a reference group of known ancestry in order to estimate the person's ethnic background. One of the limitations of this analysis, however, is that each company has their own database with which to compare the results and much of the ancestry information is self-reported. In addition, ancestry testing can tell you how much DNA you have inherited from ancestors of each ethnic group, but it cannot tell you exactly where they lived. It is also possible that a person may receive results that differ from those of a close relative, such as a brother or sister. This is because you could have inherited a segment of DNA that originated from, for example, an Italian ancestor, but your sibling inherited a different segment that originated from, for example a Northern European ancestor. Each parent does not pass on *all* of their DNA to their children — only half. And which parts a parent passes on to each child varies (unless the children are identical twins).

Another important aspect to ancestry testing that those being tested should be aware of is the possibility of obtaining unexpected information from the results. Things like undisclosed adoption and nonpaternity (having a different father than expected) can come to light from ancestry testing, and once this information is known, it cannot be taken back. Chapter 20 covers more about the ethics of direct-to-consumer testing.

Uncovering the secret social lives of animals

Gene flow can have an enormous impact on threatened and endangered species. For example, scientists in Scandinavia were studying an isolated population of gray wolves not long ago. Genetically, the population was very inbred; all the animals descended from the same pair of wolves. Heterozygosity was low and, as a consequence, so were birth rates. When the population suddenly started to grow, the scientists were shocked. Apparently, a male wolf migrated over 500 miles to join the pack and father wolf pups. Just one animal brought enough new genes to rescue the population from extinction.

Mating patterns of animals often provide biologists with surprises. Early on, scientists thought birds were similar to humans in mating patterns — that is, primarily monogamous. As it turns out, birds aren't so monogamous after all. In most species of perching birds (the group that includes pigeons and sparrows, to name two widespread types), 20 percent of all offspring are fathered by some male other than the one with whom the female spends all her time. By spreading paternity among several males, a female bird makes sure that her offspring are genetically diverse. And genetic diversity is incredibly important to help fend off stress and disease. An example of this is the fairy wren — a tiny, brilliant-blue songbird. Fairy wrens live in Australia in big groups; one female is attended by several males who help her raise her young. But none of the males attending the nest actually fathers any of the kids — female fairy wrens slip off to mate with males in distant territories.

It turns out that humans aren't the only ones who live in close association with their parents, brothers, or sisters for their entire lives. Some species of whales live in groups called *pods*. Every pod represents one family: moms, sisters, brothers, aunts, and cousins, but not dads. Different pods meet up to find mates — as in the son or brother of one pod may mate with the daughter or sister of another pod. Males father offspring in different pods but stay with their own families for their entire lives. Sadly, geneticists learned about whale family structures and mating habits by taking meat from whales that had been killed by people. Like so many of the world's creatures, whales are killed by hunters. Hopefully, though, the information that scientists gather when whales are harvested will contribute to their conservation, allowing the planet's amazing biodiversity to persist for generations to come.

Changing Forms over Time: The Genetics of Evolution

Evolution, or how organisms change over time, is a foundational principle of biology. When Charles Darwin put forth his observations about natural selection, the genetic basis for inheritance was unknown. Now, with powerful tools like DNA sequencing (which appears in Chapter 8), scientists are documenting evolutionary change in real time, as well as uncovering how species share ancestors from long ago.

When genetic variation arises (from changes in DNA sequence, which we talk about in Chapter 12), new alleles are created. Then, *natural selection* acts to make particular genetic variants more common by way of improved survival and reproductive success for some individuals over others. In this section, you discover how genetics and evolution are inextricably tied together.

Genetic variation is key

All evolutionary change occurs because genetic variation arises through changes in DNA sequence (known as *sequence variants*). Without genetic variation, evolution can't take place. While many sequence variants are decidedly bad, some sequence variants confer an advantage, such as resistance to disease.

No matter how a sequence variant arises or what consequences it causes, the change must be heritable, or passed from parent to offspring, to drive evolution. Before DNA sequencing and other molecular techniques were available, it wasn't possible to examine heritable variation directly. Instead, phenotypic variation was used as an indicator of how much genetic variation might exist. With the help of DNA sequencing, scientists have come to realize that genetic variation is vastly more complex than anyone ever imagined.

REMEMBER

Heritable genetic variation alone doesn't mean that evolution will occur. The final piece in the evolutionary puzzle is natural selection. Put simply, natural selection occurs when conditions favor individuals carrying particular traits. By favor, it's meant that those individuals reproduce and survive better than other individuals carrying a different set of traits. This success is sometimes referred to as *fitness*, which is the degree of reproductive success associated with a particular genotype. When an organism has high fitness, its genes are being passed on successfully to the next generation. Through its effects on fitness, natural selection produces *adaptations*, or sets of traits that are important for survival. The appearance of white fur of polar bears, which allow them to blend into the snowy landscape of Arctic regions, is an example of an adaptation.

Where new species come from

Probably since the dawn of time (or at least the dawn of humankind, anyhow), humans have been classifying and naming the creatures around them. The formalized species naming system, what scientists call *taxonomic classification*, has long relied on for physical differences and similarities between organisms as a means of sorting things out. For example, elephants from Asia and elephants from Africa are obviously both elephants, but they're so different in their physical characteristics that they're considered separate species. Over the past 50 years or so, the way in which species are classified has changed as scientists have gained more genetic information about various organisms.

One way of classifying species is the *biological species concept*, which bases its classification on reproductive compatibility. Organisms that can successfully reproduce together are considered to be of the same species, and those that can't reproduce together are a different species. This definition leaves a lot to be desired, because many closely related organisms can interbreed yet are clearly different enough to be separate species.

Another method of classification, one that works a bit better, says that species are groups of organisms that maintain unique identities — genetically, physically, and geographically — over time and space. A good example of this definition of species is dogs and wolves. Both dogs and wolves are in the genus *Canis*. (Sharing a genus name tells you that organisms are quite similar and very closely related.) However, their species names are different. Dogs are always *Canis familiaris*, but there are many species of wolves, all beginning with *Canis* but ending with a variety of species names to accurately describe how different they are from each other (such as gray wolves, *Canis lupus*, and red wolves, *Canis rufus*). Genetically, dogs and wolves are very distinct, but they aren't so different that they can't interbreed. Dogs and wolves occasionally mate and produce offspring, but left to their own devices, they don't interbreed.

When populations of organisms become reproductively isolated from each other (that is, they no longer interbreed), each population begins to evolve independently. Different sequence variants arise, and with natural selection, the passage of time leads to the accumulation of different adaptations. In this way, after many generations, populations may become different species.

Growing the evolutionary tree

One of the basic concepts behind evolution is that organisms have similarities because they're related by descent from a common ancestor. Genetics and DNA sequencing techniques have allowed scientists to study these evolutionary relationships, or *phylogenies*, among organisms. For example, the DNA sequence of a particular gene may be compared across many organisms. If the gene is very similar or unchanged from one species to another, the species would be considered more closely related (in an evolutionary sense) than species that have accumulated many sequence variants in the same gene.

One way to represent the evolutionary relationships is with a tree diagram. In a similar fashion to the pedigrees used to study genetics in family relationships (flip to Chapter 15 for pedigree analysis), evolutionary trees like the one in Figure 17-2 illustrate the family relationships among species. The trunk of an evolutionary tree represents the common ancestor from which all other organisms in the tree descended. The branches of the tree show the evolutionary connections between species. In general, shorter branches indicate that species are more closely related.

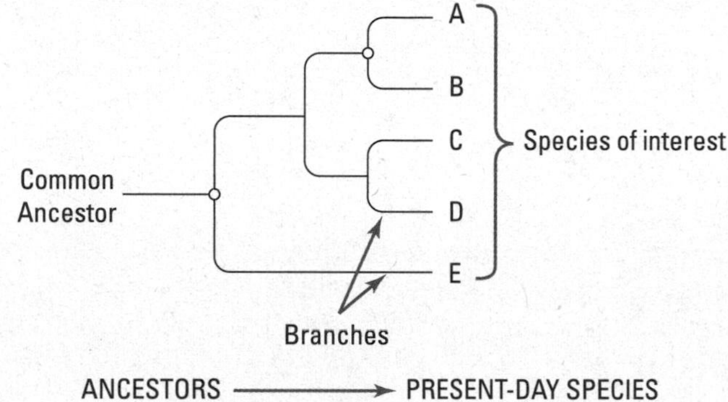

FIGURE 17-2: An evolutionary tree can show common ancestry among species.

IN THIS CHAPTER

» Generating DNA fingerprints

» Using DNA to match criminals to crimes

» Identifying persons using DNA from family members

Chapter 18 Solving Mysteries Using DNA

Forensics pops up in every cop drama and murder mystery on television these days, but what is forensics used for in the real world? Generally, *forensics* is thought of as the science used to capture and convict criminals; it includes everything from determining the source of carpet fibers and hairs to paternity testing. Technically, forensics is the application of scientific methods for legal purposes. Thus, *forensic genetics* is the exploration of DNA evidence — who is it, who did it, and who's your daddy.

Just as each person has their own unique fingerprint, every human (with the exception of identical twins) is genetically unique. *DNA fingerprinting*, also known as *DNA profiling*, is the process of uncovering the patterns within DNA. DNA fingerprinting is at the heart of forensic genetics and is often used to:

» Confirm that a person was present at a particular location.

» Determine identity (including sex).

» Assign paternity.

In this chapter, you step inside the DNA lab to discover how scientists solve forensic mysteries by identifying individuals and family relationships using genetics.

The knowledge that every human's fingerprints are unique appears to stretch back at least several thousand years. There is evidence that ancient Babylonians used their fingerprints to document business transactions. Much later, Sir Francis Galton studied fingerprints based on their patterns and was able to show convincing statistical evidence that each person's fingerprints are unique. Subsequently, in 1899, Sir Edward Henry was the first police officer to apply the patterns of loops, arches, and whorls from people's fingertips to identify individual people and match criminal to crime.

Rooting through Your DNA to Find Your Identity

It's obvious just from looking at the people around you that each one of us is unique. But getting at the *genotype* (genetic traits) behind the *phenotype* (physical traits) is tricky business, because almost all your DNA is exactly like every other human's DNA. Much of what your DNA does is provide the information to run all your body functions, and most of those functions are exactly the same from one human to another. If you were to compare your roughly 3 billion base pairs of DNA (see Chapter 5 to review how DNA is put together) with your next-door neighbor's DNA, you'd find that 99.9 percent of your DNA is exactly the same (assuming your identical twin doesn't live next door).

So what makes you look so different from the guy next door, or even from your mom and dad? Your genetic uniqueness is the result of sexual reproduction. (For more on how sexual reproduction works to make you unique, turn to Chapter 2.) Until the human genome was sequenced (see Chapter 8), the tiny differences produced by recombination and meiosis that make you genetically unique were very hard to isolate. But in 1985, a team of scientists in Britain figured out how to profile a tiny bit of DNA uniqueness into a DNA fingerprint. Surprisingly, DNA fingerprinting doesn't use the information contained in your genes that make you look unique. Instead, the process takes advantage of part of the genome that doesn't seem to do anything at all: non-coding DNA (previously referred to as *junk DNA*).

Less than 2 percent of the human genome codes for actual physical traits — that is, all your body parts and the ways they function. That's pretty astounding considering that your genome is so huge. So what's all that extra DNA doing in there? Scientists are still trying to figure that part out, but what they do know is that some non-coding DNA is very useful for identifying individual people.

Even within non-coding DNA, one human looks much like another. But short stretches of non-coding DNA vary a lot from person to person. *Short tandem repeats* (STRs) are sections of DNA arranged in back-to-back repetition (a simple sequence is repeated several times in a row). A naturally occurring non-coding DNA sequence may look something like the following example. (The spaces in these examples allow you to read the sequences more easily. Real DNA doesn't have spaces between the bases.)

TGCT AGTC AAAG TCTT CGGT TCAT

A short STR may look like this:

TCAT TCAT TCAT TCAT TCAT TCAT

The number of repeats in pairs of STRs varies from one person to another. The variations are referred to as *alleles* (different versions of a specific DNA sequence). Using two chromosome pairs from different suspects, Figure 18-1 shows how the same STRs can have different alleles. Chromosome 1 has two loci (in reality, this chromosome has thousands of loci, but we're only looking at two in this example). For the first suspect, the marker at STR Locus A is the same length on both chromosomes (5 and 5), meaning that Suspect One is homozygous for Locus A (*homozygous* means the two alleles at a particular locus are identical). At Locus B, Suspect One (S1) has alleles that are different lengths (6 and 3), meaning he's heterozygous at that locus (*heterozygous* means the two alleles differ). Now look at the STR DNA profile of Suspect Two (S2). It shows the same two loci, but the patterns are different. At Locus A, S2 is heterozygous (5 and 4) and has one allele that's different from S1's. At Locus B, S2 is homozygous (4 and 4) for a completely different allele than the ones S1 carries.

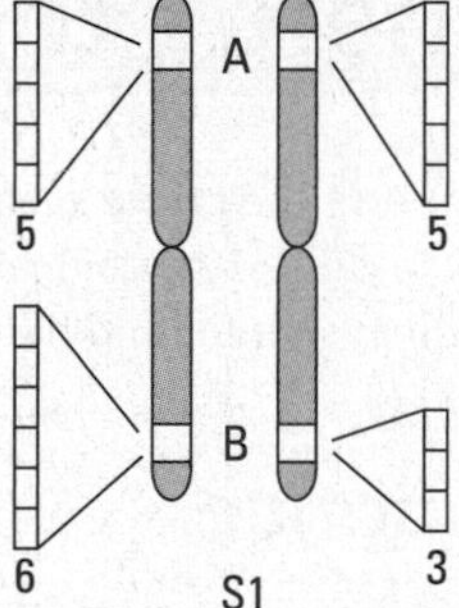

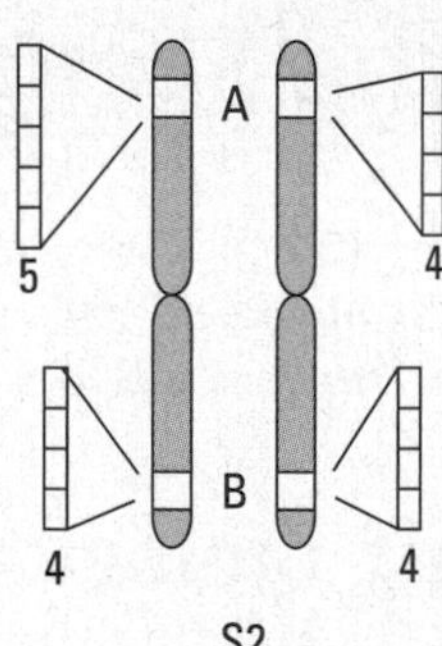

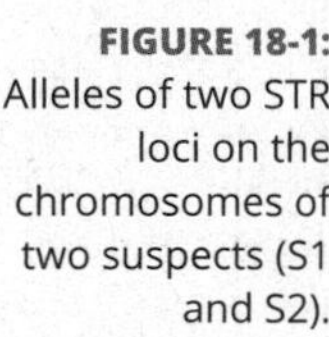
FIGURE 18-1: Alleles of two STR loci on the chromosomes of two suspects (S1 and S2).

The variation in STR alleles is called *polymorphism* (*poly-* meaning "many" and *morph-* meaning "shape or type"). Polymorphism arises from mistakes made during the DNA copying process (called *replication;* covered in Chapter 7). Normally, DNA is copied mistake-free during replication. But when the enzyme copying the DNA reaches an STR, it often gets confused by all the repeats and winds up leaving out one repeat — like one of the TCATs in the previous example — or putting an extra one in by mistake. As a result, the sequences of DNA before and after the STR are exactly the same from one person to the next, but the number of repeats *within* the STR varies. In the case of non-coding DNA, the changes in the STRs create lots of variation in how many repeats appear.

The specific STRs used in forensics are referred to as *loci* or *markers.* Loci is the plural form of *locus,* which is Latin for "place." (Genes are also referred to as loci, as described in Chapter 3.) You have hundreds of STR markers on every one of your chromosomes. These loci are named using numbers and letters, such as D5S818 or VWA.

It is expected that your STR DNA fingerprint is completely different from everyone else's on earth (except your identical twin, if you have one). The probability of anyone else having identical alleles as you at every one of his or her loci is outrageously small.

Lots of other organisms have STR DNA and, in turn, their own DNA fingerprints. Dogs, cats, horses, fish, plants — in fact, just about all eukaryotes have lots of STR DNA. (*Eukaryotes* are organisms whose cells have nuclei.) That makes STR DNA fingerprinting an extremely powerful tool to answer all kinds of biological questions.

Investigating the Scene: Where's the DNA?

When a crime occurs, the forensic geneticist and the crime scene investigator are interested in the biological evidence, because cells in biological evidence contain DNA. Biological evidence includes blood, saliva, semen, and hair.

Collecting biological evidence

Anything that started out as part of a living thing may provide useful DNA for analysis. In addition to human biological evidence (blood, saliva, semen, and hair), plant parts like seeds, leaves, and pollen, as well as hair and blood from pets, can help link a suspect and victim. (See the sidebar "Pets and plants play detective" for more on how nonhuman DNA is used to investigate crimes.)

PETS AND PLANTS PLAY DETECTIVE

DNA evidence from almost any source may provide a link between criminal and crime. For example, a particularly brutal murder in Seattle was solved entirely on the basis of DNA provided by the victim's dog. After two people and their dog were shot in their home, two suspects were arrested in the case, and blood-spattered clothing was found in their possession. The only blood on the clothing was of canine origin, and the dog's blood turned out to be the only evidence linking the suspects to the crime scene. Using markers originally designed for canine paternity analysis, investigators generated a DNA fingerprint from the dog's blood and compared it with DNA tests from the bloodstained clothing. A perfect match resulted in a conviction.

Practically any sort of biological material can provide enough DNA to match a suspect to a crime. In one murder case, the perpetrator stepped in a pile of dog feces near the scene. DNA fingerprinting matched the evidence on a suspect's shoe to the evidence at the scene, leading to a conviction. In another case, a rape victim's dog urinated on the attacker's vehicle, allowing investigators to match the pup to the truck; the suspect promptly confessed his guilt.

Even plants have a place in the DNA evidence game. The very first time DNA evidence from plants was used was in an Arizona court case in 1992. A murder victim was found near a desert tree called Paloverde. Seeds from that type of tree were found in the bed of a pickup truck that belonged to a suspect in the case, but the suspect denied ever having been in the area. The seeds in the truck were matched to the exact tree where the victim was found using DNA fingerprinting. The seeds couldn't prove the suspect's presence, but they provided a link between his truck and the tree where the body was found. The DNA evidence was convincing enough to obtain a conviction in the case.

To properly collect evidence for DNA testing, the investigator must be very, very careful, because his or her own DNA can get mixed up with DNA from the scene. Investigators wear gloves, avoid sneezing and coughing, and cover their hair (dandruff has DNA, too).

To conduct a thorough investigation, the investigator needs to collect everything at the scene (or from the suspect) that may provide evidence. DNA has been gathered from bones, teeth, hair, urine, feces, chewing gum, cigarette butts, toothbrushes, and even earwax! Blood is the most powerful evidence because even the tiniest drop of blood contains about 80,000 white blood cells, and the nucleus of every white blood cell contains a copy of the donor's entire genome and more than enough information to determine identity using a DNA fingerprint. But even one skin cell has enough DNA to make a fingerprint (see "Outlining the powerful PCR process"). That means that skin cells clinging to a cigarette butt or an envelope flap may provide the evidence needed to place a suspect at the scene.

DECOMPOSING DNA

DNA, like all biological molecules, can decompose; that process is called *degradation*. *Exonucleases,* a particular class of enzymes whose sole function is to carry out the process of DNA degradation, are practically everywhere: on your skin, on the surfaces you touch, and in bacteria. Anytime DNA is exposed to exonuclease attack, its quality rapidly deteriorates because the DNA molecule starts to get broken into smaller and smaller pieces. Degradation is bad news for evidence because DNA begins to degrade as soon as cells (like skin or blood) are separated from the living organism. To prevent DNA evidence from further degradation after it's collected, it's stored in a sterile (that is, bacteria-free) container and kept dry. As long as the sample isn't exposed to high temperatures, moisture, or strong light, DNA evidence can remain usable for more than 100 years. (Even under adverse conditions, DNA can sometimes last for centuries, as we explain in Chapter 5.)

To draw information and conclusions from the DNA evidence, the investigator needs to collect samples from the victim or victims, suspects, and witnesses for comparison. Investigators collect samples from houseplants, pets, or other living things nearby to compare those DNA fingerprints to the DNA evidence. After the investigator gets these samples, it's time to head to the lab.

Moving to the lab

Biological samples contain lots of substances besides DNA. Therefore, when an investigator gets evidence to the lab, the first thing to do is extract the DNA from the sample. (For a DNA extraction experiment using a strawberry, see Chapter 5.) There are different methods to extract DNA, but they generally follow these three basic steps:

1. Break open the cells to free the DNA from the nucleus (this is called *cell lysis*).
2. Remove the proteins (which make up most of the biological sample) by digesting them with an enzyme. This breaks down proteins only; the DNA is left intact.
3. Remove the DNA from the solution by adding ethanol (a type of alcohol).

After the DNA from the sample is isolated, it's analyzed using a process called the *polymerase chain reaction*, or PCR.

Outlining the powerful PCR process

The goal of the PCR process is to make thousands of copies of specific parts of the DNA molecule — in the case of forensic genetics, several target STR loci that are used to construct a DNA fingerprint. (Copying the entire DNA molecule would be useless because the uniqueness of each individual person is hidden among all that DNA.) Many copies of several target sequences are necessary, for two reasons:

- Current technology used in DNA fingerprinting can't detect the DNA unless large amounts are present, and to get large amounts of DNA, you need to make copies.
- Matches must be exact when it comes to DNA fingerprinting and forensic genetics; after all, people's lives are on the line. To avoid misidentifications, many STR loci from each sample must be examined.

Originally, in the US, 13 standard markers were used for matching human samples, plus one additional marker that allows determination of sex (that is, whether the sample came from a male or a female). These markers are part of CODIS, the COmbined DNA Index System, which is the US database of DNA fingerprints. As of January of 2017, there are 20 loci tested, including the original set.

Here's how PCR works, as shown in Figure 18-2:

1. To replicate DNA using PCR, you have to separate the double-stranded DNA molecule (called the *template*) into single strands. This process is called *denaturing*. When DNA is double-stranded, the bases are protected by the sugar-phosphate backbone of the double helix (you can review DNA structure in Chapter 5). DNA's complementary bases, where all the information is stored, are locked away, so to speak. To pick the lock, get at the code, and build a DNA copy, the double helix must be opened. The hydrogen bonds that hold the two DNA strands together can be broken by heating the molecule up to a temperature just short of boiling (212 degrees Fahrenheit). When heated, the two strands slowly come apart as the hydrogen bonds melt. DNA's sugar-phosphate backbone isn't damaged by heat, so the single strands stay together with the bases still in their original order.
2. When denaturing is complete, the mix is cooled slightly. Cooling allows small, complementary pieces of DNA called *primers* to attach themselves to the template DNA. The primers match up with their complements on the template strands in a process called *annealing*. Primers only attach to the template strand when the match is perfect; if no exact match is found, the next step in the PCR process doesn't occur because primers are required to start the copying process (Chapter 7 explains why primers are necessary to build strands of DNA from scratch). The primers used in PCR are marked with dyes that glow when

exposed to the right wavelength of light (think fluorescent paint under a black light). STRs of similar length (even though they may actually be on entirely different chromosomes, as in Figure 18-1) are labeled with different colors so that when the fingerprint is read, each locus shows up as a different color (see "Constructing the DNA fingerprint" later in this chapter).

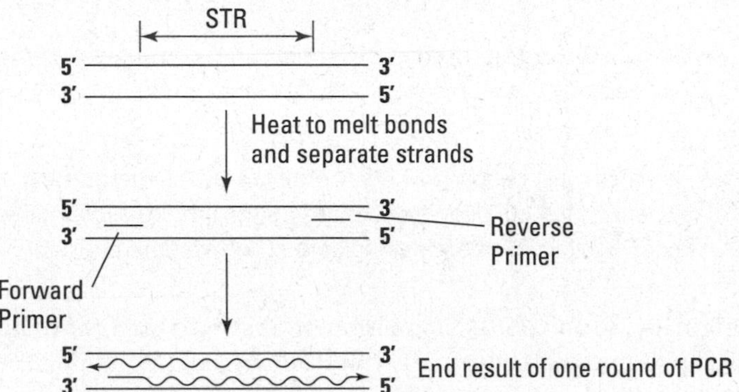

FIGURE 18-2: The process of PCR.

3. After the primers find their matches on the template strands, *Taq (pronounced "tack") polymerase* begins to do its work. Polymerases act to put things together. In this case, the thing getting put together is a DNA molecule. Taq polymerase specifically is a DNA polymerase extracted from bacteria that are able to withstand high temperatures.

   Taq polymerase starts adding bases — this stage is called *extension* — onto the 3' ends of the primers by reading the template DNA strand to determine which base belongs next. Meanwhile, on the opposite template strand at the end of the reverse primer, Taq rapidly adds complementary bases using the template as a guide. (The newly replicated DNA remains double-stranded throughout this process because the mixture isn't hot enough to melt the newly formed hydrogen bonds between the complementary bases.)

One complete round of PCR produces two identical copies of the desired STR. But two copies aren't enough to be detected by the lasers used to read the DNA fingerprints (see "Constructing the DNA fingerprint"). You need hundreds of thousands of copies of each STR, so the PCR process — denaturing, annealing, and extending — repeats over and over.

Figure 18-3 shows you how fast this copying reaction adds up — after 5 cycles (when starting with just a single strand from a single cell), you have 32 copies of the STR. Typically, a PCR reaction is repeated for 30 cycles, so with just one template strand of DNA, you end up with more than 1 billion copies of the target STR (the primers and the sequence between them). Usually, evidence samples consist of more than one cell, so it's likely that you start with 80,000 or so template strands instead of just one. With 30 rounds of PCR, this would yield . . . we'll wait while you do the math . . . okay, so it's a lot of DNA, as in trillions of copies of the target STR. That's the power of PCR. Even the tiniest drop of blood or a single hair can yield a fingerprint that may free the innocent or convict the guilty.

REMEMBER

The invention of PCR revolutionized the study of DNA. Basically, PCR is like a copier for DNA but with one big difference: A photocopier makes facsimiles; PCR makes real DNA. Before PCR came along, scientists needed large amounts of DNA directly from the evidence to make a DNA fingerprint. But DNA evidence is often found and collected in very tiny amounts. Often, the evidence that links a criminal to a crime scene is the DNA contained in a single hair! One of the biggest advantages to PCR is that a very tiny amount of DNA — even one cell's worth! — can be used to generate many exact copies of the STRs used to create a DNA fingerprint (see the section "Rooting through Your DNA to Find Your Identity" earlier in this chapter for a full explanation of STR). Chapter 21 looks at the discovery of PCR in more detail.

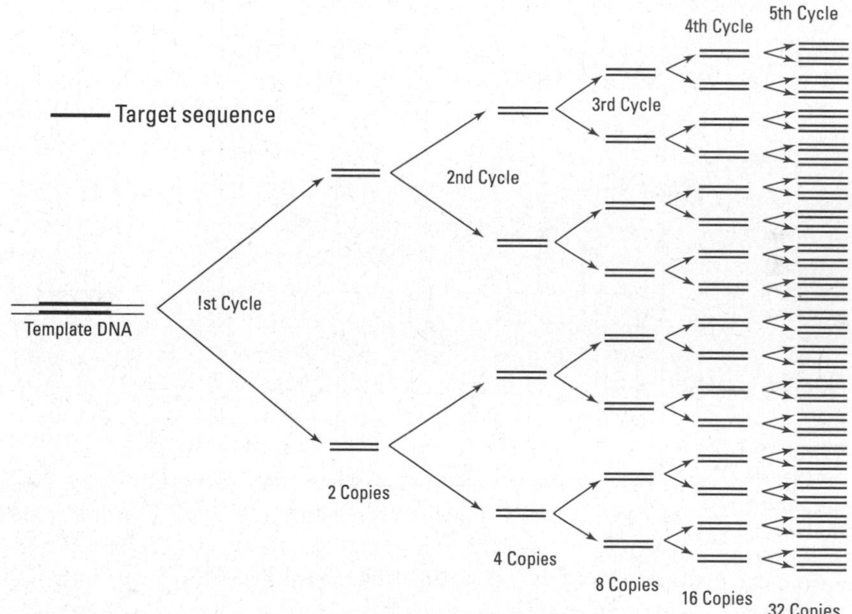

FIGURE 18-3: The number of STR copies made by five cycles of PCR.

Constructing the DNA fingerprint

For each DNA sample taken as forensic evidence and put through the process of PCR, several loci are examined. ("Several" often means either the 13 original markers or the 20 now used in the CODIS database; see the earlier "Outlining the powerful PCR process" section.) This study yields a unique pattern of colors and sizes of STRs — this is the DNA fingerprint of the individual from whom the sample came.

DNA fingerprints are "read" using a process called *electrophoresis,* which takes advantage of the fact that DNA is negatively charged. An electrical current is passed through a gelatin-like substance (referred to as a *gel*), and the completed PCR is injected into the gel. By electrical attraction, the DNA moves toward the positive pole (electrophoresis). Small STR fragments move faster than larger ones, so the STRs sort themselves according to size (see Figure 18-4). Because the fragments are tagged with dye, a computer-driven machine with a laser is used to "see" the fragments by their colors. The STR fragments show up as peaks like those shown in Figure 18-4. The results are stored in the computer for later analysis of the resulting pattern.

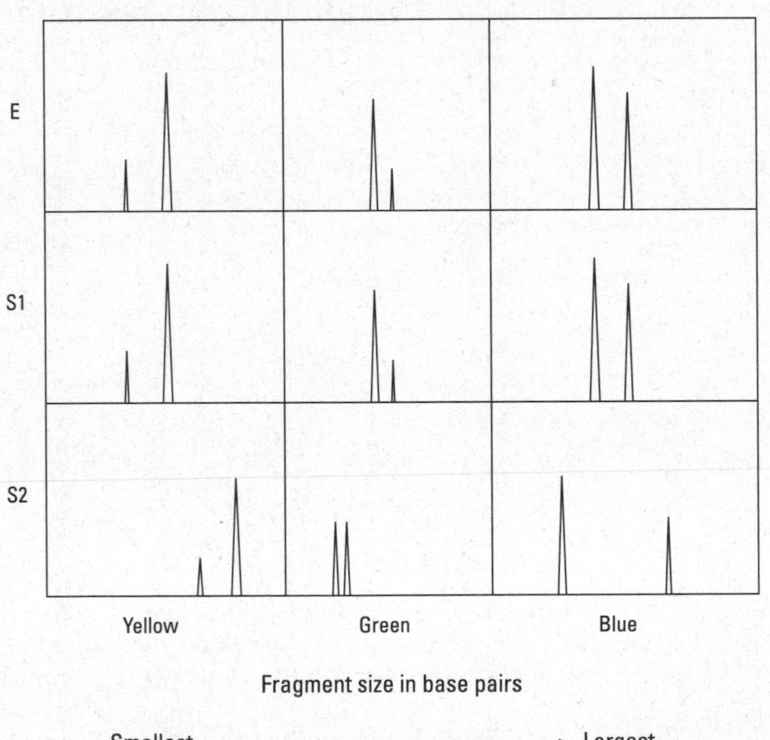

FIGURE 18-4: The DNA fingerprints of two suspects (S1 and S2) are compared with an evidence sample (E).

The technology used in DNA fingerprinting now allows the entire process, from extracting the DNA to reading the fingerprint, to be done very rapidly. If everything goes right, it takes less than 24 hours to generate one complete DNA fingerprint.

The first cases to use DNA fingerprinting appeared in the courts in 1986. Generally, legal evidence must adhere to what experts call the *Frye standard. Frye* is short for *Frye v. United States,* a court case decided in 1923. Put simply, Frye says that scientific evidence can only be used when most scientists agree that the methods and theory used to generate the evidence are well established. Following its development, DNA fingerprinting rapidly gained acceptance by the courts, and it's now considered routine. The tests used to generate DNA fingerprints have changed over the years, and STRs are now the gold standard.

Employing DNA to Catch Criminals (And Free the Innocent)

After the forensic geneticist generates DNA fingerprints from different samples, the next step is to compare the results. When it comes to getting the most information out of the fingerprints, the basic idea is to look for matches between the:

- Suspect's DNA and DNA on the victim, the victim's clothing or possessions, or the location where the victim was known to have been.
- Victim's DNA and DNA on the suspect's body, clothing, or other possessions, or a location linked to the suspect.

Matching the evidence to the bad guy

In Figure 18-4, you can see that exact matches of DNA fingerprints stand out like a sore thumb. But when you find a match between a suspect and your evidence, how do you know that no one else shares the same DNA fingerprint as that particular suspect?

With DNA fingerprints, you can't know for sure that a suspect is the culprit you're looking for, but you can calculate the odds of another person having the same pattern. Because this book isn't *Statistical Genetics For Dummies,* we'll skip the details of exactly how to calculate these odds. Instead, we'll just tell you that, in Figure 18-4, the odds of another person having the same pattern as Suspect Two is 1 in 45 for locus one, 1 in 70 for locus two, and 1 in 50 for locus three. To calculate the total odds of a match, you multiply the three probabilities: $1/45 \times 1/70 \times 1/50 = 1/157{,}500$. Based on your test using just three loci, the probability of another person having the same DNA pattern as Suspect Two is 1 in 157,500 (assuming Suspect Two doesn't have an identical twin).

TECHNICAL STUFF

When the original 13 CODIS loci are used, the probability of finding two unrelated white individuals who have the same DNA fingerprints is 1 in 575 trillion. When all 20 are used, the odds are even less. To put this in perspective, consider that the planet has only 7.5 billion people. To say the least, your lifetime odds of getting hit by lightning at some point in your life (1 in 3,000) are a lot better than this!

Of course, real life is a lot more complicated than the example in Figure 18-4. Biological evidence samples are often mixed and contain more than one person's DNA. Because humans are *diploid* (having chromosomes in pairs), mixed samples (called *admixtures* for you *CSI* fans) are easy to spot — they have three or more alleles in a single locus. By comparing samples, forensic geneticists can parse out whose DNA is whose and even determine how much DNA in a sample was contributed by each person.

TO FIND A CRIMINAL USING DNA

The FBI's CODIS system works because it contains hundreds of thousands of cataloged samples for comparison. All 50 U.S. states require DNA samples to be collected from persons convicted of sex offenses and murder. Laws vary from state to state on which other convictions require DNA sampling, but so far, CODIS has cataloged over 13 million offender samples. But what if no sample has been collected from the guilty party? What happens then?

Some law enforcement agencies have conducted mass collection efforts to obtain DNA samples for comparison. The most famous of these collection efforts occurred in Great Britain in the mid-1980s. After two teenaged girls were murdered, every male in the entire neighborhood around the crime scene was asked to donate a sample for comparison. In all, nearly 4,000 men complied with the request to donate their DNA. The actual murderer was captured after he bragged about how he had gotten someone else to volunteer a sample for him.

DNA evidence is also sometimes used to extend the statute of limitations on crimes when no arrest has been made. (The *statute of limitations* is the amount of time prosecutors have to bring charges against a suspect.) Crimes involving murder have no statute of limitations, but most states have a statute of limitations on other crimes such as rape. To allow prosecution of such crimes, DNA evidence can be used to file an arrest warrant or make an indictment against "John Doe" — the unknown person possessing the DNA fingerprint of the perpetrator. The arrest warrant extends the statute of limitations indefinitely until a suspect is captured.

But what if the evidence and the suspect's DNA don't match? The good news is that an innocent person is off the hook. The bad news is that the guilty party is still roaming free. At this point, investigators turn back to the CODIS system because it was designed not only to standardize which loci get used in DNA fingerprinting but also to provide a library of DNA fingerprints to help identify criminals and solve crimes. The FBI established the DNA fingerprint database in 1998 based on the fact that repeat offenders commit most crimes. When a person is convicted of a crime (laws vary on which convictions require DNA sampling; see the sidebar "To find a criminal using DNA"), his or her DNA is sampled — often by using a cotton swab to collect a few skin cells from the inside of the mouth. As of June 2019, the CODIS database had provided over 470,000 matches and assisted in over 460,000 investigations. If no match is found in CODIS, the evidence is added to the database. If the perpetrator is ever found, then a match can be made to other crimes he or she may have committed.

Taking a second look at guilty verdicts

Not all persons convicted of crimes are guilty. One study estimates that up to 10,000 persons are wrongfully convicted each year in the U.S. alone. The reasons behind wrongful conviction are varied, but the fact remains that innocent persons shouldn't be jailed for crimes they didn't commit.

In 1992, Barry Scheck and Peter Neufeld founded the Innocence Project in an effort to exonerate innocent men and women. The project relies on DNA evidence, and services are free of charge to all who qualify.

Walter D. Smith is one of the 365 persons in the U.S. exonerated by DNA evidence as of August 2019. In 1985, Smith was wrongfully accused of raping three women. Despite his claims of innocence, eyewitness testimony brought about a conviction, and Smith received a sentence of 78 to 190 years in prison. During his 11 years of incarceration, Smith earned a degree in business and conquered drug addiction. In 1996, the Innocence Project conducted DNA testing that ultimately proved his innocence and set him free.

It's unclear how many criminal cases have been subjected to postconviction DNA testing, and the success rates for such cases are unreported. Surprisingly, many states have opposed postconviction testing, but laws are being passed to allow or require such testing when circumstances warrant.

It's All Relative: Finding Family

Family relationships are important in forensic genetics when it comes to paternity for court cases or determining the identities of persons killed in mass disasters. Individuals who are related to one another have sections of their DNA in common. Within a family tree, the amount of genetic relatedness, or *kinship,* among individuals is very predictable. Assuming that Mom and Dad are unrelated to each other, full siblings have roughly half their DNA in common because they each inherit all their DNA from the same parents.

Paternity testing

Recent studies suggest that the rate of nonpaternity (when the presumed father of a child is not the biological father) is much lower than initially believed. It is now reported that between 1 and 5 percent of children are fathered by someone other than the father listed on the birth certificate, depending on the population studied. Given that nonpaternity is not exactly uncommon, tests to determine what male fathered what child are of considerable interest. Paternity testing is used in divorce and custody cases, determination of rightful inheritance, and a variety of other legal and social situations.

Paternity testing using STR techniques has become very common and relatively inexpensive. The methods are exactly the same as those used in evidence testing (see the earlier section "Outlining the powerful PCR process"). The only difference is the way the matches are interpreted. Because the STR alleles are on chromosomes (see the earlier section "Rooting through Your DNA to Find Your Identity"), a mother contributes half the STR alleles possessed by a child, and the father contributes the other half. Figure 18-5 shows what these contributions may look like in a DNA fingerprint. (M is the mother, C is the child, and F1 and F2 are the possible fathers.) Alleles are depicted here as peaks, and arrows indicate the maternal alleles. Assuming that mother and father are unrelated, half the child's alleles came from F2, indicating that F2 is likely the father.

Two values are often reported in paternity tests conducted with DNA fingerprinting:

- **Paternity index:** A value that indicates the strength of the evidence. The higher the paternity index value, the more likely it is that the alleged father is the actual genetic father. The paternity index is a more accurate estimate than probability of paternity.
- **Probability of paternity:** The probability that a particular person could have contributed the same pattern shown by the DNA fingerprint. The odds calculations for probability of paternity are more complicated than multiplying

simple probabilities (see "Matching the evidence to the bad guy") because an individual who's heterozygous at a particular locus has an equal probability of contributing either allele. The probability of a particular male being the father also depends on how often the various alleles at a locus show up in the population at large (which is also true for the estimates of odds shown in "Matching the evidence to the bad guy"; see Chapter 17 for the lowdown on how population genetics works).

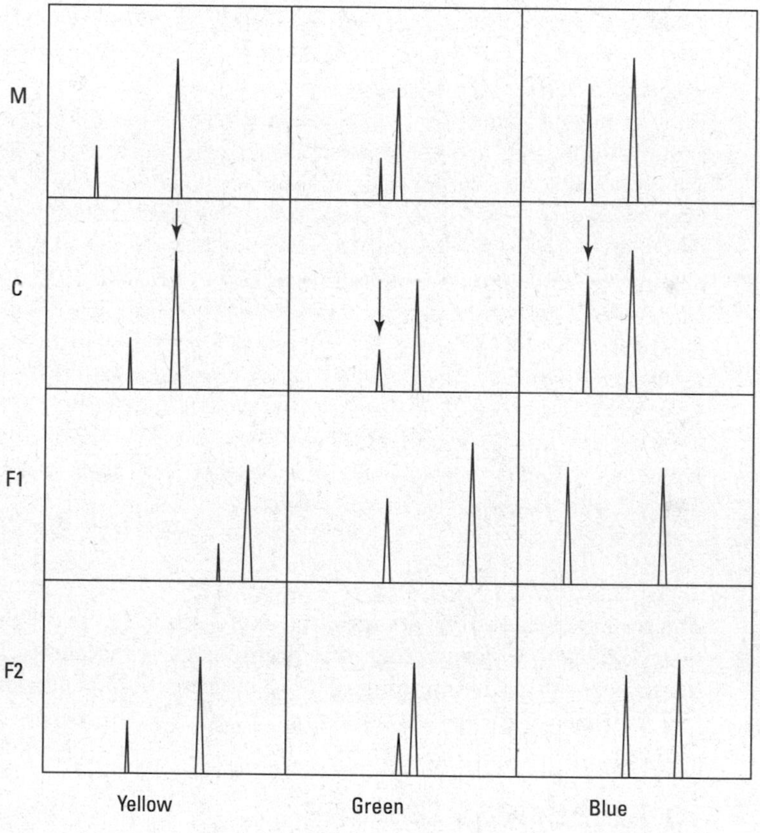

FIGURE 18-5: Paternity testing using STR loci.

THOMAS JEFFERSON'S SON

Male children receive their one and only Y chromosome from their fathers (see Chapter 6). Thus, paternity of male children can be resolved by using DNA markers on the Y chromosome. The discovery of this testing option led to the unusual resolution of a long-term mystery involving the second U.S. president, Thomas Jefferson.

In 1802, Jefferson was accused of fathering a son by one of his slaves, Sally Hemings. Jefferson's only acknowledged offspring to survive into adulthood were daughters, but Jefferson's paternal uncle has surviving male relatives who are descended in an unbroken male line. Thus, the Y chromosome DNA from these Jefferson family members was expected to be essentially identical to the Y chromosome DNA that Jefferson inherited from his paternal grandfather — DNA he would have contributed to a son. Five men known to have descended from Jefferson's uncle agreed to contribute DNA samples for comparison with the only remaining male descendant of Sally Hemings's youngest son. In all, 19 samples were examined. These samples included descendants of other potential fathers along with unrelated persons for comparison. A total of 19 markers found only on the Y chromosome were used. (None of the CODIS markers is on the Y chromosome; they'd be useless for females if they were.) The Jefferson and Hemings descendants matched at all 19 markers. Since the publication of the genetic analysis, historical records have been examined to provide additional evidence that Jefferson fathered Sally Hemings's son, Eston. For example, Jefferson was the only male of his family present at the time Eston was conceived. Interestingly, examination of the historical records seems to indicate that Jefferson is likely the father of all of Sally Hemings's six children; however, this conclusion remains controversial.

REMEMBER

The results of paternity tests are often expressed in terms of "proof" of paternity or lack thereof. Unfortunately, this terminology is inaccurate. Genetic paternity testing doesn't *prove* anything. It only indicates a high likelihood that a given interpretation of the data is correct.

Relatedness testing

Paternity analysis isn't the only time that DNA fingerprinting is used to determine family relationships. Historical investigations (like the Jefferson-Hemings case we explain in the sidebar "Thomas Jefferson's son") may also use patterns inherited within the DNA to show how closely related people are and to identify remains. Mass fatality incidents such as plane crashes and the World Trade Center disaster of September 11, 2001, rely on DNA technologies to identify deceased persons. Several methods are used under such circumstances, including STR DNA fingerprinting, mitochondrial DNA analysis (see Chapter 5 for the details on mitochondrial DNA), and Y-chromosome analysis.

Several conditions complicate DNA identification of victims of mass fatality incidents. Bodies are often badly mutilated and fragmented, and decomposition damages what DNA remains in the tissues. Furthermore, reference samples of the deceased person's DNA often don't exist, making it necessary to make inferences from persons closely related to the deceased.

Reconstructing individual genotypes

Much of what forensic geneticists know about identifying victims of mass fatalities comes from airplane crashes. In 1998, Swissair Flight 111 crashed into the Atlantic Ocean just off the coast of Halifax, Nova Scotia, Canada. This disaster sparked an unusually comprehensive DNA typing effort that now serves as the model for forensic scientists the world over dealing with similar cases.

In all, 1,200 samples from 229 persons were recovered from Swissair Flight 111. Only one body could be identified by appearance alone, so investigators obtained 397 reference samples either from personal items belonging to victims (like toothbrushes) or from family members. Because most reference samples from the victims themselves were lost in the crash, 93 percent of identifications depended on samples from parents, children, and siblings of the deceased. The number of alleles shared by family members is fairly predictable, allowing investigators to conduct parentage analysis based on the expected rate of matching alleles. In the Swissair case, 43 family groups (including 6 families of both parents and some or all of their children) were among the victims, so the analyses were complicated by kinship among the victims.

The initial DNA fingerprinting of remains revealed 228 unique genotypes (including one pair of twins). The 13 CODIS loci were tested using PCR. All the data from DNA fingerprinting was entered into a computer program specifically designed to compare large numbers of DNA fingerprints. The program searched for several kinds of matches:

- A perfect match between a victim and a reference sample from a personal item.
- Matches between victims that would identify family groups (parents and children, and siblings).
- Matches between samples from living family members.

The computer then generated reports for all matches within given samples. Two investigators independently reviewed every report and only declared identifications when the probability of a correct identification was greater than a million to one. Altogether, over 180,000 comparisons were made to determine the identities of the 229 victims.

Forty-seven persons were identified based on matches with personal items. The remaining 182 persons were identified by comparing victims' genotypes with those of living family members. The power of PCR, combined with many loci and computer software, led to rapid comparisons and the positive identification of all the victims.

Bringing closure in times of tragedy

On September 11, 2001, two jetliners crashed into the twin towers of the World Trade Center in New York City. The enormous fires resulting from the crashes caused both buildings to collapse. Roughly 2,700 persons died in the disaster. Over 20,000 body parts were recovered from the rubble; therefore, the task of forensic geneticists was two-fold: determine the identity of each deceased person and collect the remains of particular individuals for interment. Unlike Swissair Flight 111, few victims of the WTC tragedy were related to each other. However, other issues complicated the task of identifying the victims. Many bodies were subjected to extreme heat, and others were recovered weeks after the disaster, as rubble was removed. Thus, many victim samples had very little remaining DNA for analysis.

DNA reference samples from missing persons were collected from personal effects such as toothbrushes, razors, and hairbrushes. Skin cells clinging to toothbrushes accounted for almost 80 percent of the reference samples obtained for comparison. These samples were DNA fingerprinted using PCR with the standard 13 CODIS loci. By July 2002, roughly 300 identifications were made using these direct reference samples. An additional 200 identifications were made by comparing victim samples to samples from living relatives using the methods we describe for the Swissair crash (see the preceding section, "Reconstructing individual genotypes").

By July 2004, a total of 1,500 victims had been positively identified, but subsequent progress was slow. The remaining samples were so damaged that the DNA was in very short pieces, too short to support STR analysis. Two avenues for additional identifications remained:

- **Mitochondrial DNA (mtDNA)** is useful for two reasons:
  - It's multicopy DNA, meaning that each cell has many mitochondria, and each mitochondrion has its own molecule of mtDNA.
  - It's circular, making it somewhat more resistant to decomposition because the nucleases that destroy DNA often start at the end of the molecule (see the section "Collecting biological evidence," earlier in this chapter), and a circle has no end, so to speak.

  mtDNA is inherited directly from mother to child; therefore, only maternal relatives can provide matching DNA. Unlike STR markers, mtDNA is usually analyzed by comparing the sequences of nucleotides from various samples (see Chapter 8 to find out how DNA sequences are generated and analyzed). Because sequence comparison is more complicated than STR marker comparison, the analyses take longer to perform.

- **Single nucleotide polymorphism (SNP analysis)** (pronounced *"snip"*; discussed in Chapter 17) relies on the fact that DNA tolerates some kinds of changes without harming the organism. SNPs occur when one base replaces another. These tiny changes occur often (estimated to occur in about one in every 1000 bases), and when many SNPs are compared, the changes can create a unique DNA profile similar to a DNA fingerprint.

The downside to SNP analysis is that the base changes don't create obvious size differences that traditional DNA fingerprinting can detect. Therefore, sequencing or gene chips (see Chapter 22 for more on gene chips) must be used to detect the SNP profile of various individuals. Because SNP analysis can be conducted on very small fragments of DNA, it allowed investigators to make more identifications than were possible otherwise. The most recent identification was made in June of 2019.

IN THIS CHAPTER

» Understanding how transgenics works

» Weighing the pros and cons of transgenic crops

» Learning about gene editing

Chapter 19

Genetic Makeovers: Using Genetic Engineering to Change the Genome

One of the most controversial applications of genetics technology is *genetic modification* (GM). Some forms of genetic modification don't involve the direct, mechanical manipulation of the genome; instead, they involve things like selective breeding or inducing mutations by radiation or other mutagen. Other forms of genetic modification, those that tend to be the most contentious, utilize genetic engineering techniques to make actual changes in an organism's DNA. The most common type of genetic modification is called *transgenics,* which involves transferring genes from one organism to another. More recently, technologies have been discovered to directly alter the genome of an organism by changing a specific sequence in the genome in a very targeted manner. This is referred to as *gene editing* or *genome editing.* In this chapter, you discover how scientists move DNA around or change DNA to endow plants, animals, bacteria, and insects with new combinations of genes and traits.

Genetically Modified Organisms Are Everywhere

News items about genetically modified this, that, and the other appear practically every day, and most of this news seems to revolve around protests, bans, and lawsuits. Despite all the brouhaha, genetically modified "stuff" is neither rare nor wholly dangerous. In fact, most processed foods that you eat are likely to contain one or more transgenic ingredients. You'll see proclamations of "No GMO," which is meant to reassure you that no *transgenes* — genes that have been artificially introduced using recombinant DNA methods (described in the section "Making a transgene using recombinant DNA technology") — were present in the plants or animals used to make the product in question. However, there are other ways to genetically modify an organism.

In truth, you can't avoid genetically modified organisms in your everyday life. Genetic modification by humans by selective breeding has been around for thousands of years. Humans started domesticating plants and animals many centuries ago (take a look at the sidebar "Amazing maize" to see how corn made the transition from grass to gracing your table).

Historically, farmers preferentially grew certain types of plants to increase the frequency of desirable traits, such as sweeter grapes and more kernels per stalk of wheat. Many, if not all, of the cereal grains humans depend on, such as wheat, rice, and barley, are the result of selective hybridization events that created polyploid organisms (organisms with more than two sets of chromosomes; see Chapter 13). When plants become polyploid, their fruits get substantially larger. Fruits from polyploids are more commercially valuable. They're also better tasting — try a wild strawberry if you're unconvinced.

In addition to domestication and selective breeding, new plant breeds have been created by purposefully induced, albeit random, gene changes. In essence, plants are exposed to radiation (such as X-rays, gamma rays, and neutrons) or chemicals to produce altered alleles (different versions of genes) aimed at producing desired traits (see Chapter 12 for how radiation damages DNA to cause changes in DNA sequence). Plants that have received radiation and chemical treatment have resulted in new varieties of food crops and ornamental flowers.

The ability to move genes from one species to another also isn't new — viruses and bacteria do it all the time. It may be unclear as to why transgenesis is less acceptable than induced mutagenesis and artificial selection, but no matter what you call it, it's all genetic modification.

AMAZING MAIZE

Plants depend on a variety of helpers to spread their seeds around: The wind, birds, animals, and waterways all carry seeds from one place to another. Most plants get along just fine without humans. Not so with corn. Corn depends *entirely* on humans to spread its seeds; archeological evidence confirms that corn has traveled only where humans have taken it. What's striking about this story is that modern geneticists have pinpointed the mutations that humans took advantage of to create one of the world's most widely used crops.

Primitive corn (called *maize*) put in its first appearance around 9,000 years ago. The predecessor of maize is a grass called *teosinte*. You need a good imagination to see an ear of corn when you look at the seed heads of teosinte; there's only a vague resemblance, and unlike corn, teosinte is only barely edible — it has a few rock-hard kernels per stalk. Yet corn and teosinte (going by the scientific name of *Zea mays*) are the same species.

The changes in five genes that turned teosinte into maize popped up naturally and changed several things about teosinte to make it a more palatable food source:

- One gene controls where cobs appear on the plant stalk: Maize has its cobs along the entire stem instead of on long branches like teosinte.
- Three genes control sugar and starch storage in the kernels: Maize is easier to digest and better tasting than teosinte.
- One gene controls the size and position of kernels on the cob: Unlike teosinte, maize has an appearance normally associated with modern corn.

Humans apparently used teosinte for food before it acquired its mutational makeover, so it's likely that people caught on quickly to the change that developed. The changes in the five genes were cemented into the genome by selective harvest and planting of the new variety. People grew the altered plants on purpose, and the only reason corn is so common now is because humans made it that way. The first true maize crops were planted in Mexico 6,250 years ago, and, as a popular addition to the diets of people in the area, its cultivation spread rapidly. Archeological sites in the United States bear evidence of maize cultivation as early as 3,200 years ago. By the time Europeans arrived, most native peoples in the New World grew maize to supplement their diets.

Old Genes in New Places

If genetic modification is so ubiquitous, what's the problem with transgenic organisms? After all, humans have been at this whole genetic modification thing for centuries, right? Well, historically, humans have modified organisms by controlling matings between animals and plants with preexisting genetic compatibility, not by introducing sequences from different species.

REMEMBER

Transgenics are often endowed with genes from very different species. (The bacterial gene that's been popped into corn to make it resistant to attack by plant-eating insects is a good example.) Therefore, transgenic organisms wind up with genes that never could have moved from one organism to another without considerable help (or massive luck; see the "Traveling genes" sidebar for more about natural gene transfer events).

After these "foreign" genes get into an organism, they don't necessarily stay put. One of the biggest issues with transgenic plants, for example, is uncontrolled gene transfer to other, unintended species. Another controversial aspect of transgenic organisms has to do with gene expression; many people worry that transgenes will be expressed in agricultural products in unwanted or unexpected ways, making food harmful to eat.

REMEMBER

To understand the promises and pitfalls of transgenics, you first need to know how transgenes are transferred and why. *Recombinant DNA technology* is the set of methods used for all transgenic applications. This set of techniques also falls under the classification of *genetic engineering*. Genetic engineering refers to the directed manipulation of genes to alter phenotype in a particular way. Thus, genetic engineering is also used in gene therapy to bring in healthy genes to counteract the effects of mutations.

TRAVELING GENES

Movement of genes from one organism to another usually occurs through mitosis or meiosis, the normal mechanisms of inheritance. With *horizontal gene transfer*, genes can move from one species to another without mating or cell division. Bacteria and viruses accomplish this task with ease; they can slip their genes into the genomes of their hosts to alter the functions of host genes or supply the hosts with new, sometimes unwanted ones. This movement of genes isn't merely scientific fiction or a rare event, either. The appearance of antibiotic-resistant genes in various species of bacteria is due to horizontal gene transfers. Horizontal transfer also occurs in multicellular organisms (various species of fruit flies have shared their genes this way). Indeed, your own genome may owe some of its size and genetic complexity to genes acquired from bacteria. The possibility of genes turning up in unexpected places is real.

While different vectors, different source DNA, and different cell populations can be used, this basic process that underlies the creation of *transgenes* is essentially the same in the different types of organisms – animals, insects, bacteria, or plants.

Following the transgenesis process

In general, developing transgenic organisms involves three major steps:

1. Find (and potentially alter) the gene that controls the trait(s) of interest.
2. Slip the transgene into an appropriate delivery vehicle (a *vector*).
3. Create fully transgenic organisms that pass the new gene on to future generations.

The process of finding and mapping genes is pretty similar from one organism to another and has been made much easier now that so many different organisms have had their genomes sequenced. After scientists identify the gene they want to transfer, they man need to alter the gene so that it works properly outside the original organism.

Making a transgene using recombinant DNA technology

Recombinant DNA technology involves taking DNA from two different sources, cutting it, and then putting the pieces back together in different combinations. This process has been used for decades in order to clone genes or other segments of DNA. When we say "clone genes," we mean to isolate the gene (various methods exist for doing this), insert it into a vector (such as a bacterial plasmid — a circular piece of DNA from a bacterium), and replicate it in appropriate host cells, in order to create exact copies of the gene.

In order to clone a gene, the DNA first needs to be cut. To do this, scientists use restriction enzymes. *Restriction enzymes* are naturally made by bacteria, which use these proteins as a defense for viral infection. More than 3000 restriction enzymes have been characterized, with each one cutting DNA at a very specific sequence. For example, a common restriction enzyme known as *Eco*RI (which is made by the bacterium *E. coli*) cuts specifically at the sequence 5′-GAATTC-3′. *Eco*RI is also known to cut DNA so that sticky ends result — that is, it leaves short single-stranded DNA sequences at the ends of the cut fragments. Figure 19-1 shows how a restriction enzyme can cut DNA to leave sticky ends. For more about the 5′ to 3′ orientation of DNA, see Chapter 5.

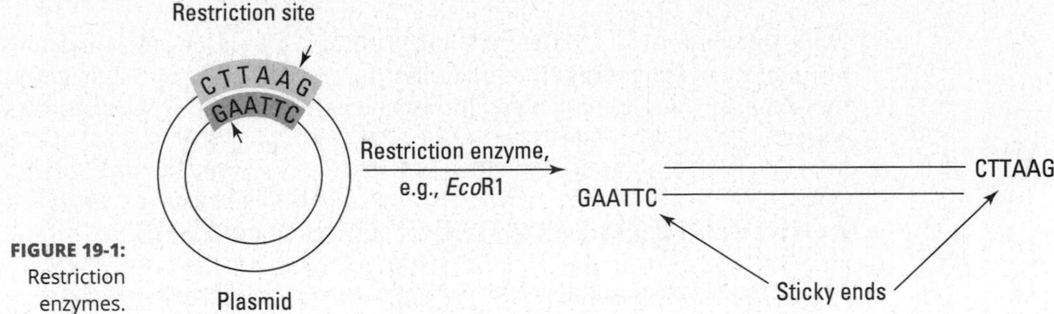

FIGURE 19-1: Restriction enzymes.

The AATT overhang allows another fragment with a complementary end to come in and bind here. So, if you cut a fragment of human DNA with *Eco*RI and you cut a bacterial plasmid with *Eco*RI, both would end up having the same sticky ends. When you combine them, the fragment of human DNA can then be inserted between the two sticky ends of the bacterial plasmid (see Figure 19-2).

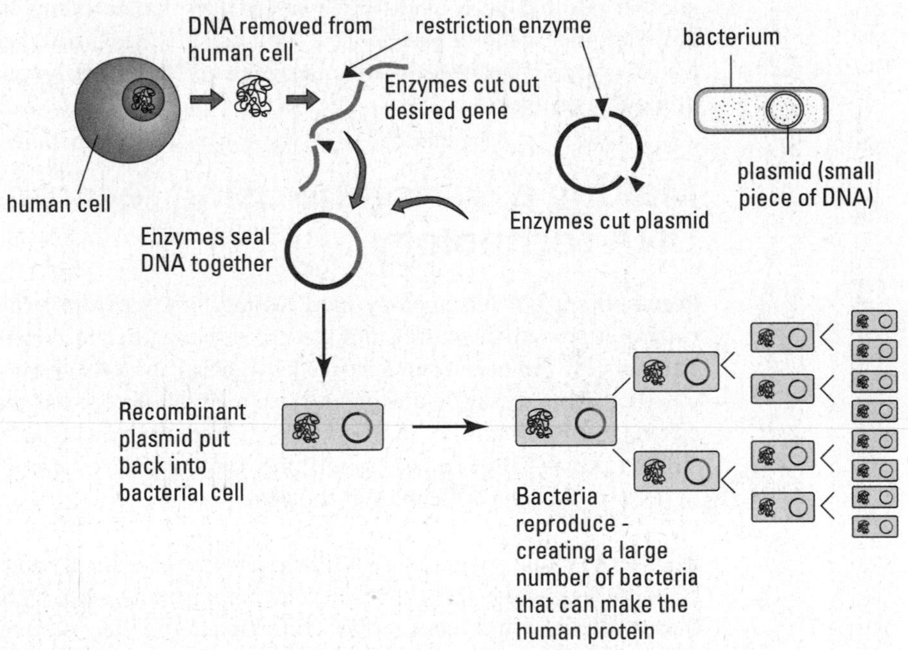

FIGURE 19-2: Cloning a gene.

The plasmid can then be introduced into bacterial cells, which are grown in culture. In every new bacterial cell, there will be a copy of the plasmid vector containing the gene of interest. The gene is considered cloned after it is placed in the vector and the vector has been reproduced.

Modifying the gene to reside in its new home

All genes must have *promoter sequences*, the genetic landmarks that identify the start of a gene, to allow transcription to occur (for the scoop on transcription, see Chapter 9). When it comes to creating a transgenic organism, the promoter sequence in the original organism may not be very useful in the new host; as a result, a new promoter sequence is needed to make sure the gene gets turned on when and where it's wanted. Using the same recombinant DNA techniques used in cloning a gene, an appropriate promoter can be inserted upstream of the gene to be introduced, such that the gene would now be under the control of the designated promoter. Some promoters that are used are set to keep the gene on continually in all the cells of the organism. Other promoters are used because they only turn on (express the gene) in certain cell types or at a certain point in development. When more precise regulation is needed, genetic engineers can use promoters that respond to conditions in the environment (see Chapter 11 for more about how cues in the environment can control genes).

In addition to the promoter, genetic engineers must also find a good companion gene — called a *marker gene* — to accompany the transgene. The marker gene provides a strong and reliable signal indicating that the whole unit (marker and transgene) is in place and working. Common markers include genes that convey resistance to antibiotics. With these kinds of markers, geneticists grow transgenic cells in a medium that contains the antibiotic. Only those that have resistance (conveyed by the marker gene) survive, providing a quick and easy way to tell which cells have the transgene (alive) and which don't (dead). Other markers or reported genes that can be used are those that allow for visualization of transgene expression, such as the gene for green fluorescent protein (GFP; see the sidebar "Transgenic pets: Glow-in-the-dark fish" for more on this gene).

Looking at the GMO Menagerie

Transgenic critters are all over the place. Animals, insects, and bacteria have all gotten in on the fun. In this section, you take a trip to the transgenic zoo.

Transgenic animals

Mice were the organisms of choice in the development of transgenic methods. One of the ways that transgenic mice can be created is by inserting a transgene into a mouse's genome during the process of fertilization. When a sperm enters an egg, there's a brief period before the two sets of DNA (maternal and paternal) fuse to become one. The two sets of DNA existing during this intermission are called *pronuclei.* Geneticists discovered that by injecting many copies of the transgene (with its promoter and sometimes with a marker gene, too; see the earlier section "Modifying the gene to reside in its new home") directly into the paternal pronucleus (see Figure 19-3), the transgene was sometimes integrated into the embryo's chromosomes.

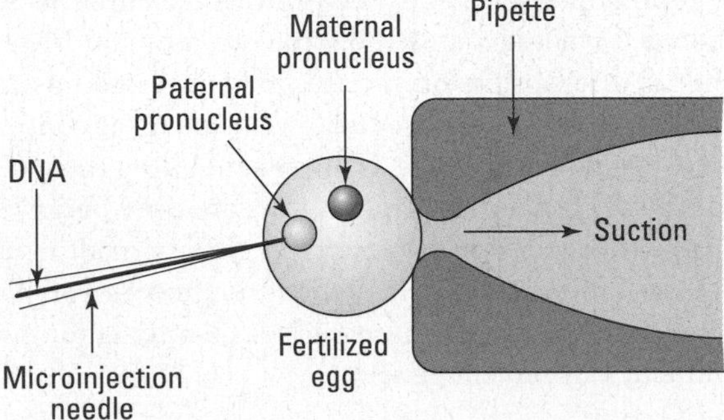

FIGURE 19-3: Researchers introduce transgenes into mouse embryos before fertilization occurs.

TECHNICAL STUFF

If the transgene integrates into one of the mouse chromosomes when still at the one-cell stage, it will end up in all the cells of the mouse's body. If it integrates after several rounds of cell division, then not all the embryo's cells will contain the transgene. The cells that do have the transgene often have multiple copies (oddly, these end up together in a head-to-tail arrangement), and the transgenes are inserted into the mouse's chromosomes at random. The resulting, partly transgenic mouse is *mosaic.* Mosaicism is when there are two populations of cells in the animal. In this case, there is one that contains and expresses the transgene and one that does not. To get a fully transgenic animal, many mosaic animals are mated in the hope that non-mosaic transgenic offspring will be produced from one or more matings.

One of the first applications of the highly successful mouse transgenesis method used growth hormone genes. When introduced into the mouse genome, rat, human, and bovine growth hormone genes all produced mice that were much larger than normal. The result encouraged the idea that growth hormone genes engineered into meat animals would allow faster production of larger, leaner animals. However, transgenic pigs with the human growth hormone gene didn't fare very well; in studies, they grew faster than their nontransgenic counterparts but only when fed large amounts of protein. And female transgenic pigs turned out to be sterile. All the pigs showed muscle weakness, and many developed arthritis and ulcers. Unfortunately, cows didn't fare any better. In contrast, fish do swimmingly with transgenes (see the sidebar "Transgenic pets: Glow-in-the-dark fish" for one application of transgenics in fish) and transgenic salmon with the growth hormone gene grow six times faster than their nontransgenic cousins and convert their food to body weight much more efficiently, meaning that less food makes a bigger fish.

Primates have also been targeted for transgenesis as a way to study human disorders including aging, neurological diseases, and immune disorders. The first transgenic monkey was born in 2000. This rhesus monkey was endowed with a simple marker gene because the purpose of the study was simply to determine whether transgenesis in monkeys was possible. Since then, transgenic primates that model human disease, such as a transgenic monkey that can be used to model Huntington disease (described in Chapters 4 and 12), have been created and are showing great promise as a tool for studying treatments for these conditions.

TRANSGENIC PETS: GLOW-IN-THE-DARK FISH

Ever have one of those groovy posters that glows under a black light? Well, move that black light over to the aquarium — there's a new fish in town. Originally derived from zebrafish, a tiny, black-and-white–striped native of India's Ganges River, these glowing versions bear a gene that makes them fluorescent. The little, glow-in-the-dark wonders (referred to as GloFish) are the first commercially available transgenic pets.

Zebrafish are tried-and-true laboratory veterans — they even have their own scientific journal! Developmental biologists love zebrafish because their transparent eggs make it simple to observe development. Geneticists use zebrafish to study the functions of all sorts of genes, many of which have direct counterparts in other organisms, including humans. And genetic engineers have taken advantage of these easy-to-keep fish, too; scientists in Singapore saw the potential to use zebrafish as little pollution indicators. The Singapore geneticists used a gene from jellyfish to make their zebrafish glow in the dark. The action of the fluorescent gene is set up to respond to cues in the environment (like hormones, toxins, or temperature; see Chapter 11 for how environmental cues turn on your genes). The transgenic zebrafish then provide a quick and easy to read signal: If they glow, a pollutant is present.

Of course, glowing fish are so unique that some enterprising soul couldn't let lab scientists have all the fun. Thus, these made-over zebrafish have hit the market. Currently, GloFish are available in more than 10 different colors, including green, red, orange, pink, purple, or blue! Initially, when GloFish were introduced in 2003, the state of California banned their sale outright. However, they changed their decision in 2015, after GloFish sales were approved by the FDA, the US Fish and Wildlife Service, and a variety of state-level regulators.

Transgenic insects

A number of uses for transgenic insects appear to be on the horizon. Malaria and other mosquito-borne diseases are a major health problem worldwide, but the use of pesticides to combat mosquito populations is problematic because resistant populations rapidly replace susceptible ones. And in fact, the problem isn't really the mosquitoes themselves (despite what you may think when you're being buzzed and bitten). The problem is the parasites and viruses the mosquitoes carry and transmit through their bites. In response to these problems, at least one research group has developed a transgenic mosquito in which expression of the transgene results in an early death. Research trials with these mosquitos in Brazil were designed to see if they may help decrease the population of mosquitoes that spread dengue fever or the Zika virus. However, it is still unclear whether the release of these transgenic mosquitos is decreasing the population without the unintended consequence of creating hybrid mosquitos.

Another potential use for transgenic insects would be to release millions of transgenically infertile bugs that attract the mating attentions of fertile ones. The matings result in infertile eggs, reducing the reproduction of the target insect population. This is an especially appealing idea when used to combat invasive species that can sweep through crops with economically devastating results.

Transgenic bacteria

Bacteria are extremely amenable to transgenesis. Unlike other transgenic organisms, genes can be inserted into bacteria with great precision, making expression far easier to control. As a result, many products can be produced using bacteria, which can be grown under highly controlled conditions, essentially eliminating the danger of transgene escape.

REMEMBER

Many important drugs are produced by recombinant bacteria, such as insulin for treatment of diabetes, clotting factors for the treatment of hemophilia, and human growth hormone for the treatment of some forms of short stature. These sorts of medical advances can have important side benefits as well:

- Transgenic bacteria can produce much greater volumes of proteins than traditional methods.
- Transgenic bacteria are safer than animal substitutes, such as pig insulin, which are slightly different from the human version and may therefore cause allergic reactions.

To use bacteria to make human proteins for the treatment of disease, the gene for the human protein must be isolated. For example, to make human insulin, DNA is obtained from an insulin-producing pancreatic cell (see Figure 19-4). The insulin gene is isolated and inserted into a bacterial plasmid. The plasmid containing the human gene is then introduced into *E. coli*, which then acts as a factory for making the insulin protein. Transgenic bacteria are used to produce large quantities of insulin, which can then be purified and used to treat patients with diabetes.

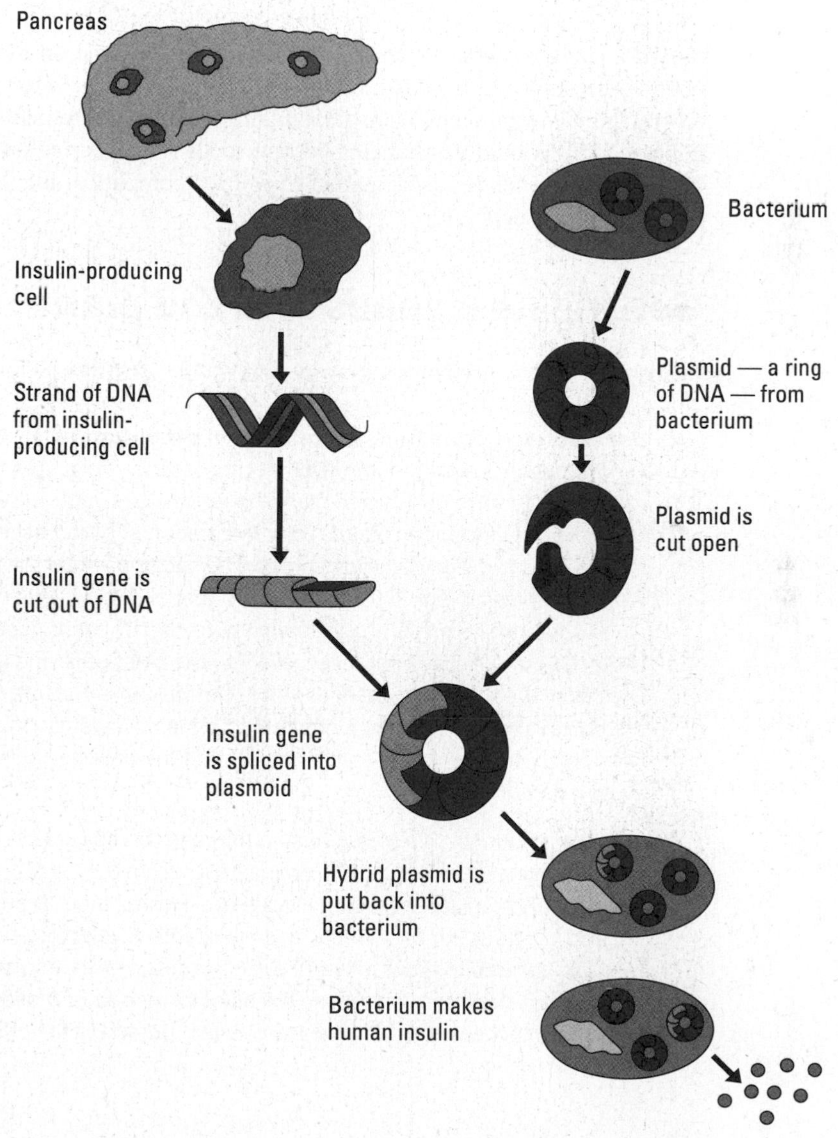

FIGURE 19-4: The use of transgenic bacteria to make human insulin.

Transgenic plants

Plants are really different from animals, but not in the way you may think. Plant cells are *totipotent*, meaning that practically any plant cell can eventually give rise to every sort of plant tissue: roots, leaves, and seeds. When animal cells differentiate during embryo development, they lose their totipotency forever (but the DNA in every cell retains the potential to be totipotent). For genetic engineers, the totipotency of plant cells reveals vast possibilities for genetic manipulation.

Much of the transgenic revolution in plants has focused on moving genes from one plant to another, from bacteria to plants, or even from animals to plants. Like all transgenic organisms, transgenic plants are created to achieve various ends, including nutritionally enhancing certain foods (such as rice) or altering crops to resist either herbicides used against unwanted competitor plants or the attack of plant-eating insects.

Getting new genes into the plant

To put new genes into plants, genetic engineers can either:

- **Use a vector system from a common soil bacterium called *Agrobacterium*.** *Agrobacterium* is a plant pathogen that causes *galls* — big, ugly, tumor-like growths — to form on infected plants. In Figure 19-5, you can see what a gall looks like. Gall formation results from integration of bacterial genes directly into the infected plant's chromosomes. The bacteria enters the plant from a wound such as a break in the plant's stem that allows bacteria to get past the woody defense cells that protect the plant from pathogens (just as your skin protects you). The bacterial cells move into the plant cells, and once inside, DNA from the bacteria's plasmids — the circular DNAs that are separate from the bacterial chromosome — integrate into the host plant's DNA. The bacterial DNA pops itself in more or less randomly and then hijacks the plant cell to allow it to replicate.

  Like the geneticists using virus vectors for gene therapy (see Chapter 16), genetic engineers snip out gall-forming genes from the *Agrobacterium* plasmids and replace them with transgenes. Host plant cells are grown in the lab and infected with the *Agrobacterium.* Because these cells are totipotent, they can be used to grow an entire plant — roots, leaves, and all — and every cell contains the transgene. When the plant forms seeds, those contain the transgene, too, ensuring that the transgene is passed to the offspring.

» **Shoot plants with a *gene gun* so that microscopic particles of gold or other metals carry the transgene unit into the plant nucleus by brute force.** Gene guns are a bit less dependable than *Agrobacterium* as a method for getting transgenes into plant cells. However, some plants are resistant to *Agrobacterium,* thus making the gene gun a viable alternative. With gene guns, the idea is to coat microscopic pellets with many copies of a transgene and by brute force (provided by compressed air) shove the pellets directly into the cell nuclei. By chance, some of the transgenes are inserted into the plant chromosomes.

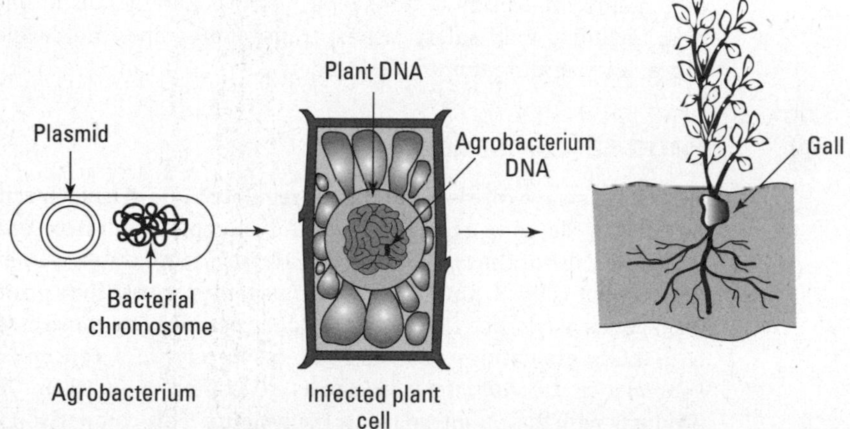

FIGURE 19-5: *Agrobacterium* inserts its genes into plant cells to cause gall formation.

Exploring commercial applications

Transgenic plants have made quite a splash in the world of agriculture. So far, the main applications of this technology have addressed two primary threats to crops:

» **Weeds:** The addition of herbicide-resistant genes make crop plants immune to the effects of weed-killing chemicals, allowing farmers to spread herbicides over their entire fields without worrying about killing their crops. Weeds compete with crop plants for water and nutrients, reducing yields considerably. Soybeans, cotton, and canola (a seed that produces cooking oil) are only a few of the crop plants that have been genetically altered to tolerate certain herbicides.

» **Bugs:** The addition of transgenes that confer pest-killing properties to plants effectively reduces crop losses to plant-eating bugs. Geneticists provide pest-protection traits using the genes from *Bacillus thuringiensis* (otherwise known as *Bt*). Organic gardeners discovered the pesticide qualities of *Bt,* a soil bacterium, years ago. *Bt* produces a protein called *Cry.* When an insect eats the soil bacteria, digestion of *Cry* releases a toxin that kills the insect shortly after its meal. However, the *Cry* toxin is not toxic to animals and has been deemed safe for consumption by the FDA. Transgenic corn and cotton were the first to carry the *Bt Cry* gene. Others now include potatoes, eggplants, soybeans, and tomatoes, although not all are commercially available yet.

Weighing points of contention

Few genetic issues have excited the almost frenzied response met by transgenic crop plants. Opposition to transgenic plants generally falls into four basic categories, including food safety issues, transgene escape, the development of resistance, and harming unintended targets.

Food safety issues

Normally, gene expression is highly regulated and tissue-specific, meaning that proteins produced in a plant's leaves, for example, don't necessarily show up in its fruits. Because of the way transgenes are inserted, however, their expression isn't under tight control. Opponents to transgenics worry that proteins produced by transgenes may prove toxic, making foods produced by those crop plants unsafe to eat. Safety evaluations of transgenic crops rely on a concept called substantial equivalence. *Substantial equivalence* is a detailed comparison of transgenic crop products with their nontransgenic equivalents. This comparison involves chemical and nutritional analyses, including tests for toxic substances. If the transgenic product has some detectable difference, that trait is targeted for further evaluation. Thus, substantial equivalence is based on the assumptions that any ingredient or component of the nontransgenic product is already deemed safe and that only new differences found in the transgenic version are worth investigating. For example, in the case of transgenic potatoes, unmodified potatoes are thought to be safe, so only the Bt that had been introduced was slated for further tests.

Escaped transgenes

The escape of transgenes into other hosts is a widely reported fear of transgenics opponents. Canola, a common oil-seed crop, provides one good example of how quickly transgenes can get around. Herbicide-resistant canola was marketed in Canada in 1996, or so. By 1998, wild canola plants in fields where no transgenic crop had ever been grown already had not one but *two* different transgenes for herbicide resistance. This finding was quite a surprise because no commercially available transgenic canola came equipped with both transgenes. It's likely that the accidental transgenic acquired its new genes via pollination. In 2002, several companies in the United States failed to take adequate precautions mandated by law to prevent the escape of corn transgenes via pollination or the accidental germination of untended transgenic seeds. These lapses resulted in fines — and the release of transgenes into unintended crops.

Developing resistance

The third major point of opposition to transgenics — the development of resistance to transgene effects — is connected to the widespread movement of transgenes. The point of developing most of these transgenic crops is to make controlling weeds or insect pests easier. Additionally, transgenic crops (particularly transgenic cotton) have the potential to significantly reduce chemical use, which is a huge environmental plus. However, when weeds or insects acquire resistance to transgene effects, the chemicals that transgenics are designed to replace are rendered obsolete. Full-blown resistance development depends on artificial selection supplied by the herbicide or the plant itself. Resistance develops and spreads when insects that are susceptible to the pesticide transgene being used are all killed. The only insects that survive and reproduce are, you guessed it, able to tolerate the pesticide transgene. Insects produce hundreds of thousands of offspring, so it doesn't take long to replace susceptible populations with resistant ones.

Damaging unintended targets

The argument against transgenic plants is that nontarget organisms may suffer ill effects. For example, when Bt corn was introduced (see the section "Exploring commercial applications," earlier in this chapter), controversy arose surrounding the corn's toxicity to beneficial insects (that is, bugs that eat other bugs) and desirable creatures like butterflies. Indeed, Bt is toxic to some of these insects, but it's unclear how much damage these natural populations sustain from Bt plants. The biggest threat to migratory monarch butterflies is likely habitat destruction in their overwintering sites in Mexico, not Bt corn.

Changing the Blueprint: Gene Editing

One of the hottest topics these days is gene editing (also known as genome editing). *Gene editing* is a group of genetic engineering technologies that allow scientists to change a specific sequence within the genome. Each of these technologies involves an engineered enzyme called a *nuclease*, which can cut DNA, along with some kind of guide to lead the enzyme to the right place in the genome. Subsequently, the cell's own DNA repair machinery can repair the break, inserting a new segment of DNA if a fragment of DNA with the desired sequence is provided.

Gene editing could be used in a couple of different ways. First, it could be used to replace a sequence with a disease-causing mutation with the normal sequence, thereby correcting the underlying cause of a patient's condition. Another use could be to disrupt a gene that is being expressed and turn it off. This would be helpful for conditions in which there is an excess of expression that contributes to the disorder, such as oncogenes in cancer (see Chapter 14).

Technologies to perform gene editing have been around for more than 20 years; however, the different tools vary with respect to how well they work. Gene editing tools that have been developed include the following:

- **Zinc finger nucleases (ZFNs):** ZFNs use a specific type of DNA binding domain (referred to as a zinc finger domain) to recognize the target sequence. While they have been successfully used to modify various plants, they are not really used anymore because they frequently affected the wrong sequence.
- **Transcription activator-like effector nucleases (TALENs):** TALENs use a transcription activator-like effector DNA binding domain that can be customized to target a specific sequence. TALENs have a higher efficiency than ZFNs and have been used to alter the genomes of a variety of different organisms. It has also been used experimentally to correct mutations that cause human disease.
- **CRISPR-Cas9 (pronounced "*crisper cass* 9"):** The CRISPR-Cas9 system (which is described in the following section) is the most recently introduced and appears to have the highest accuracy of any of the methods. The system uses a customizable RNA guide to target the sequence to be edited, along with a CRISPR-associated protein (Cas). Because of its efficiency and accuracy, this method is rapidly becoming the most widely used approach for performing gene editing.

The CRISPR-Cas9 gene editing system is also the method that has been hitting the headlines. The basics of this system are described in the following section. In addition, since the system could be used in either somatic cells (the cells of the body other than the eggs or sperm) or in the germline (the sex cells – eggs and sperm), the importance of this distinction is discussed. The promise that the CRISPR-Cas9 system holds for the future treatment of genetic diseases has also raised some significant concerns, which are introduced in the section "Discussing the ethics of gene editing."

CRISPR-Cas9 gene editing

CRISPR stands for clustered regularly interspersed short palindromic repeats. *Cas9* is the CRISPR-associated protein number 9. CRISPR-Cas9 is based on a defense system that occurs naturally in some bacteria.

Scientists discovered that the DNA in these bacteria contained many short palindromic sequences. Palindromes are words or sequences that are the same both forward and backward, such as the word *racecar* or the DNA sequence *CATAATAC*.

In the bacteria, these short palindromic sequences flanked segments of DNA from a virus that had previously infected the cell. Basically, every time a bacterium was infected with a virus (and survived the infection), the bacterium saved a bit of the viral DNA by inserting it between the palindromic DNA sequences (see Figure 19-6). It would then make a small piece of RNA (see Chapter 7 for more on RNA) from this sequence, which would then be attached to a Cas protein that is able to cut DNA into pieces. If the bacterium was infected again by the same virus, the armed Cas protein would recognize the viral DNA and destroy it immediately.

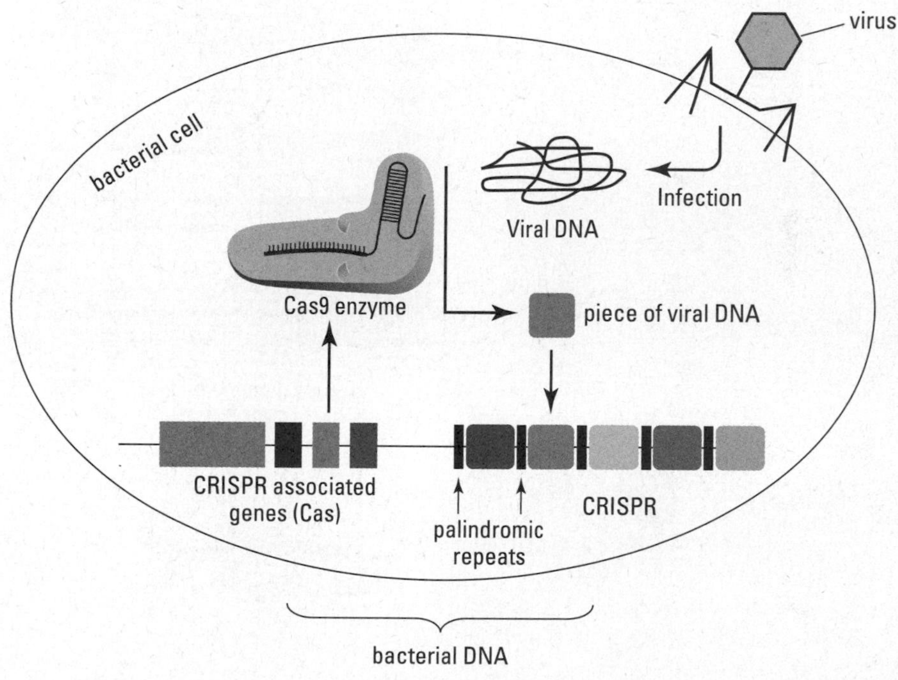

FIGURE 19-6: The CRISPR-Cas9 system in bacteria.

The CRISPR-Cas9 system for gene editing uses a short RNA guide sequence that can bind to a specific target sequence in the genome (see Figure 19-7). This guide RNA also binds to the Cas9 protein. To use this system for correcting a gene known to carry a mutation, the Cas9 protein with the guide RNA attached could be introduced into a cell with the defective gene, along with fragments containing the normal gene sequence. Once the guide RNA-Cas9 complex binds to its target sequence in the genome, the Cas9 protein would cut the DNA at the target location. The cells' own DNA repair machinery could then be used to replace the defective copy of the gene with a correctly functioning version (without any mutation). Another potential use for this system would be to disrupt genes that are causing problems. For example, if a gene is producing a growth factor that is contributing to cancer cell division and proliferation, the guide sequence could target the gene and the attached Cas9 protein could cut it and stop it from making the growth factor.

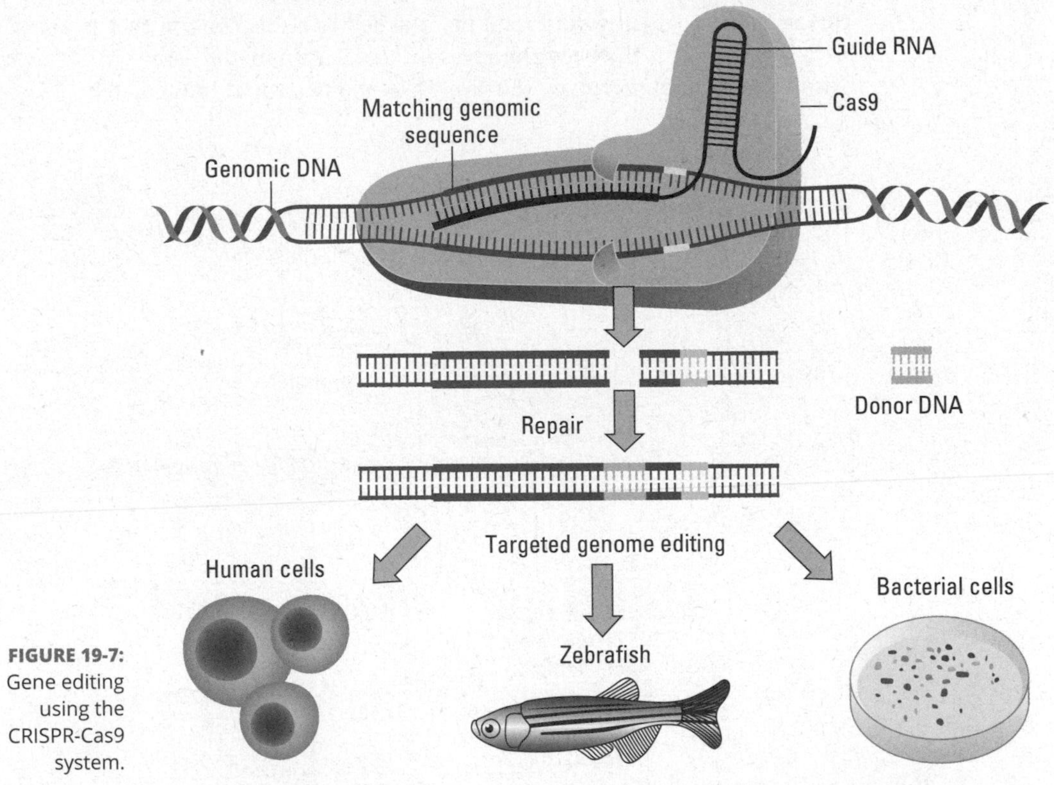

FIGURE 19-7: Gene editing using the CRISPR-Cas9 system.

Germline versus somatic gene editing

Gene editing could be performed in a somatic cell (a body cell other than an egg or sperm), such that the edited sequence ends up only in the tissue of interest. In this case, the genetic change would not be passed on to subsequent generations. Alternatively, gene editing could be performed in a manner that results in the altered sequence being in all the cells of the body. In this case, the editing would need to occur in a germ cell (such as an egg or sperm) or in a one-cell zygote (fertilized egg). This would also lead to a genome alteration what would be passed onto future generations, something associated with a unique set of ethical issues.

Discussing the ethics of gene editing

Both germline and somatic gene editing raise significant ethical concerns. In both cases, there are concerns about off-target editing (that is, when the wrong sequence gets changed, perhaps with devastating consequences). And like any new technology, there are definite concerns regarding the overall safety of such a procedure and the health implications for the organism in which the editing was performed (and potentially future offspring if germline editing is carried out).

In addition, like all other genetic engineering technologies and some genetic testing approaches, there are worries regarding the potential to use these techniques to enhance one's genetics, as opposed to treating a genetic condition.

Informed consent (described in detail in Chapter 20) also triggers some very real apprehensions, since modifying someone's germline involves performing the change before the person is born and before they can provide consent.

From a societal perspective, concerns have been raised regarding the equal access to such technologies and concerns related to how genetic diseases or traits are viewed. If therapies based on gene editing become clinically available, will everyone be able to afford them? Will this lead to a negative view of certain genetic disorders or traits, and an attempt to remove them from society?

Most of these concerns have been brought to the forefront since the first report demonstrating that genome editing in a human embryo is possible, which was published by a Chinese research group in 2015. This study, however, utilized an embryo that was not capable of developing to term. Later in 2015, there was a call for a voluntary worldwide moratorium on using CRISPR-Cas9 to alter the germline genome in humans. Despite this, a later report from China in 2018, described the use of gene editing to make twins who were immune to HIV infection. However, the genetic status of these individuals has not been verified.

At the time this book was written, a search on the clinicaltrials.gov website (which lists future, ongoing, and completed clinical trials) identified 19 different trials designed to study gene editing in somatic cells using CRISPR-Cas9. This included studies for the treatment of sickle cell disease, beta-thalassemia, and blood cell cancers (such as leukemia), and at least three of these trials have study sites in the United States. Little has been reported from these studies, but one researcher has stated that the first beta-thalassemia patient he treated hadn't needed a blood transfusion in the four months since treatment was initiated. No trials were identified that involved modification of the human germline.

IN THIS CHAPTER

» Learning about eugenics

» Exploring the possibility of designer babies

» Examining the ethical issues of genetic testing

» Treating genetic disorders safely

Chapter 20
Giving Ethical Considerations Their Due

The field of genetics grows and changes constantly. When it comes to genetics, the amount of information is bewildering, and the possibilities are endless. If you've already read many of the chapters in this book, you have a taste of the many choices and debates created by the burgeoning technology surrounding our genes.

With such a fast-growing and far-reaching field as genetics, ethical questions and issues arise around every corner and are interconnected with the applications and procedures. In this chapter, you find out how genetics has been misunderstood, misinterpreted, and misused to cause people harm based on their racial, ethnic, or socioeconomic status. It also dispels the myth of designer babies and discusses the ethical issues that surround genetic testing, including the testing of minors and direct-to-consumer testing. You also discover how information you give out and receive can be used for and against you.

Profiling Genetic Discrimination

One of the biggest hot button issues of all time must be *eugenics*. In a nutshell, eugenics is the idea that humans should practice selective reproduction in an effort to "improve" the species. If you read Chapter 19, which explains how organisms can be genetically engineered, you probably already have some idea of what eugenics in the modern age may entail (transgenic, made-to-order babies, perhaps?). Historically, the most blatant examples of eugenics are genocidal activities the world over. (Perhaps the most infamous example occurred in Nazi Germany during the 1930s and 1940s.)

HISTORICAL STUFF

The story of eugenics begins with the otherwise laudable Francis Galton, who coined the term in 1883. (Galton is best remembered for his contribution to law enforcement: He invented the process used to identify persons by their fingerprints. Check out Chapter 18 for more on the genetic version of fingerprinting.) In direct and vocal opposition to the U.S. Constitution, Galton was quite sure that all men were *not* created equal (we emphasize here that he was particularly fixated on men; women were of no consequence in his day). Instead, Galton believed that some men were quite superior to others. To this end, he attempted to prove that "genius" is inherited. The view that superior intelligence is heritable is still widely held despite abundant evidence to the contrary. For example, twin studies conducted as far back as the 1930s show that genetically identical persons are not intellectually identical.

Galton, who was one of Charles Darwin's cousins, gave eugenics its name, but his ideas weren't unique or revolutionary. During the early 20th century, as understanding of Mendelian genetics (see Chapter 3) gathered steam, many people viewed eugenics as a highly admirable field of study. Charles Davenport was one such person. Davenport holds the dubious distinction of being the father of the American eugenics movement (one of his eugenics texts is subtitled "The science of human improvement by better breeding"). The basis of Davenport's idea is that "degenerate" people shouldn't reproduce. This notion arose from something called *degeneracy theory,* which posits that "unfit" humans acquire certain undesirable traits because of "bad environments" and then pass on these traits genetically. To these eugenicists, unfit included "shiftlessness," "feeblemindedness," and poverty, among other things.

While the British, including Galton, advocated perpetuating good breeding (along with wealth and privilege), many American eugenicists focused their attention on preventing *cacogenics,* which is the erosion of genetic quality. Therefore, they advocated forcibly sterilizing people judged undesirable or merely inconvenient. Shockingly, the forcible sterilization laws of this era have never been overturned, and until the 1970s, it was still common practice to sterilize mentally ill persons without their consent — an estimated 60,000 people in the United States suffered this atrocity. Some societies have taken this sick idea a step further and *murdered* the "unfit" in an effort to remove them and their genes permanently.

REMEMBER

Sadly, violent forms of eugenics, such as genocide and forced sterilization, are still advocated and practiced all over the world. But not all forms of eugenics are as easy to recognize as these extreme examples. To some degree, eugenics lies at the heart of most of the other ethical quandaries we address in this chapter. In addition, it only requires a little imagination to see how gene therapy (covered in Chapter 16), gene transfer (covered in Chapter 19), or other genetic technologies can be abused to advance the cause of eugenics.

Ordering Up Designer Babies

One of the more contentious issues with a root in eugenics stems from a combination of genetic technologies and the fantasy of the perfect child to create a truly extreme makeover — designer babies. In theory, a designer baby may be made-to-order according to a parent's desire for a particular sex, hair and eye color, and maybe even athletic ability. However, in reality, the technologies are not really there yet for "designing" the perfect child.

REMEMBER

The fantasy of the designer baby rests on the fallacy of *biological determinism*. Biological determinism assumes that genes are expressed in precise, repeatable ways — in other words, genetics is identity is genetics. However, this assumption isn't true. Gene expression is highly dependent on environment, among other things (see Chapter 11 for more about how gene expression works).

So, where does the myth of designer babies come from? A lot of it started with the development of *preimplantation genetic diagnosis,* or PGD (which was introduced in Chapter 15). PGD is genetic testing that is performed before an embryo is implanted in the womb, after *in vitro* fertilization is performed. Although it's true that PGD opens the possibility of creating transgenic humans or humans with edited genomes (see Chapter 19 for details), the likelihood of PGD becoming commonplace is extremely remote.

First, the *in vitro* fertilization process (the fertilization of an egg by sperm in the laboratory) that plays a role in current-day applications of the science in question isn't always easy — just ask any couple who's gone through it in an effort to get pregnant. *In vitro* procedures are extremely expensive, invasive, and painful, and women must take large quantities of strong and potentially dangerous fertility medications to produce a sufficient number of eggs. And in the end, many fertilizations don't result in pregnancies.

TECHNICAL STUFF

Second, the process of PGD can be technologically complicated. After *in vitro* fertilization is performed and the fertilized eggs are allowed to grow for a few days, the embryos are then biopsied and tested for specific gene mutations or other genetic variations for which the baby is at risk. No matter why PGD is performed, the decision about what's done with the fertilized eggs that don't meet the desired requirements fosters ethical concerns. And for those that are chosen and inserted into the mother's uterus, many will fail to implant and would thus not survive. Although lack of implantation is also true when conception occurs naturally, it's still a very tough call when someone is deciding the fate of each embryo. For those not used, options include donation to other couples, donation for research purposes, or destruction.

Like PGD, the decision to use prenatal diagnosis after a pregnancy is started (which was introduced in Chapter 15) and the decision of what to do with the results isn't easy or straightforward. Without getting too philosophical, suffering is a highly personal experience; that is, what is okay to one person may not be for someone else. One example of relative suffering that comes up a lot is hereditary deafness. If a deaf couple chooses prenatal diagnosis, what's the most desirable outcome? On one hand, a deaf child shares the worldview of his or her parents. On the other hand, a hearing child fits into the world of nondeaf people more easily. By now, you see how complex the issues surrounding prenatal diagnosis are. It seems clear that right answers, if there are any, will be very hard to come by.

PGD and other forms of prenatal diagnosis may give parents the choice of whether or not to have a child with a particular genetic condition. They do not, however, provide parents with the means to choose all the traits their children may have. They may be able to choose sex in some cases (like when a couple chooses to implant only female embryos in order to minimize the chance of having a child with an X-linked disorder that is running through the family), but they cannot pick the child that has high intelligence, brown hair, and blue eyes. The technology for that just doesn't exist (and even if it did, most genetics professionals and other clinicians would be seriously opposed and unwilling to do it).

Ethical Issues Surrounding Genetic Testing

Most people have had some type of medical test at some point in their life —an X-ray, blood work, a urine test, or other test. Most medical tests provide information about the patient taking the test at a certain point in their life. In contrast, genetic testing can provide results not only about the patient, but also about his or her family. In addition, the results of a genetic test can not only provide information relevant to the current management of a patient, it can provide information about his or her future. Another key difference about genetic testing is the potential complexity of the results. The results of most medical tests are relatively straightforward; the tests have been around for a long time, the benefits and limitations have been well established, and physicians have been trained to understand them. The results of a genetic test are not always so straightforward. Often a result is obtained that has never been reported before, and the implications of such a result may be unknown.

Another unique feature of genetic testing is that it can glean massive amounts of information about a person — from an individual's sex to his or her racial and ethnic makeup — from even a very tiny sample of tissue. Of course, the procedure can also detect the presence of mutations for inherited disorders. But given that your DNA has so much personal information stored in it, shouldn't you have complete control over the data it contains? The answer to this question is becoming more and more contentious as the definitions of, and limits to, informed consent are explored. The rights of persons both living and dead are at stake. For example, the descendants of Thomas Jefferson consented to genetic testing in 1998 to settle a long-standing controversy about Jefferson's relationship with one of his slaves, Sally Hemings (see Chapter 18 for the full story). In the Jefferson case, the matter was more than just academic curiosity because the right to burial in the family cemetery at Monticello was at stake.

Because of differences such as those described above, there are many ethical issues surrounding genetic testing to contend with. Several of the biggest issues are informed consent, patient autonomy, privacy and confidentiality, genetic discrimination, and the possibility of unexpected or unclear results. In addition, both the testing of minor and direct-to-consumer genetic testing bring up some serious concerns.

Informed Consent

Informed consent is an important ethical and legal issue. Basically, the idea is that a person can only truly make an informed decision about having a procedure when he or she is fully apprised of all the facts, risks, and benefits. Informed consent can only be given by the person receiving the procedure or by that person's legal guardian. Generally, guardianship is established in cases where the recipient of the procedure is too young to make decisions for themselves or is deemed legally incapacitated in some way; presumably, guardians have the best interests of their wards at heart. In order for informed consent to be considered valid, an individual needs to be adequately informed, legally competent to make decisions, and able to make the relevant decision voluntarily and without coercion.

Several major issues exist in the debate over informed consent:

- Many genetic tests are quite broad in nature (like whole exome sequencing, which tests all the coding regions of the entire genome; see Chapter 22), and the possible results are extensive and complex. Unexpected results are a definite possibility.
- The results of genetic tests have implications not only for the patient, but also for their family members, who may or may not want the information obtained by the test.
- Genetic testing can be performed on embryos and fetuses, obtaining genetic information about them without them ever having given consent (and they may not want to have had this information when they are adults).
- After DNA samples are obtained and genetic testing is done, information storage and privacy assurance could be problematic.

The issue of informed consent, or lack thereof, is complicated by the ability to store tissue for long periods of time. In some cases, patients or their guardians gave informed consent for certain tests but didn't include tests that hadn't yet been developed. Some institutions routinely practice long-term tissue storage, making informed consent a frequent point of contention. For example, a children's hospital in Britain was taken to task over storage of organs that were obtained during autopsies but weren't returned for internment with the rest of the body. Parents of the affected deceased gave consent for the autopsies but not for the retention of tissues.

Biologists also use stored tissue to create *cell lines*, living tissues that grow in culture for research purposes. The original cell donors are often dead, usually from the disease under study. Cell lines aren't that hard to make and maintain (if you know what you're doing), but the creation of cell lines raises the question of whether the original donor has ownership rights to cells descended from his or her tissue.

One of the areas in which a lack of informed consent is a huge issue is genetic testing in minors. Genetic testing in children who show symptoms of a genetic disorder (in hopes of obtaining a diagnosis) is one thing; performing predictive genetic testing in healthy children is another. Legally, children under the age of 18 cannot provide consent for medical procedures or testing. This must be done by their parents or other legal guardian. In general, the field of genetics recommends *against* performing genetic testing in healthy children at risk for later-onset conditions.

Testing children in this situation basically takes away the child's right to make that decision when they are adults. And the decision to do predictive or disease susceptibility testing (described in Chapter 15) is a very personal one. Some people might want to know what their future risks are, but others will not.

The situation gets to be a bit unclear with regard to doing PGD or prenatal diagnosis for later-onset conditions in order to avoid having a child with whatever condition is running through the family, such as Huntington disease (an adult-onset, progressive neurological condition that is always fatal). Some see this as the same thing as testing a minor, in that you are taking the future offspring's right to decide out of their hands.

Others see this as preventing the future offspring from suffering through whatever their family members had to go through. Once again, this is a very personal decision and what is right for one family may not be right for another.

Patient Autonomy

Patient autonomy is the right for patients to make their own decisions, without coercion from anyone else, including the patient's healthcare provider or family members. In medicine, autonomy and informed consent go hand in hand. And respect for autonomy is a key tenet of genetic counseling — genetic counselors are trained specifically to be nondirective. Genetic counselors provide information, discuss risks and benefits, and help individuals make difficult decisions. However, they do not tell patients what they should or should not do.

Problems dealing with autonomy can arise due to the fact that genetic testing not only provides information about the patient, it can reveal a lot about family members as well.

Consider this scenario: a woman with a strong family history of breast and ovarian cancer comes in for genetic counseling. Her mother died of breast cancer in her forties, and she is extremely concerned about the same thing happening to her. She would like to have genetic testing to see if she carries a mutation that increases her risk of breast cancer. Seems relatively straightforward, right?

Well, the patient then reveals that she is an identical twin and her sister is adamant that she does not want to know her risk. The results of genetic testing have direct implications for the sister's twin, since they have the same DNA. If the patient finds out she carries a mutation, that means the twin does as well. Can the patient keep this information from her sister? Will her sister figure it out based on the medical decisions the patient makes (such as having a prophylactic mastectomy)? What impact would testing have on the sisters' relationship? While respecting the autonomy for one person, it could take away the autonomy for another.

Situations like these are definitely problematic and, in many cases, there is no "right" answer. Sometimes the best thing a counselor can do is thoroughly counsel the family about what all the possible results mean for each individual and help the family reach a decision that is best for everyone. Unfortunately, that is not always possible.

Privacy and Confidentiality

Another issue with genetic testing relates to privacy and confidentiality. When genetic tests are conducted, the data recorded often includes detailed medical histories and other personal information, all of which aids researchers or physicians in the interpretation of the genetic data obtained. So far, so good. But what happens to all that information? Who sees it? Where's it stored? And for how long?

Privacy is a big deal, particularly in American culture. Laws exist to protect one's private medical information, financial status, and juvenile criminal records (if any). Individuals are protected from unwarranted searches and surveillance, and they have the right to exclude unwanted persons from their private property. Genetic information falls under existing medical privacy laws (in that it is protected health information), but there's one twist: Genetic information contains an element of the future, not just the past.

REMEMBER

When you carry a mutation for susceptibility to breast cancer, you have a greater likelihood of developing breast cancer than someone who doesn't have a mutation (see Chapter 14). A breast cancer allele doesn't guarantee you'll develop the cancer, though; it just increases one's chances. If you were to be tested for a breast cancer mutation and found to have one, that information would become part of your medical record. Besides your doctor and appropriate medical personnel, who might learn about your condition?

As a patient, knowing that you have a genetic mutation is a really good thing, because the condition may be treatable, or an early detection screening may help you prevent more serious complications. For example, cancers that are caught early have far better prognoses than those diagnosed in later stages. However, there is concern that someone's genetic information could lead to unfair treatment by insurance companies and employers.

Fortunately, in 2008, former U.S. President George W. Bush signed the Genetic Information Nondiscrimination Act (GINA), which prohibits both health insurance companies and employers from discriminating against someone based on information obtained from genetic testing. GINA states that health insurance companies cannot use someone's genetic information (like the fact that they have a mutation that predisposes them to cancer or other disease) to determine eligibility, premiums, or coverage. It also states that employers who have more than 15 employees cannot use a person's genetic information for hiring, firing, pay decisions, promotions, or job assignments. GINA, however, does not apply to life insurance, long-term care insurance, or disability insurance; companies that provide these types of policies are currently able to use genetic information in their policy decisions.

Incidental Findings

Historically, when patients went to a geneticist because they were showing symptoms of a genetic syndrome, they underwent a complete physical examination, a review of their family history, and often additional medical evaluations (such as X-rays or bloodwork). In some cases, this could lead to a diagnosis. In other cases, genetic testing would be recommended. In general, the specific test(s) recommended would be based on the specific features of the patient. As a result, many patients would end up on a diagnostic odyssey undergoing one genetic test and then another until hopefully a diagnosis was reached. Unfortunately, in many cases, patients never received a specific diagnosis and are said to have an "unknown genetic syndrome" — likely the result of a genetic change in a gene we didn't have a test for.

Today, many genetic tests that are performed are quite broad in nature. For example, whole exome sequencing involves determining the order of nucleotides in all the coding exons of the genome (all the parts of the genes that provide instructions for the protein products). One of the most common reasons whole exome sequencing is now being performed for patients is to diagnose them with rare genetic conditions, some of whom have already been on that diagnostic odyssey we just mentioned. Well, with whole exome sequencing you are essentially testing all the genes at the same time. While this can definitely shorten the time from initial evaluation to diagnosis, it also brings up new and very important issues. And unfortunately, it still doesn't guarantee a diagnosis.

One of the main issues with this type of test is the possibility of receiving information one wasn't expecting and is completely unrelated to why the testing was performed. These are referred to as *incidental findings*. For example, whole exome sequencing for a child with developmental delays and poor growth could reveal that the child carries a mutation that significantly increases his risk for developing cancer as an adult. So, inadvertently, a child was tested for (and diagnosed with) an adult-onset genetic condition, something that most genetics professionals and other clinicians *strongly* recommended against.

In order to help deal with incidental findings, the American College of Medical Genetics and Genomics created a list of 59 genes for which they recommend that "medically actionable" changes be reported to patients if they are identified on whole exome or whole genome sequencing. Changes in each of these genes would lead to a condition for which disease surveillance should be performed or there is treatment that could be provided, even in childhood.

However, there is still much debate over the return of *any* incidental finding, especially since some individuals with changes in these genes will *never* develop the corresponding condition. This is likely to be an issue that is debated for a long time to come and it is unlikely that everyone will reach a consensus about the "right" thing to do.

Another issue with genome-wide genetic tests and incidental findings is how one makes sure that the patient (or parent/guardian) is adequately informed about all the possible outcomes of the test. It would be impossible to counsel someone about all the possible genetic conditions that could be diagnosed from taking the test. In this case, the informed consent may be quite broad and discuss only generally what the possible outcomes could be.

Direct-to-Consumer Testing

As genetic testing became easier and less expensive, it was only a matter of time before it hit the direct-to-consumer market. And it has done exactly that. Everywhere you turn, there is an advertisement for some direct-to-consumer test. Some of these tests are designed to give you an idea where your ancestors came from. Others provide you with estimates of your risks for certain health conditions or genetic traits. Some do all these things. Regardless of the type of test, direct-to-consumer testing has definite limitations and potential risks to the consumer.

While direct-to-consumer genetic tests can provide you with some interesting information, they are not designed for making medical decisions or for establishing a diagnosis. There is no physician involved in the direct-to-consumer testing process and these tests are not regulated in the same manner as clinical genetic tests. In addition, one study has shown that the raw data that consumers can download after testing (which can then be submitted to a third-party interpretation service) had a very high false positive rate. Specifically, 40 percent of the disease-associated variants that were reported to be in this raw data were not really there.

Direct-to-consumer testing also raises concerns about the privacy of consumers' genetic information. Who owns the information after testing has been performed? Will the company use this information for other purposes, like research or advertising? Will the company destroy the person's sample after testing is complete or will they store it? There are many different questions one should ask before deciding on a direct-to-consumer genetic test of any type.

Individuals who choose to undergo direct-to-consumer testing also need to be aware of the possibility of receiving unexpected findings. Cases have been reported in which individuals have found out that the person they thought was their father was not in fact their biological father. Adults have also discovered that they were adopted through testing. The psychological impact of unexpected findings and the implications for family relationships can be quite devastating.

Practicing Safe Genetic Treatments

If you've ever had to sign a consent for treatment form, you know it can be a sobering experience. Almost all such forms include some phrase that communicates the possibility of death. With a gulp, most of us sign off and hope for the best. For routine procedures and treatments, our faith is usually repaid with survival. Experimental treatments are harder to gauge, though, and fully informing someone about possible outcomes is very difficult.

The 1999 case of Jesse Gelsinger (covered in Chapter 16) brought the problem of informed consent and experimental treatment into a glaring, harsh light. Jesse died after receiving an experimental treatment for a hereditary disorder that, by itself, wasn't likely to kill him. His treatment took place as part of a clinical trial designed to assess the effects of a particular therapy in relatively healthy patients and to work out any difficulties before initiating treatments on patients for whom the disease would, without a doubt, be fatal (in this case, infants homozygous for the allele). What researchers knew about all the possible outcomes and what the Gelsinger family was told before treatment began is debatable.

Almost every article on gene therapy published since the Gelsinger case makes mention of it. In fact, most researchers in the field divide the development of gene therapies into two categories: before and after Gelsinger. Sadly, Gelsinger's death probably contributed very little to the broader understanding of gene therapy. Instead, the impacts of the Gelsinger case are that clinical trials are now harder to initiate, criteria for patient inclusion and exclusion are heightened, and disclosure and reporting requirements are far more stringent.

These changes are basically a double-edged sword: New regulations protect patients' rights and simultaneously decrease the likelihood that researchers will develop treatments to help those who desperately need them. Like so many ethical issues, a safe and effective solution may prove elusive.

Genetic Property Rights

Another news-worthy topic related to genetics is gene patenting. According to U.S. law, a patent gives the owner of the patent exclusive rights to manufacture and sell his or her invention for a certain length of time. Patents are not only designed to protect one's inventions, they are also designed to protect intellectual property rights. Intellectual property is a type of property in which the product is an intangible creation of the human intellect, such as a method or process.

Well, in 2001, after discovering the *BRCA1* breast cancer gene (see Chapter 14 to learn more about this gene), an American company was granted a patent for this gene. Mutations in the *BRCA1* gene lead to a significantly increased risk of developing breast and ovarian cancer. So, why would a company want to patent the *gene*? Well, while the company held the patent to that gene, no other clinical laboratory was able to offer *BRCA1* gene testing to patients. That meant that any physician who wanted to order *BRCA1* gene testing for a patient had to do so through that one laboratory (they also later patented the *BRCA2* gene).

In 2013, however, after having a monopoly on such testing for more than a decade, the patent was invalidated after the Supreme Court decided that genes could not be patented. The Supreme Court decided that genes are unpatentable because they are a "product of nature" and that, since nothing new was physically created in the process of discovering a gene, there were no intellectual property rights to consider. This decision did not only affect the breast cancer genes; at the time of the ruling, patents had been issued for more than 4300 different genes. The Supreme Court's decision made each and every one of these genes accessible to researchers and clinical laboratories. The court did, however, grant that DNA that had been manipulated in the laboratory (such as by genetic engineering technologies) could be patented, since this DNA was not found in nature.

5 The Part of Tens

IN THIS PART . . .

Learn about the ten most important people and events that shaped what genetics is today.

Discover ten of the hottest topics in genetics right now.

IN THIS CHAPTER

» Appreciating the history of genetics

» Highlighting the people behind great discoveries

Chapter 21
Ten Defining Events in Genetics

Many milestones define the history of genetics. This chapter focuses on nine that we don't cover in other chapters of the book and one that we do (the Human Genome Project is so important that we cover it in Chapter 8 and here, too). The events listed here appear roughly in order of historical occurrence.

The Publication of Darwin's "The Origin of Species"

Earthquakes have aftershocks — little mini-earthquakes that rattle around after the main quake. Events in history sometimes cause aftershocks, too. The publication of one man's life's work is such an event. From the moment it hit the shelves in 1856, Charles Darwin's *The Origin of Species* was deeply controversial (and still is).

The basis of evolution is elegantly simple: Individual organisms vary in their ability to survive and reproduce. For example, a sudden cold snap occurs, and most individuals of a certain bird species die because they can't tolerate the rapid drop in temperature. But individuals of the same species that can tolerate the unexpected freeze survive and reproduce. As long as the ability to deal with rapid temperature drops is heritable, the trait is passed to future generations, and more and more individuals inherit it. When groups of individuals are isolated from each other, they wind up being subjected to different sorts of events (such as weather patterns). After many, many years, stepwise changes in the kinds of traits that individuals inherit based on events like a sudden freeze accumulate to the point that populations with common ancestors become separate species.

Darwin concluded that all life on earth is related by inheritance in this fashion and thus has a common origin. Darwin arrived at his conclusions after years of studying plants and animals all over the world. What he lacked was a convincing explanation for how individuals inherit advantageous traits. Yet the explanation was literally at his fingertips. Gregor Mendel figured out the laws of inheritance at about the same time that Darwin was working on his book (see Chapter 3). Apparently, Darwin failed to read Mendel's paper — he scrawled notes on the papers immediately preceding and following Mendel's paper but left Mendel's unmarked. Darwin's copious notes show no evidence that he was even aware of Mendel's work.

REMEMBER

Even without knowledge of how inheritance works, Darwin accurately summarized three principles that are confirmed by genetics:

- **Variation is random and unpredictable.** Studies of mutation confirm this principle (see Chapter 12).
- **Variation is *heritable* (it can be passed on from one generation to the next).** Mendel's research — and thousands of studies over the past century — confirms heritability.
- **Variation changes in frequency over the course of time.** For decades, genetic studies have confirmed that genetic variation within populations changes because of things like mutation, accidents, and geographic isolation (to name only a few causes).

Regardless of how you view it, the publication of Darwin's *The Origin of Species* is pivotal in the history of genetics. If no genetic variation existed, all life on earth would be precisely identical. Variation gives the world its rich texture and complexity, and it's what makes you wonderfully unique.

The Rediscovery of Mendel's Work

In 1866, Gregor Mendel wrote a summary of the results of his gardening experiments with peas (which we detail in Chapter 3). His work was published in the scientific journal *Versuche Pflanzen Hybriden*, where it gathered dust for nearly

40 years. Although Mendel wasn't big on self-promotion, he sent copies of his paper to two well-known scientists of his time. One copy remains missing; the other was found in what amounts to an unopened envelope — the pages were never cut. (Old printing practices resulted in pages being folded together; the only way to read the paper was to cut the pages apart.) Thus, despite the fact that his findings were published and distributed (though limitedly), his peers didn't grasp the magnitude of Mendel's discovery.

Mendel's work went unnoticed until three botanists — Hugo de Vries, Erich von Tschermak, and Carl Correns — all reinvented Mendel's wheel, so to speak. These three men conducted experiments that were very similar to Mendel's. Their conclusions were identical — all three "discovered" the laws of heredity. De Vries found Mendel's work referenced in a paper published in 1881. (De Vries coined the term *mutation*, by the way.) The author of the 1881 paper, a man by the name of Focke, summarized Mendel's findings but didn't have a clue as to their significance. De Vries correctly interpreted Mendel's work and cited it in his own paper, which was published in 1900. Shortly thereafter, Tschermak and Correns also discovered Mendel's publication through de Vries's published works and indicated that their own independent findings confirmed Mendel's conclusions as well.

William Bateson is perhaps the great hero of this story. He was already incredibly influential by the time he read de Vries's paper citing Mendel, and unlike many around him, he recognized that Mendel's laws of inheritance were revolutionary and absolutely correct. Bateson became an ardent voice spreading the word. He coined the terms *genetics, allele* (shortened from the original *allelomorph*), *homozygote*, and *heterozygote.* Bateson was also responsible for the discovery of linkage (see Chapter 4), which was experimentally confirmed later by Morgan and Bridges.

DNA Transformation

Frederick Griffith wasn't working to discover DNA. The year was 1928, and the memory of the deadly flu epidemic of 1918 was still fresh in everyone's mind. Griffith was studying pneumonia in an effort to prevent future epidemics. He was particularly interested in why some strains of bacteria cause illness and other seemingly identical strains do not. To get to the bottom of the issue, he conducted a series of experiments using two strains of the same species of bacteria, *Streptococcus pneumonia.* The two strains looked very different when grown in a Petri dish, because one grew a smooth carpet and the other a lumpy one (he called it "rough"). When Griffith injected smooth bacteria into mice, they died; rough bacteria, on the other hand, were harmless.

To figure out why one strain of bacteria was deadly and the other wasn't, Griffith conducted a series of experiments. He injected some mice with heat-killed smooth bacteria (which turned out to be harmless) and others with heat-killed smooth in combination with living rough bacteria. This combo proved deadly to the mice. Griffith quickly figured out that something in the smooth bacteria *transformed* rough bacteria into a killer. But what? For lack of anything better, he called the responsible factor the *transforming principle*.

Oswald Avery, Maclyn McCarty, and Colin MacLeod teamed up in the 1940s to discover that Griffith's transforming principle was actually DNA. This trio made the discovery by a dogged process of elimination. They showed that fats and proteins don't do the trick; only the DNA of smooth bacteria provides live rough bacteria with the needed ingredient to become a killer. Their results were published in 1944, and like Mendel's work nearly a century before, their findings were largely rejected.

It wasn't until Erwin Chargaff came along that the transforming principle started to get the appreciation it deserved. Chargaff was so impressed that he changed his entire research focus to DNA. Chargaff eventually determined the ratios of bases in DNA that helped lead to Watson and Crick's momentous discovery of DNA's double helix structure (flip to Chapter 5 for all the details).

The Discovery of Jumping Genes

By all accounts, Barbara McClintock was both brilliant and a little odd; a friend once described her as "not fooled or foolable." McClintock was unorthodox in both her research and her outlook, as she lived and worked alone for most of her life. Her career began in the early 1930s and took her into a man's world — very few women worked in the sciences in her day.

HISTORICAL STUFF

In 1931, McClintock collaborated with another woman, Harriet Creighton, to demonstrate that genes are located on chromosomes. This fact sounds so self-evident now, but back then, it was a revolutionary idea. McClintock's contribution to genetics goes beyond locating genes on chromosomes, though. She also discovered traveling bits of DNA, sometimes known as jumping genes. In 1948, McClintock, working independently, published her results demonstrating that certain genes in corn could hop around from one chromosome to another *without* translocation. Her announcement triggered little reaction at first. It's not that people thought McClintock was wrong; she was just so far ahead of the curve that her fellow geneticists couldn't comprehend her findings. Alfred Sturtevant (who was responsible for the discovery of gene mapping) once said, "I didn't understand one word she said, but if she says it is so, it must be so!"

Now referred to as *transposable elements* or *transposons*, scientists have since discovered that there are many different types. It is also now known that up to half of the human genome contains sequences from transposable elements. However, many of these are basically ancient relics that are no longer able to jump. And many others are rendered silent by certain genetic mechanisms. For those that are still active and able to move, the effect depends on where they land. If they land within a gene, they can cause a mutation that results in disease. In other cases, their movement can contribute to genetic diversity and evolution of the species. The frequency with which transposons move depends on the species and the type of transposon.

It took nearly 40 years before the genetics world caught up with Barbara McClintock and awarded her the Nobel Prize in Physiology or Medicine in 1983. By then, jumping genes had been discovered in many organisms. Feisty to the end, this grand dame of genetics passed away in 1992 at the age of 90.

The Birth of DNA Sequencing

So many events in the history of genetics lay a foundation for other events to follow. Frederick Sanger's invention of chain-reaction DNA sequencing (which we cover in Chapter 8) is one of those foundational events. In 1980, Sanger shared the Nobel Prize in Chemistry with Walter Gilbert for their work on DNA.

HISTORICAL STUFF

Sanger figured out how to use the characteristics of DNA and of DNA replication to determine the sequence of DNA. *Chain-termination sequencing*, as Sanger's method is called, uses the same mechanics as replication in your cells (see Chapter 7 for a rundown of replication). Sanger figured out that he could control the DNA building process by snipping off one oxygen molecule from the building blocks of DNA. The resulting method allowed identification of every base, in order, along a DNA strand, sparking a revolution in the understanding of how your genes work. This process is responsible for the Human Genome Project, DNA fingerprinting (see Chapter 18), genetic engineering (see Chapter 19), and gene therapy (see Chapter 16).

The Invention of PCR

In 1985, while driving along a California highway in the middle of the night, Kary Mullis had a brainstorm about how to carry out DNA replication in a tube (see Chapter 7 for the scoop on replication). His idea led to the invention of *polymerase chain reaction* (PCR), a pivotal point in the history of genetics.

We detail the entire process of how PCR is used in DNA fingerprinting in Chapter 18. In essence, PCR acts like a copier for DNA. Even the tiniest snippet of DNA can be copied. Scientists need many copies of a DNA molecule before enough is present for them to examine. Without PCR, large amounts of DNA are needed to generate a DNA fingerprint. However, at many crime scenes, only tiny amounts of DNA are present. PCR is the powerful tool that every crime lab in the country now uses to detect the DNA left behind at crime scenes and to generate DNA fingerprints.

Mullis's bright idea turned into a billion-dollar industry. Although he reportedly was paid a paltry $10,000 for his invention (from the lab where he worked), he received the Nobel Prize for Chemistry in 1993 (a sort of consolation prize).

The Development of Recombinant DNA Technology

HISTORICAL STUFF

In 1970, Hamilton O. Smith discovered *restriction enzymes,* which act as chemical cleavers to chop DNA into pieces at very specific sequences (see Chapter 19 to learn more about these enzymes). As part of other research, Smith put bacteria and a bacteria-attacking virus together. The bacteria didn't go down without a fight — instead, it produced an enzyme that chopped the viral DNA into pieces, effectively destroying the invading virus altogether. Smith determined that the enzyme, now known as *HindII* (named for the bacteria *Haemophilus influenzae Rd*), cuts DNA every time it finds certain bases all in a row and cuts between the same two bases every time.

This fortuitous (and completely accidental!) discovery was just what was needed to spark a revolution in the study of DNA. Some restriction enzymes make offset cuts in DNA, leaving single-stranded ends. The single-strand bits of DNA allow geneticists to "cut-and-paste" pieces of DNA together in novel ways, forming the entire basis of what's now known as *recombinant DNA technology.*

REMEMBER

Gene therapy (see Chapter 16), the creation of genetically engineered organisms (see Chapter 19), and practically every other advance in the field of genetics these days all depend on the ability to cut DNA into pieces without disabling the genes and then to put the genes into new places — a feat made possible thanks to restriction enzymes.

Researchers have used thousands of restriction enzymes to help map genes on chromosomes, study gene function, and manipulate DNA for diagnosis and treatment of disease. Smith shared the Nobel Prize in Physiology or Medicine in 1978 with two other geneticists, Dan Nathans and Werner Arber, for their joint contributions to the discovery of restriction enzymes.

The Invention of DNA Fingerprinting

Sir Alec Jeffreys has put thousands of wrongdoers behind bars. Almost single-handedly, he's also set hundreds of innocent people free from prison. Not bad for a guy who spent most of his time in the genetics lab.

Jeffreys invented DNA fingerprinting in 1985. By examining the patterns made by human DNA after it was diced up by restriction enzymes, Jeffreys realized that every person's DNA produces a slightly different number of various sized fragments (which number in the thousands).

Jeffreys's invention has seen a number of refinements since its inception. PCR and the use of STRs (*short tandem repeats*; see Chapter 18) have replaced the use of restriction enzymes. Modern methods of DNA fingerprinting are highly repeatable and extremely accurate, meaning that a DNA fingerprint can be stored much like a fingerprint impression from your fingertip. Crime laboratories all over the United States make use of the methods that Jeffreys pioneered, and the information that these labs generate is housed in a huge database hosted by the FBI. This data can then be accessed by police departments in order to help match criminals to crimes. In 1994, Queen Elizabeth II knighted Jeffreys for his contributions to law enforcement and his accomplishments in genetics.

The Birth of Developmental Genetics

Every cell in your body has a full set of genetic instructions to make all of you. The master plan of how an entire organism is built from genetic instructions remained a mystery until 1980, when Christiane Nüsslein-Volhard and Eric Wieschaus identified the genes that control the whole body plan during fly development.

Fruit flies and other insects are constructed of interlocking pieces, or segments. A group of genes (collectively called *segmentation genes*) tells the cells which body segments go where. These genes, along with others, give directions like top and bottom and front and back, as well as the order of body regions in between. Nüsslein-Volhard and Wieschaus made their discovery by mutating genes and looking for the effects of the "broken" genes. When segmentation genes get mutated, the fly ends up lacking whole sections of important body parts or certain pairs of organs.

A different set of genes (called *homeotic genes*) controls the placement of all the fly's organs and appendages, such as wings, legs, eyes, and so on. One such gene is *eyeless*. Contrary to what would seem logical, *eyeless* actually codes for normal eye development. Using the same recombinant DNA techniques made possible by restriction enzymes, Nüsslein-Volhard and Wieschaus moved *eyeless* to different chromosomes where it could be turned on in cells in which it was normally turned off. The resulting flies grew eyes in all sorts of strange locations — on their wings, legs, butts, you name it. This research showed that, working together, segmentation and homeotic genes put all the parts in all the right places. Humans have versions of these genes, too; your body-plan genes were discovered by comparing fruit fly genes to human DNA.

The Work of Francis Collins and the Human Genome Project

In 1989, Francis Collins and Lap-Chee Tsui identified the single gene responsible for cystic fibrosis. The very next year, the Human Genome Project (HGP) officially got underway. Collins, who has both a medical degree and a PhD in physical chemistry, later replaced James Watson as the head of the National Human Genome Research Institute in the United States and supervised the race to sequence the entire human genome from start to finish. In 2009, Collins became the director of the U.S. National Institutes of Health.

Collins is one of the true heroes of modern genetics. He kept the HGP ahead of schedule and under budget. He continues to champion the right to free access to all the HGP data, making him a courageous opponent of gene patents and other practices that restrict access to discovery and healthcare, and he's a staunch defender of genetic privacy (see Chapter 20 for more on these subjects). The human genome project wouldn't have been a success without the tireless work of Collins.

IN THIS CHAPTER

» Understanding the basics of direct-to-consumer genetic testing

» Tracking potential advances in genetic testing

» Continuing research on stem cells and antibiotic resistance

» Learning about the follow-ups to the Human Genome Project

Chapter 22
Ten Hot Issues in Genetics

Genetics is a field that grows and changes with every passing day. This chapter shines the spotlight on ten of the hottest topics and next big things in this ever-changing scientific landscape.

Direct-to-Consumer Genetic Testing

Not too long ago, genetic testing was uncommon and was reserved for visits to a geneticist, genetic counselor, or other specialized healthcare provider. Now, a trip to the drug store, a little spit in a tube that you put in the mail, and that's it. A few weeks or months later and you get to learn all about yourself! Of course, it's not really that straightforward.

Direct-to-consumer tests come in several varieties. The most common of these tests is genetic ancestry testing. Genetic ancestry testing involves testing individuals for sequence variations throughout the genome. The frequency of certain variants differs among different ancestral populations. Therefore, an analysis of the variants that a person carries and a comparison with previously tested individuals of known ancestry can give us an idea of where that person's ancestors came from. To learn more about genetic ancestry testing, you can flip back to Chapter 17.

Popular direct-to-consumer genetic tests also include testing for health risks and common traits. The companies that offer these tests provide risk assessments for certain health conditions based on the particular sequence variants a person carries. For example, a specific variant in the gene for apoliprotein E (*APOE*) has been associated with an increased risk for developing late-onset Alzheimer disease (AD). Carrying the *APOE* e4 allele appears to increase one's risk for developing AD late in life. Carrying two copies of the allele raises the risk even higher.

It is important to note that this is *not* a predictive test. It cannot tell you if you will develop AD or not, only whether you have an increased risk based on the versions of the *APOE* gene that you carry. Similar testing can tell you about your *risk* for conditions such as celiac disease, Parkinson's disease, and diabetes, among others, but it cannot tell you whether or not you will get them.

At least one company offers testing for three variants in the breast cancer genes *BRCA1* and *BRCA2* (described in more detail in Chapter 14). The three variants that are a part of this test are found much more often in individuals of Ashkenazi (Eastern European) Jewish ancestry. Carriers of one of these three gene changes have a significantly increased risk for breast and ovarian cancer.

One of the things that is hard for many patients to understand about the *BRCA1* and *BRCA2* testing that is performed is that they will not be tested for the vast majority of mutations in these genes (thousands have been reported). So if the person tested does not carry one of the three mutations included in the test, it does not rule out the possibility that they carry another mutation in one of these genes. And if the person is not of Ashkenazi Jewish ancestry, this portion of the test is of little value, since it is unlikely they would carry one of the three mutations anyway.

Whole Exome Sequencing

The genome is a complete set of chromosomes from an organism, including all coding and non-coding DNA. The *exome* contains all of the coding sequences found in the genome — all the exons in all of the genes. The exome, which is estimated to account for less than two percent of the genome, provides the instructions for creating the proteins in the body. Most mutations that cause genetic disorders are found in the exome.

With advances in DNA sequencing technologies (described in Chapter 8), it is now relatively straightforward (and not crazy expensive) to sequence the exomes of individuals in search of a genetic diagnosis. Consequently, whole exome sequencing has now entered mainstream genetics and is increasingly being used for patients who are suspected of having a genetic syndrome that has yet to be diagnosed. In many cases, patients have been on a long diagnostic odyssey with numerous medical evaluations, chromosome studies, and single gene tests, with no success. Whole exome sequencing allows for the testing of all known genes at the same time, regardless of the specific clinical features the person demonstrates.

Not only is whole exome sequencing providing diagnoses for those who were previously diagnosed with an "unknown genetic syndrome" (and many answers to the many questions of frustrated parents), it is providing a lot of information about the spectrum of features that can be associated with any particular disorder. We are finding out that the symptoms of genetic disorders can be quite varied. This means that someone may have a condition that would not have previously been suspected because they show features that are quite different than what is typically seen in individuals who have been diagnosed with the disorder.

One of the limitations of whole exome sequencing is the identification of incidental findings (described in Chapter 20) or genetic changes of unclear consequence. Incidental findings are those that are unrelated to why the testing was performed, such as identifying a gene change that increases the risk for Alzheimer disease in a child who was tested because of intellectual disabilities. Genetic changes for which the effect is unclear are typically referred to as variants of unknown significance. It's possible these changes could be the cause of some disorder, but it is also possible that they are completely harmless. The problem is there is not enough information about them yet to know which is true, and it is not always clear whether someone's medical care should be altered based on the finding of a variant for which the effect is uncertain.

Whole exome sequencing is also being performed in large groups of individuals without genetic syndromes in order to determine how common different variants are in the population. The Exome Aggregation Consortium (ExAC) and the 100,000 Genomes Project are two collaborations that are collecting exome and/or genome data from a large number of individuals from a variety of ethnic backgrounds. Because of data like this, variants that were once classified as pathogenic (that is, as a disease-causing mutation) are now being reclassified as benign because we now know these variants are quite common in the general population (and therefore, highly unlikely to cause any problems).

Whole Genome Sequencing

While whole exome sequencing involves determining the order of the nucleotides in all the exons in the genome, *whole genome sequencing* involves determining the sequence of the nucleotides in the *entire genome*. Whole genome sequencing was what was done during the Human Genome Project. And in the time since the Human Genome Project was completed, the ease with which sequencing can be performed and the significantly lowered cost have led to whole genome sequencing entering the clinic (just as with whole exome sequencing).

The idea with whole genome sequencing is that it should theoretically be able to identify the cause of rare genetic conditions that cannot be diagnosed using more traditional methods, or even whole exome sequencing. Scientists estimate that about 85 percent of disease-causing mutations involve the coding sequences of the genome. The remaining 15 percent are expected to be located in non-coding sequences that are not covered in other types of tests. Whole genome sequencing could identify the previously unidentifiable mutations in those with undiagnosed genetic syndrome. Indeed, a project in the United Kingdom is allowing all children with serious illnesses to have whole genome testing. So far, the program has found that about 1 in 4 seriously ill children had an underlying genetic syndrome. One case was a girl with a rare and severe form of epilepsy, and once her diagnosis was made, it was realized that the child had been taking medication that was known to aggravate that type of epilepsy. The mutation that caused this condition could have been found using more traditional approaches, but whole genome sequencing was able to provide a result in several weeks without having to undergo a diagnostic odyssey.

As with whole exome sequencing, two of the main concerns with whole genome sequencing are incidental findings and the identification of variants of unknown clinical significance (described in the previous section and in Chapter 20).

Stem Cell Research

Stem cells may hold the key to curing brain and spinal cord injuries. They may be part of the cure for cancer. These little wonders may be *the* magic bullet to solving all sorts of medical problems, but they're at the center of controversies so big that their potential remains unknown.

You've probably guessed (or already knew) that one of the sources of stem cells for research is embryonic tissue — and therein lies the rub. However, in 2006, scientists figured out a way to turn multipotent stem cells (cells that can give rise to just a few, related cell types) from an adult organism into pluripotent stem cells

(cells that can give rise to any of the cell types in the body of an organism). These stem cells are referred to as *induced pluripotent stem cells.* This discovery meant that it was now possible to collect stem cells from a patient, modify them, and return them to the patient, eliminating the chance of tissue rejection and without having to use embryonic stem cells.

Currently, the most common use of stem cells is in hematopoietic stem cell therapy, which can be used to treat certain types of cancer and immune conditions. In hematopoietic stem cell therapy, the stem cells can be obtained from bone marrow, from the bloodstream, or from blood in the umbilical cord. This type of stem cell is a multipotent stem cell, which can give rise to any of the blood-related cell types (such as red blood cells and white blood cells).

Other uses of stem cells are being evaluated, with hopes that stem cell therapy will be able to help those with a number of different conditions, including neurodegenerative disorders and blindness. In fact, in 2018, researchers from England reported the results of a phase I clinical trial in which two patients with age-related macular degeneration were treated with stem cells. In age-related macular degeneration, specific light-sensitive cells in the retina die off, resulting in a progressive loss of central vision. In the trial, the eyes of both patients were injected with a patch of stem cells. Over the course of the following year, both patients showed dramatic improvement in their eyesight.

While stem cells in one form or another have yet to find their way into everyday medicine, ongoing studies are promising, and it is likely just a matter of time.

REMEMBER

When considering new or experimental medical therapies, it is important that the patient is aware of the history and experience of the provider and medical practice offering the treatment. Anyone considering medical treatment for a chronic condition should consult with a licensed and reputable healthcare provider.

The ENCODE Project

The goal of the Human Genome Project was to sequence the entire human genome. In 2001, a draft of the human genome sequence was published. And after about 13 years from start to finish (2 years ahead of schedule), the project was deemed complete. As a result of the project, we now know that there are approximately 22,000 genes (compared to the original prediction of 100,000 genes) and that coding sequences (the sequences that actually code for the protein products made by the genes) account for less than 2 percent of the entire human genome. We also have many new and improved tools for the analysis of genetic data. What we didn't know is the function of the more than 98 percent of sequences — non-coding sequences that were once referred to as *junk DNA.*

The Encyclopedia of DNA Elements (ENCODE) Project is a follow-up to the Human Genome Project. This project involves more than 30 research groups and more than 400 scientists from across the globe. The main goal of this project is to explore and determine the function of the non-coding sequences in the human genome. Research to date suggests that more than 80 percent of the non-coding sequences play a role in the regulation of gene expression, which we discuss in Chapter 11.

Research is also attempting to determine whether differences in the expression of certain genes (as opposed to difference in gene sequence) can be linked with the development of specific genetic disorders. While most disease-causing mutations are found in the protein-coding portions of genes (the exons), some are found in non-coding regions of the genes (such as in the promoters or introns, which are described in Chapter 9).

Scientists also expect that sequence variants located in other non-coding regions, including regions located distant to the gene itself (such as in enhancers or silencers, which are also reviewed in Chapter 9), could affect gene expression and result in the development of genetic conditions.

Proteomics

Genomics is the study of whole genomes. *Proteomics* is the study of all the proteins an organism makes. Proteins do all the work in your body. They carry out all the functions that genes encode, so when a gene mutation occurs, the protein is what winds up being altered (or goes missing altogether). Given the link between genes and proteins, the study of proteins may end up telling researchers more about genes than the genes themselves!

Proteins are three-dimensional. Proteins not only get folded into complex shapes but also get hooked up with other proteins and decorated with other elements such as metals. (See Chapter 10 for more on how proteins are folded and modified from plain amino acid chains in order to do their jobs.) Scientists can't just look at a protein and tell what its function is. However, if researchers can better understand each protein and what role it may play, proteins may be a big deal in the development of medicines and other treatments, because medications act upon the proteins in your system.

Cataloging all the proteins in your proteome isn't easy, because researchers have to sample every tissue to find them all. Nonetheless, the rewards of discovering new drugs and treatments for previously untreatable diseases may make the effort worthwhile. Proteomics hasn't made a big splash in clinical settings just yet — complexities and technological setbacks have slowed progress. However, like most things in genetics, it's likely just a matter of time.

Gene Chips

Technology is at the heart of modern genetics, and one of the most useful developments in genetic technology is the *gene chip.* Also known as *microarrays,* gene chips allow researchers to quickly determine which genes are at work (that is, being expressed) in a given cell (see Chapter 11 for a full rundown on how your genes do their jobs).

TECHNICAL STUFF

Gene expression depends on messenger RNA (mRNA), which is produced through transcription (see Chapter 9 to review transcription). The mRNAs get tidied up and sent out into the cell cytoplasm to be translated into proteins (see Chapter 10 for how translation works to make proteins). The various mRNAs in each cell tell exactly which of the thousands of genes are at work at any given moment. In addition, the number of copies of each mRNA conveys an index of the strength of gene expression. The more copies of a particular mRNA, the stronger the action of the gene that produced it.

Gene chips are grids composed of bits of DNA that are complementary to the mRNAs the geneticist expects to find in a cell. It works like this: The bits of DNA are attached to a glass slide. All the mRNAs from a cell are passed over the gene chip, and the mRNAs bind to their DNA complements on the slide. Geneticists measure how many copies of a given mRNA attach themselves to any given spot on the slide to determine which genes are active and what their strength is.

Gene chips are relatively inexpensive to make and can each test hundreds of different mRNAs, making them a valuable tool. One thing this screening can be used for is to compare mRNAs from normal cells to those from diseased cells (such as cancer). By comparing the genes that are turned on or off in the two cell types, geneticists can determine what's gone wrong and how the disease may be treated. Scientists are also using microarrays to screen thousands of genes rapidly to identify specific mutations that cause disease, or to test patients for certain types of chromosome abnormalities (we describe this type of microarray in Chapter 13).

Evolution of Antibiotic Resistance

Unfortunately, not all "next big things" are good. Antibiotics are used to fight diseases caused by bacteria. When penicillin (a common antibiotic) was developed, it was a wonder drug that saved many, many lives. However, many antibiotics are nearly useless now because of the evolution of *antibiotic resistance.*

REMEMBER

Bacteria don't have sex, but they still pass their genes around. They achieve this feat by passing around little circular bits of DNA called *plasmids.* Almost any species of bacteria can pass its plasmids on to any other species. Thus, when bacteria that are resistant to a particular antibiotic run into bacteria that aren't resistant, the exchange of DNA endows the formerly susceptible bacteria with antibiotic resistance. Antibacterial soaps and the overprescribing of antibiotics make the situation worse by killing off all the nonresistant bacteria, leaving only the resistant kind behind. This can make illnesses that result from bacterial infections very difficult to treat. Consequently, scientists must continually work to develop new, more powerful antibiotics in an effort to stay one step ahead of the bacteria.

Circumventing Mother Nature

While you get your nuclear DNA (your autosomes and sex chromosomes) from both your mother and your father, your mitochondrial DNA (mtDNA) comes solely from your mother. So, conditions that are the result of changes in the mtDNA can only be inherited from mom. And if mom carries a mutation in her mtDNA, each of her children have a risk of inheriting it and developing a mitochondrial disorder, many of which are quite severe and potentially lethal. Well, a relatively new technique for preventing the inheritance of a mitochrondrial disorders is *mitochondrial transfer* (or *mitochondrial replacement therapy*).

To minimize the chance of a child inheriting his mother's mtDNA mutation and developing a mitochondrial disorder, scientists developed a procedure where the mother's mitochondria could be replaced with the mitochondria from a donor, so the child would inherit someone else's mitochondrial DNA but her mother's nuclear DNA. In 2016, the birth of a child that resulted from this procedure was reported. The family had already had two children who had died from the mitochondrial disorder Leigh syndrome. The nucleus of one of the mother's eggs (containing the nuclear DNA) was transferred into a donor egg that had the nucleus removed. The egg was then fertilized by sperm from the father and implanted in the mother. The couple had a healthy son who had very low levels of the mother's mtDNA mutation and no clinical symptoms of Leigh syndrome.

Genetics from Afar

For many, a referral to a geneticist or genetic counselor means traveling several hours (potentially to another state) and often having to navigate a large, unfamiliar city. It may also mean a very long wait, from the time of referral to the time of the actual appointment. In the entire state of Alaska, there is currently a single genetics clinic with a single clinical geneticist. And in some states, the demand far exceeds the availability; consequently, the wait for an appointment can be as long as 12 to 18 months. To improve access to genetics services, there has been a significant increase in the use of telegenetics.

Telegenetics involves using either the telephone or videoconferencing to connect a patient with a genetics provider. In a virtual patient visit with a clinical geneticist, the patient attends a clinic at a local physician's office. The visit requires a computer, high-speed internet, a good camera, a few specialized medical tools, and a provider onsite (such as a nurse or genetic counselor) to assist. The geneticist can perform a thorough evaluation while being hundreds of miles away. Even more common these days are telephone genetic counseling sessions. Genetic counselors can provide all aspects of genetic counseling (including pre-test and post-test counseling) while the patient enjoys the comforts of home.

Glossary

acrocentric chromosome: A chromosome with its centromere located toward one end of the chromosome.

adenine: Purine base found in DNA and RNA.

allele: Alternative version of a gene.

amino acid: Unit composed of an amino group, a carboxyl group, and a radical group; amino acids link together in chains to form polypeptides (proteins).

anaphase: Stage of cell division in mitosis when replicated chromosomes (as chromatids) separate. In meiosis, homologous chromosomes separate during anaphase I, and replicated chromosomes (as chromatids) separate during anaphase II.

aneuploidy: Increase or decrease in the number of chromosomes; a deviation from an exact multiple of the haploid number of chromosomes.

anticipation: Increasing severity or decreasing age of onset of a genetic trait or disorder with successive generations.

anticodon: The three nucleotides in a tRNA (transfer RNA) complementary to a corresponding codon of mRNA.

antiparallel: Parallel but running in opposite directions; orientation of two complementary strands of DNA.

apoptosis: Normal process of regulated cell death.

autosome: A non-sex chromosome.

backcross: Cross between an individual with an F1 (first generation) genotype and an individual with one of the parental (P) genotypes.

bacteriophage: Virus that infects bacterial cells.

base: One of the three components of a nucleotide; DNA and RNA have four bases.

carrier: An individual who is heterozygous for a recessive gene mutation; carriers are typically unaffected.

cell cycle: Repeated process of cell growth, DNA replication, mitosis, and cytokinesis (cell division).

centromere: Region at the center of a chromosome that appears pinched during metaphase; where spindle fibers attach during mitosis and meiosis.

chloroplast: Organelle in plants and algae that processes sunlight into energy.

chromatid: One half of a replicated chromosome; separate during anaphase of mitosis or meiosis.

chromosome: Linear or circular strand of DNA that contains genes.

codominance: When heterozygotes express both alleles equally.

codon: Combination of three nucleotides in an mRNA that correspond to an amino acid.

complementary: Specific matching of base pairs in DNA (adenine-thymine and cytosine-guanine) or RNA (adenine-uracil and cytosine-guanine).

compound heterozygote: Individual with two different alleles of a given gene or locus.

consanguineous: Mating by related individuals.

crossing-over: Equal exchange of DNA between homologous chromosomes during meiosis; also called recombination.

cytokinesis: Cell division.

cytosine: A pyrimidine base found in DNA and RNA.

ddNTP: Dideoxyribonucleotide; identical to dNTP but lacking an oxygen at the 3' site. Used in DNA sequencing.

deamination: When a base loses an amino group.

degenerate: Property of the genetic code whereby some amino acids are encoded by more than one codon.

deletion: The loss of one or more nucleotides from a DNA sequence.

denaturation: Melting bonds between DNA strands, thereby separating the double helix into single strands.

deoxyribose: Ribose sugar that has lost one oxygen atom; a component of DNA.

depurination: When a nucleotide loses a purine base.

dihybrid cross: Cross between two individuals who differ at two traits or loci.

diploid: Possessing two copies of each chromosome.

DNA: Deoxyribonucleic acid; the molecule that carries genetic information.

dNTP: Deoxyribonucleotide; the basic building block of DNA used during DNA replication consisting of a deoxyribose sugar, three phosphate molecules, and one of four nitrogenous bases.

dominant: An allele or phenotype that completely masks another allele or phenotype. For traits or disorders inherited in a dominant manner, only one copy of the causative gene change is needed to develop the phenotype.

epigenetics: Changes in gene expression and phenotype caused by characteristics of DNA outside the genetic code itself (such as the addition of chemical groups to the nucleotides in DNA).

epistasis: Gene interaction in which one gene hides the action of another.

eukaryote: Organism with a complex cell structure and a cell nucleus.

euploid: Organism possessing an exact multiple of the haploid number of chromosomes.

exon: Coding part of a gene; the sequence that codes for the protein product.

expressivity: Variation in the strength of traits.

F1 generation: First generation offspring of a specific cross.

F2 generation: Offspring of the F1 generation.

gamete: Reproductive cell; sperm or egg cell.

gene: Fundamental unit of heredity; a specific section of DNA within a chromosome.

genome: A particular organism's full set of chromosomes.

genotype: The genetic makeup of an individual; the allele(s) possessed at a given locus.

germline: Present in the sex cells of an organism (the sperm or eggs); can be passed from parent to offspring; opposite of *somatic*.

guanine: Purine base found in DNA and RNA.

gyrase: Enzyme that acts during DNA replication to prevent tangles from forming in the DNA strand.

haploid: Possessing one copy of each chromosome.

helicase: Enzyme that acts during DNA replication to open the double helix.

heterozygote: Individual with a single copy of an allele at a given gene or locus; has a different allele on the homologous chromosome.

homologous chromosomes: Two chromosomes that are identical in shape and structure and carry the same genes. Diploid organisms inherit one homologous chromosome from each parent.

homozygote: Individual with two identical alleles of a given gene or locus.

incomplete dominance: When the phenotype of a dominant allele is not fully expressed.

imprinting: When the sex of the parent who contributes an allele determines whether it is expressed.

insertion: The addition of one or more nucleotides to a DNA sequence.

interphase: Period of cell growth between divisions.

intron: Non-coding part of a gene. Intervening sequences that interrupt exons.

ligase: Enzyme that acts during replication to seal gaps created by lagging strand DNA synthesis.

linkage: Inheriting genes located close together on chromosomes as a unit.

locus: A specific location on a chromosome (plural: loci; pronounced "*low*-sigh").

meiosis: Cell division in sexually reproducing organisms that reduces amount of genetic information by half.

metacentric chromosome: A chromosome with its centromere located in the middle of the chromosome.

metaphase: Stage of cell division when chromosomes align along the equator of the dividing cell.

metastasis: The spread of cancer cells to other parts of the body.

mitochondria: Organelle in plants, algae, and animals that generates energy for cellular functions.

mitosis: Simple cell division without a reduction in chromosome number.

mutagen: Any factor that causes an increase in the rate of DNA sequence changes.

mutation: A change in DNA sequence that leads to a functional change in the protein and/or has been associated with disease.

nondisjunction: The failure of chromosomes to separate properly during cell division; leads to too many or too few chromosomes (that is, aneuploidy).

nucleotide: Building block of DNA; composed of a deoxyribose sugar, a phosphate, and one of four nitrogenous bases.

oocyte: A haploid egg cell; can be fertilized by a sperm in order to create a zygote.

oogonia: A diploid egg cell; precursor to an oocyte.

p arm: The shorter arm of a chromosome, or the arm above the centromere.

P generation: Parental generation in a genetic cross.

pathogenic: Disease-causing.

penetrance: Percentage of individuals with a particular genotype that express the associated trait or disorder.

phenotype: Physical characteristics of an individual.

phosphodiester bond: The type of bond found between the phosphate and two sugar molecules in DNA or RNA.

pleiotropy: When a gene or allele controls more than one phenotype or trait.

polar body: Small cells that result from female meiosis (in addition to a mature egg cell); polar bodies lack cytoplasm and cannot be fertilized.

polymorphism: A change in DNA sequence that does not cause a change in protein function; considered a *neutral* or *benign* change in the DNA.

polypeptide: Chain of amino acids that form a protein.

polyploid: Having more than two complete sets of chromosomes.

prokaryote: Organism with a simple cell structure and no cell nucleus.

prophase: Stage of cell division when chromosomes contract and become visible and nuclear membrane begins to break down. In meiosis, crossing-over takes place during prophase.

proto-oncogene: A gene whose protein product stimulate the cell to grow and divide.

purine: Compound composed of two rings.

pyrimidine: Chemicals that have a single, six-sided ring structure.

q arm: The longer arm of a chromosome, or the arm below the centromere.

recessive: A phenotype or allele that is masked by one that is dominant. For traits or disorders inherited in a recessive manner, both copies of the gene need to be changed in order to develop the associated trait or phenotype.

replication: Process of making an exact copy of a DNA molecule.

RNA: Ribonucleic acid; the single-stranded molecule that transfers information carried by DNA to the protein-manufacturing part of the cell.

somatic: Present in the non-sex cells of an organism (cells other than the sperm or eggs); cannot be passed from parent to offspring; opposite of *germline*.

spermatid: An immature sperm cell that contains a single copy of each chromosome.

spermatocyte: Precursor to a spermatid that contains two copies of each chromosome in the form of chromatids.

spermatogonia: Precursor to a spermatocyte that contains two copies of each chromosome as individual chromosomes.

submetacentric chromosome: A chromosome with its centromere located somewhere between the end of the chromosome and the middle of the chromosome.

telomere: End of a chromosome.

telophase: Stage of cell division when chromosomes relax and the nuclear membrane re-forms.

thymine: Pyrimidine base found in DNA but not RNA.

totipotent: A cell that can develop into any type of cell.

transcription: The process by which a gene is "read" and a messenger RNA is created.

translation: The process by which a messenger RNA is "read" and a polypeptide chain (protein) is created.

tumor suppressor gene: A gene whose protein product acts to stop cell growth and tell the cell when its normal life span has ended.

uracil: Pyrimidine base found in RNA but not DNA.

vector: A DNA molecule that is used in DNA cloning and as a delivery system for a gene in gene therapy; vectors may be derived from bacteria, viruses, or other sources.

zygote: Fertilized egg resulting from the fusion of a sperm and egg cell.

Index

Numbers

A

B

C

D

E

F

G

H

I

J

K

L

M

N

Q

R

S

T

U

V

W

X

Y

Z

About the Authors

Tara Rodden Robinson, PhD, grew up in Monroe, Louisiana. She earned a degree in nursing from the University of Southern Mississippi and worked as a nurse for six years. She left nursing to pursue graduate studies at the University of Illinois, Urbana-Champaign. Her dissertation work, conducted in the Republic of Panama, was on social behavior of Song Wrens. She received post-doctoral training at the University of Miami and held a teaching fellowship in genetics at Auburn University. She moved to Oregon State University in 2002 and taught genetics via distance ed for ten years.

After leaving academia in 2015, Tara became a professional artist. She is an accomplished painter and illustrator. Her work focuses on birds and nature. You can learn more about her at `www.tararobinson.com`.

Lisa Cushman Spock, PhD, CGC, is a Hoosier native who graduated from Clay High School in South Bend, Indiana. Lisa attended college at Indiana University in Bloomington where she earned a Bachelor of Science degree in biology, with a minor in psychology. Lisa continued her education at the University of Michigan, obtaining a PhD in Human Genetics in the laboratory of Dr. Sally Camper. Lisa's dissertation focused on the molecular genetics of pituitary gland development, which she continued studying after graduation with Dr. Simon Rhodes at Indiana University-Purdue University at Indianapolis. Lisa then decided to return to graduate school and earned a master's degree from the Indiana University Genetic Counseling Program.

After completing her education, Lisa worked as a clinical genetic counselor at the Indiana University School of Medicine, as well as an instructor and assistant director of the genetic counseling program from which she graduated. After relocating, Lisa worked as a medical research analyst for Hayes, Inc., evaluating the validity and utility of genetic and genomic tests. She also remotely counseled subjects participating in the Parkinson's Progression Markers Initiative study, which was sponsored by the Michael J. Fox Foundation. Lisa then worked as clinical genomics scientist for Myriad Women's Health (formerly Counsyl, Inc.), specializing in genetic testing and sequence variant classification. Currently, Lisa works as a freelance medical writer who focuses on molecular genetics, genetic counseling, genetic and genomic testing, and related medical specialties. This allows her time to spend with her husband Mike and her daughter Emma, volunteer in her community, and do all the other things she loves.

Authors' Acknowledgments

Lisa Cushman Spock, PhD, CGC

I would like to thank everyone at Wiley for making this edition of *Genetics For Dummies* possible. Many thanks to my editors Colleen Diamond and Lindsay Lefevere for all their help during the writing and editing process. Thank you to Mary Corder and Kristie Pyles, and production editor Magesh Elangovan and the entire production team at SPi, for their assistance with the illustrations and overall production of the book. A special thank you to Dr. Nikki Plaster for her technical review of this edition; your attention to detail and feedback were incredibly helpful.

I would also like to thank all the wonderful supervisors and mentors I have had over the years, including Sally Camper, Simon Rhodes, Paula Delk, David Weaver, Wilfredo Torres, Laurence Walsh, Diane Allingham-Hawkins, Tatiana Foroud, Krista Moyer, and Liz Collins. I wouldn't be where I am today without these amazing people. I also want to acknowledge all my genetic counseling colleagues who have provided support, guidance, and knowledge over the years, including insight into the hottest topics in genetics right now.

Finally, I would like to extend my deepest gratitude to my wonderful family, including my husband Mike (my resident "dummy" who is actually quite brilliant), my daughter Emma (whose love, support, and hugs get me through each day), my sister Jackie (the best sister in the world), and my parents Joe and Jill (who have always been encouraging and supportive beyond measure).

Dedication

Lisa Cushman Spock, PhD, CGC:

To my husband Mike, for his never-ending support and encouragement, and my daughter Emma, who I think is the perfect combination of our genes.

In honor of my brother Jeff, the person who got me into biology in the first place.

Publisher's Acknowledgments

Acquisitions Editor: Lindsay Lefevere
Project Manager: Colleen Diamond
Development Editor: Colleen Diamond
Copy Editor: Colleen Diamond
Technical Editor: Nikki Plaster, PhD

Editorial Assistant: Matt Lowe
Sr. Editorial Assistant: Cherie Case
Production Editor: Magesh Elangovan
Cover Photo: © MEHAU KULYK/SCIENCE PHOTO LIBRARY/Getty Images